Fritz Oberhettinger

Tables of Fourier Transforms and Fourier Transforms of Distributions

Springer-Verlag Berlin Heidelberg New York
London Paris Tokyo Hong Kong

Fritz Oberhettinger
Professor emeritus, Oregon State University
P.O.Box 84, Seal Rock, OR 97376/USA

Mathematics Subject Classification (1980): 42A38, 44A10, 44A15

ISBN-13: 978-3-540-50630-0   e-ISBN-13: 978-3-642-74349-8
DOI: 10.1007/ 978-3-642-74349-8

Library of Congress Cataloging-in-Publication Data
Oberhettinger, Fritz. (Tabellen zur Fourier Transformation. English) Tables of Fourier
transforms and Fourier transforms of distribution/Fritz Oberhettinger. p. cm.
Rev. and enl. translation of: Tabellen zur Fourier Transformation. 1957.

1. Fourier transformations. 2. Mathematics-Tables. I. Title. QA404.0213   1990
515'.723-dc20   90-9507

2141/3140-5 4 3 2 1 0 – Printed on acid-free paper

Preface

These tables represent a new, revised and enlarged version of the
previously published book by this author, entitled "Tabellen zur Fourier
Transformation" (Springer Verlag 1957). Known errors have been correc-
ted, apart from the addition of a considerable number of new results,
which involve almost exclusively higher functions. Again, the following tables contain a collection of integrals of the form

$$\text{(A)} \quad g_c(y) = \int_0^\infty f(x)\cos(xy)\,dx \quad \text{Fourier Cosine Transform}$$

$$\text{(B)} \quad g_s(y) = \int_0^\infty f(x)\sin(xy)\,dx \quad \text{Fourier Sine Transform}$$

$$\text{(C)} \quad g_e(y) = \int_{-\infty}^\infty f(x)e^{ixy}\,dx \quad \text{Exponential Fourier Transform}$$

Clearly, (A) and (B) are special cases of (C) if $f(x)$ is respectively an even or an odd function. The transform parameter y in (A)
and (B) is assumed to be positive, while in (C) negative values are
also included. A possible analytic continuation to complex parameters
y* should present no difficulties. In some cases the result function
$g(y)$ is given over a partial range of y only. This means that $g(y)$
for the remaining part of y cannot be given in a reasonably simple
form. Under certain conditions the following inversion formulas for
(A), (B), (C) hold:

$$\text{(A')} \quad f(x) = \frac{2}{\pi} \int_0^\infty g_c(y)\cos(xy)\,dy$$

$$\text{(B')} \quad f(x) = \frac{2}{\pi} \int_0^\infty g_s(y)\sin(xy)\,dy$$

$$\text{(C')} \quad f(x) = (2\pi)^{-1} \int_{-\infty}^\infty g_e(y)e^{-ixy}\,dy$$

In the following parts I, II, III tables for the transforms (A),
(B) and (C) are given. The parts I and II are subdivided into 23 sec-
tions each involving the same class of functions. The first and the
second column (in parenthesis) refers to the location of the corre-
spondent page number for the cosine- and sine transform respectively.

---

*The domain of analyticity is the strip in the direction of the real
axis of the complex y plane.

Compared with the before-mentioned previous edition, a new part IV
titled "Fourier Transforms of Distributions" has been added.   In this,
those functions f(x) occuring in the parts I-III have been singled
out which represent so-called probability density (or frequency distribution) functions.   The corresponding normalization factors are
likewise listed.

The author wishes to express his thanks for the expertise of
Mrs. Jolan Eröss in the completion of this book.

Seal Rock, January 1990   Fritz Oberhettinger

Contents

Part I
Fourier Cosine Transforms
(Tables I)

1.1 Algebraic Functions

	$f(x)$		$g_c(y)$
1.1	1	$x<a$	$y^{-1}\sin(ay)$
	0	$x>a$	
1.2	x	$a<x<b$	$y^{-1}\left[b\,\sin(by)-a\,\sin(ay)\right]$
	0	$x>b$	$+\,y^{-2}\left[\cos(by)-\cos(ay)\right]$
1.3	x	$x<a$	$4y^{-2}\cos(ay)\sin^2(\tfrac{1}{2}ay)$
	$2a-x$	$a<x<2a$	
	0	$x>2a$	
1.4	0	$x<a$	$-\mathrm{Ci}(ay)$
	x^{-1}	$x>a$	
1.5	$x^{-\frac{1}{2}}$		$(\tfrac{1}{2}\pi/y)^{\frac{1}{2}}$
1.6	$x^{-\frac{1}{2}}$	$x<a$	$(2\pi/y)^{\frac{1}{2}}C(ay)$
	0	$x>a$	
1.7	0	$x<a$	$(2\pi/y)^{\frac{1}{2}}\{\tfrac{1}{2}-C(ay)\}$
	$x^{-\frac{1}{2}}$	$x>a$	
1.8	$(a+x)^{-1}$	$x<b$	$\cos(ay)\left[\mathrm{Ci}(ay+by)-\mathrm{Ci}(ay)\right]$
	0	$x>b$	$+\,\sin(ay)\left[\mathrm{si}(ay+by)-\mathrm{si}(ay)\right]$
1.9	0	$x<b$	$-\sin(ay)\mathrm{si}(ay+by)+\cos(ay)\mathrm{Ci}(ay+by)$
	$(a+x)^{-1}$	$x>b$	
1.10	$(a-x)^{-1}$	$x<b$	$\cos(ay)\left[\mathrm{Ci}(ay)-\mathrm{Ci}(ay-by)\right]$
	0	$x>b$	$+\,\sin(ay)\left[\mathrm{si}(ay)-\mathrm{si}(ay-by)\right]$
	$a>b$		
1.11	0	$x<b$	$a^{-1}\left[\cos(ay)\mathrm{Ci}(ay+by)\right.$
	$x^{-1}(a+x)^{-1}$	$x>b$	$\left.+\,\sin(ay)\mathrm{si}(ay+by)-\mathrm{Ci}(by)\right]$

	$f(x)$	$g_c(y)$
1.12	$(a+x)^{-n}$ $n=2,3,\ldots$	$\dfrac{(-y)^{n-1}}{(n-1)!}\,\big(\cos(ay+\tfrac{1}{2}\pi n)\,\text{si}(ay)$ $\qquad -\sin(ay+\tfrac{1}{2}\pi n)\,\text{Ci}(ay)\big)$ $+\displaystyle\sum_{m=1}^{n-1}\left(\dfrac{(m-1)!}{(n-1)!}\right)a^{-m}\sin(\tfrac{1}{2}\pi n-\tfrac{1}{2}\pi m)(-y)^{n-m-1}$
1.13	$0 \qquad\quad x<b$ $(a+x)^{-n} \quad x>b$ $n=2,3,\ldots$	$\dfrac{(-y)^{n-1}}{(n-1)!}\,\big(\cos(ay+\tfrac{1}{2}\pi n)\,\text{si}(ay+by)$ $\qquad -\sin(ay+\tfrac{1}{2}\pi n)\,\text{Ci}(ay+by)\big)$ $+\displaystyle\sum_{m=1}^{n-1}\dfrac{(m-1)!}{(n-1)!}\,(a+b)^{-m}\sin(\tfrac{1}{2}\pi n-\tfrac{1}{2}\pi m-by)(-y)^{n-m-1}$
1.14	$x^{\frac{1}{2}}(a+x)^{-1}$	$(2y/\pi)^{-\frac{1}{2}}-\pi a^{\frac{1}{2}}\cos(ay)\{1-C(ay)-S(ay)\}$ $\qquad -\pi a^{\frac{1}{2}}\sin(ay)\{C(ay)-S(ay)\}$
1.15	$x^{-\frac{1}{2}}(a+x)^{-1}$	$\pi a^{-\frac{1}{2}}\big(\cos(ay)\{1-C(ay)-S(ay)\}$ $\qquad +\sin(ay)\{C(ay)-S(ay)\}\big)$
1.16	$(a+x)^{-\frac{1}{2}} \quad x<b$ $0 \qquad\qquad x>b$	$(2\pi/y)^{\frac{1}{2}}\big(\cos(ay)\{C(ay+by)-C(ay)\}$ $\qquad +\sin(ay)\{S(ay+by)-S(ay)\}\big)$
1.17	$0 \qquad\qquad x<b$ $(a+x)^{-\frac{1}{2}} \quad x>b$	$(2\pi/y)^{\frac{1}{2}}\big(\sin(ay)\{\tfrac{1}{2}-S(ay+by)\}$ $\qquad +\cos(ay)\{\tfrac{1}{2}-C)ay+by)\}\big)$
1.18	$(a+x)^{-3/2}$	$2a^{-\frac{1}{2}}-(2\pi y)^{\frac{1}{2}}\big(\cos(ay)\{1-2S(ay)\}$ $\qquad -\sin z\{1-2C(ay)\}\big)$
1.19	$0 \qquad\qquad x<a$ $(x-a)^{-\frac{1}{2}} \quad x>a$	$(2y/\pi)^{-\frac{1}{2}}\{\cos(ay)-\sin(ay)\}$
1.20	$x^{-1}(x-a)^{\frac{1}{2}} \quad x>a$ $0 \qquad\qquad\;\; x<a$	$(\tfrac{1}{2}\pi/y)^{\frac{1}{2}}\{\cos(ay)-\sin(ay)\}-\pi a^{\frac{1}{2}}\{1-C(ay)-S(ay)\}$
1.21	$(a-x)^{-1}$	$\cos(ay)\,\text{Ci}(ay)+\sin(ay)\{\tfrac{1}{2}\pi+\text{Si}(ay)\}$ (Cauchy principal value)
1.22	$0 \qquad\qquad\quad x<a$ $x^{-1}(x-a)^{-\frac{1}{2}} \quad x>a$	$\pi a^{-\frac{1}{2}}\{1-C(ay)-S(ay)\}$

	$f(x)$	$g_c(y)$
1.23	$(a-x)^{-\frac{1}{2}}$ $x<a$ 0 $x>a$	$(2\pi/y)^{\frac{1}{2}}\{\cos(ay)C(ay)$ $+\sin(ay)S(ay)\}$
1.24	0 $x<b$ $(x-b)^{-\frac{1}{2}}(x+a)^{-1}$ $x>b$	$\pi(a+b)^{-\frac{1}{2}}(\cos(ay)\{1-C(ay+by)-S(ay+by)\}$ $+\sin(ay)\{C(ay+by)-S(ay+by)\})$
1.25	$(a^2+x^2)^{-1}$	$\frac{1}{2}\pi a^{-1}e^{-ay}$
1.26	$x(a^2+x^2)^{-1}$	$-\frac{1}{2}(e^{-ay}\overline{\mathrm{Ei}}(ay)+e^{ay}\mathrm{Ei}(-ay))$
1.27	$(a^2-x^2)^{-1}$ $x<b$ 0 $x>b$ $a>b$	$(2a)^{-1}(\cos(ay)\{\mathrm{Ci}(ay+by)-\mathrm{Ci}(ay-by)\}$ $+\sin(ay)\{\mathrm{si}(ay+by)-\mathrm{si}(ay-by)\})$
1.28	$(x^2-a^2)^{-1}$ $x<b$ 0 $x>b$ $b>a$	$(2a)^{-1}(\sin(ay)\{\mathrm{si}(by-ay)+\mathrm{si}(by+ay)\}$ $-\cos(ay)\{\mathrm{Ci}(by-ay)-\mathrm{Ci}(by+ay)\})$
1.29	$(a^2-x^2)^{-1}$	$\frac{1}{2}\pi a^{-1}\sin(ay)$ (Cauchy principal value)
1.30	$x(a^2-x^2)^{-1}$	$\cos(ay)\mathrm{Ci}(ay)+\sin(ay)\mathrm{Si}(ay)$ (Cauchy principal value)
1.31	$x^{-\frac{1}{2}}(a^2-x^2)^{-1}$	$\frac{1}{2}\pi a^{-3/2}\sin(ay)+(\frac{1}{2}\pi y)^{\frac{1}{2}}a^{-1}S_{0,\frac{1}{2}}(ay)$ (Cauchy principal value)
1.32	$(a^2+x^2)^{-\frac{1}{2}}$	$K_0(ay)$
1.33	$x^{-\frac{1}{2}}(a^2+x^2)^{-\frac{1}{2}}$	$(\frac{1}{2}\pi y)^{\frac{1}{2}}I_{-\frac{1}{4}}(\frac{1}{2}ay)K_{\frac{1}{4}}(\frac{1}{2}ay)$
1.34	$((a^2+x^2)(b^2+x^2))^{-1}$	$\frac{1}{2}\pi(a^2-b^2)^{-1}(b^{-1}e^{-by}-a^{-1}e^{-ay})$
1.35	$(x+(a^2+x^2)^{\frac{1}{2}})^{-1}$	$(a/y)^2-(ay)^{-1}K_1(ay)$
1.36	$\{b^2+(a-x)^2\}^{-1}$ $+\{b^2+(a+x)^2\}^{-1}$	$\pi b^{-1}e^{-by}\cos(ay)$

	$f(x)$	$g_c(y)$
1.37	$(a+x)(b^2+a+x)^{-1}$ $+(a-x)(b^2+a-x)^{-1}$	$\pi e^{-by}\sin(ay)$
1.38	$r^{-1}(r+a)^{\frac{1}{2}}$	$(\tfrac{1}{2}\pi/y)^{\frac{1}{2}}e^{-ay}$
1.39	$r^{-1}(r+a)^{-\frac{1}{2}}$	$\pi(2a)^{-\frac{1}{2}}\mathrm{Erfc}\{(ay)^{\frac{1}{2}}\}$
1.40	$x^{-\frac{1}{2}}r^{-1}(r+x)^{-\frac{1}{2}}$	$2^{-\frac{1}{2}}\pi a^{-1}e^{-\frac{1}{2}ay}I_0(\tfrac{1}{2}ay)$
1.41	$x^{-\frac{1}{2}}r^{-1}(r+x)^{-3/2}$	$2^{-\frac{1}{2}}a^{-2}\sinh(\tfrac{1}{2}ay)\,K_1(\tfrac{1}{2}ay)$
1.42	$x^{-\frac{1}{2}}\{a+x+(2ax)^{\frac{1}{2}}\}^{-1}$	$\pi(2a)^{-\frac{1}{2}}e^{ay}\mathrm{Erfc}\{(ay)^{\frac{1}{2}}\}$
1.43	$(a^2-x^2)^{-\frac{1}{2}}\quad x<a$ $0\qquad\qquad x>a$	$\tfrac{1}{2}\pi J_0(ay)$
1.44	$0\qquad\qquad x<b$ $(a^2-x^2)^{-\frac{1}{2}}\ b<x<a$ $0\qquad\qquad x>a$	$\pi\sum\limits_0^\infty J_{n+\frac{1}{2}}(\tfrac{1}{2}ay-\tfrac{1}{2}by)J_{-n-\frac{1}{2}}(\tfrac{1}{2}ay+\tfrac{1}{2}by)$
1.45	$x(a^2-x^2)^{-\frac{1}{2}}\quad x<a$ $0\qquad\qquad x>a$	$a\{1-\tfrac{1}{2}\pi\,\mathbf{H}_1(ay)\}$
1.46	$x^{-\frac{1}{2}}(a^2-x^2)^{-\frac{1}{2}}\quad x<a$ $0\qquad\qquad x>a$	$(\tfrac{1}{2}\pi)^{3/2}y^{\frac{1}{2}}\left(J_{-\frac{1}{4}}(\tfrac{1}{2}ay)\right)^2$
1.47	$0\qquad\qquad x<a$ $(x^2-a^2)^{-\frac{1}{2}}\quad x>a$	$-\tfrac{1}{2}\pi Y_0(ay)$
1.48	$0\qquad\qquad x<a$ $x^{-1}(x^2-a^2)^{-\frac{1}{2}}\quad x>a$	$\tfrac{1}{2}\pi a^{-1}\big(1-\tfrac{1}{2}\pi ay\{J_0(ay)\,\mathbf{H}_{-1}(ay)$ $+\,\mathbf{H}_0(ay)J_1(ay)\}\big)$
1.49	$0\qquad\qquad x<a$ $(x^2-a^2)^{-\frac{1}{2}}$ $\cdot\left((x+(x^2-a^2)^{\frac{1}{2}}\right)^{-n}\ x>a$ $n=1,2,3,\dots$	$-\tfrac{1}{2}\pi a^{-n}\big(\sin(\tfrac{1}{2}n\pi)J_n(ay)$ $+\cos(\tfrac{1}{2}n\pi)Y_n(ay)\big)$ $-\sum\limits_{k=1}^n\left(\frac{k!(n+k-1)!}{(2k)!(n-k)!}(\tfrac{1}{2}ay)^{-k}\sin(ay+\tfrac{1}{2}\pi k)\right)$

$r=(a^2+x^2)^{\frac{1}{2}}$

	$f(x)$	$g_c(y)$
1.50	$\begin{array}{ll} 0 & x<a \\ x^{-\frac{1}{2}}(x^2-a^2)^{-\frac{1}{2}} & x>a \end{array}$	$-(\tfrac{1}{2}\pi)^{3/2}y^{\frac{1}{2}}J_{-\frac{1}{4}}(\tfrac{1}{2}ay)Y_{-\frac{1}{4}}(\tfrac{1}{2}ay)$
1.51	$(a^4+x^4)^{-1}$	$\tfrac{1}{2}\pi a^{-3}\exp(-2^{-\frac{1}{2}}ay\sin(\tfrac{1}{4}\pi+2^{-\frac{1}{2}}ay)$
1.52	$(x^4+2a^2x^2\cos b+a^4)^{-1}$ $-\pi<b<\pi$	$\tfrac{1}{2}\pi a^{-3}\exp(-ay\cos\tfrac{1}{2}b)\sin(\tfrac{1}{2}b+ay\sin\tfrac{1}{2}b\ \text{cosec } b$
1.53	x^2 $\cdot(x^4+2a^2x^2\cos b+a^4)^{-1}$ $-\pi<b<\pi$	$-\tfrac{1}{2}\pi a^{-1}\exp(-ay\cos\tfrac{1}{2}b)\sin(\tfrac{1}{2}b-ay\sin\tfrac{1}{2}b)\text{cosec} b$
1.54	$\dfrac{x^{\frac{1}{2}}}{A_1A_2}\left(\dfrac{A_1+A_2}{A_2-A_1}\right)^{\frac{1}{2}}$ $A_1=\left(a^2+(b-x)^2\right)^{\frac{1}{2}}$ $A_2=\left(a^2+(b+x)^2\right)^{\frac{1}{2}}$	$b^{-\frac{1}{2}}\cos(by)K_0(ay)$
1.55	$x^{2m}(x^2+z)^{-n-1}$ $n+1>m>0$	$\tfrac{1}{2}(-1)^{m+n}\pi(n!)^{-1}\dfrac{d^n}{dz^n}(z^{m-\frac{1}{2}}e^{-yz^{\frac{1}{2}}}$
1.56	$x^{2m}(a^{2n}+x^{2n})^{-1}$ $n,m=1,2,3,\ldots$ $2m\ 2n-1$	$\tfrac{1}{2}\pi n^{-1}a^{2m-2n+1}\sum\limits_{k=1}^{n}\sin\left((2m+1)a_k+ay\cos(a_k)\right)$ $\cdot\exp\left(-ay\sin(a_k)\right),\ a_k=(k-\tfrac{1}{2})\pi/n$
1.57	$(a^2+x^2)^{-m-1}$ $m=0,1,2,\ldots$	$\pi(2a)^{-m-1}e^{-ay}\dfrac{y^m}{m!}\sum\limits_{k=0}^{m}\dfrac{(m+k)!}{(m-k)!}(2ay)^{-k}$

1.2   Arbitrary Powers

	$f(x)$	$g_c(y)$
2.1	$(a+x)^\nu$ $\operatorname{Re}\nu<0$	$\tfrac12 iy^{-\nu-1}\Big(\exp(\tfrac12 i\pi\nu-iay)\,\Gamma(1+\nu,-iay)$ $\quad -\exp(iay-\tfrac12 i\pi\nu)\,\Gamma(1+\nu,iay)\Big]$
2.2	$(a+x)^\nu$ $\qquad x<b$ 0 $\qquad\qquad x>b$	$\tfrac12 iy^{-\nu-1}\Big(\exp(\tfrac12 i\pi\nu-iay)\,\{\gamma(1+\nu,-iay-iby)$ $\quad -\gamma(1+\nu,-iay)\}$ $\quad -\exp(iay-\tfrac12 i\pi\nu)\,\{\gamma(1+\nu,iay+iby)$ $\quad -\gamma(1+\nu,iay)\}\Big]$
2.3	$x^\nu(a+x)^{-1},-1\ \operatorname{Re}\nu<1$	$\tfrac12 a^\nu\Gamma(1+\nu)\,\{e^{iay}\Gamma(-\nu,iay)+e^{-iay}\Gamma(-\nu,-iay)\}$
2.4	$(a-x)^\nu$ $\qquad x<a$ 0 $\qquad\qquad x>a$ $\operatorname{Re}\nu>-1$	$\tfrac12 iy^{-\nu-1}\Big(\exp(\tfrac12 i\pi\nu+iay)\,\gamma(1+\nu,iay)$ $\quad -\exp(-iay-\tfrac12 i\pi\nu)\,\gamma(1+\nu-iay)\Big]$
2.5	$x^\nu(a-x)^\mu$ $\qquad x<a$ 0 $\qquad\qquad x>a$ $\operatorname{Re}(\nu,\mu)>-1$	$\tfrac12 B(\nu+1,\mu+1)\,a^{\nu+\mu+1}$ $\Big(\,_1F_1(1+\nu;2+\nu+\mu;iay)+\,_1F_1(1+\nu;2+\nu+\mu;-iay)\Big]$
2.6	0 $\qquad\qquad x<a$ $x^{-1}(x-a)^\nu$ $\quad x>a$ $-1<\operatorname{Re}\nu<1$	$\tfrac12 a^\nu\Gamma(1+\nu)\,\{\Gamma(-\nu,iay)+\Gamma(-\nu,-iay)\}$
2.7	$(a^2+x^2)^{-\nu-\frac12},\operatorname{Re}\nu>-\frac12$	$\pi^{\frac12}(\tfrac12 y/a)^\nu\{\Gamma(\tfrac12+\nu)\}^{-1}K_\nu(ay)$
2.8	$(a^2-x^2)^{\nu-\frac12}$ $\quad x<a$ 0 $\qquad\qquad x>a$ $\operatorname{Re}\nu>-\frac12$	$\tfrac12\pi^{\frac12}(\tfrac12 y/a)^{-\nu}\Gamma(\tfrac12+\nu)J_\nu(ay)$
2.9	$(x^2-a^2)^{-\nu-\frac12}$ $\quad x>a$ 0 $\qquad\qquad x<a$ $-\tfrac12<\operatorname{Re}\nu<\tfrac12$	$-\tfrac12\pi^{\frac12}(\tfrac12 y/a)^\nu\Gamma(\tfrac12-\nu)Y_\nu(ay)$
2.10	$x(a^2-x^2)^{\nu-\frac12},$ $\quad x<a$ 0 $\qquad\qquad x>a$ $\operatorname{Re}\nu>-\frac12$	$-a(a/y)^\nu s_{\nu-1,\nu+1}(ay)$
2.11	0 $\qquad\qquad x<a$ $x^{-1}(x^2-a^2)^{-\nu-\frac12},x>a$ $-1<\operatorname{Re}\nu<\tfrac12$	$\tfrac12\pi\sec(\pi\nu)a^{-2\nu-1}\Big\{1-\tfrac12\pi ay\Big(J_\nu(ay)\mathbf{H}_{\nu-1}(ay)$ $\quad -\mathbf{H}_\nu(ay)J_{\nu-1}(ay)\Big]\Big\}$

	$f(x)$	$g_c(y)$
2.12	$0 \qquad x<2a$ $(x^2-2ax)^{-\nu-\frac{1}{2}} \quad x>2a$ $-\frac{1}{2}<\mathrm{Re}\,\nu<\frac{1}{2}$	$-\frac{1}{2}\pi^{\frac{1}{2}}\Gamma(\frac{1}{2}-\nu)(2a/y)^{-\nu}$ $\cdot\left(J_\nu(ay)\sin(ay)+Y_\nu(ay)\cos(ay)\right)$
2.13	$(x^2+2ax)^{-\nu-\frac{1}{2}}$ $-\frac{1}{2}<\mathrm{Re}\,\nu<\frac{1}{2}$	$-\frac{1}{2}\pi^{\frac{1}{2}}\Gamma(\frac{1}{2}-\nu)(2a/y)^{-\nu}$ $\cdot\left(Y_\nu(ay)\cos(ay)-J_\nu(ay)\sin(ay)\right)$
2.14	$(2ax-x^2)^{\nu-\frac{1}{2}} \quad x<2a$ $0 \qquad x>2a$ $\mathrm{Re}\,\nu>-\frac{1}{2}$	$\pi^{\frac{1}{2}}\Gamma(\frac{1}{2}+\nu)(2a/y)^\nu\cos(ay)J_\nu(ay)$
2.15	$(2ax-x^2)^{\nu-\frac{1}{2}} \quad x<a$ $0 \qquad x>a$ $\mathrm{Re}\,\nu>-\frac{1}{2}$	$\frac{1}{2}\pi^{\frac{1}{2}}\Gamma(\frac{1}{2}+\nu)(2a/y)^\nu$ $\cdot\left(J_\nu(ay)\cos(ay)+\mathbf{H}_\nu(ay)\sin(ay)\right)$
2.16	$x^{-\frac{1}{2}}r^{-1}(r+x)^\nu$ $\mathrm{Re}\,\nu<\frac{3}{2}$	$(\frac{1}{2}\pi y)^{\frac{1}{2}}a^\nu I_{-\frac{1}{2}\nu-\frac{1}{4}}(\frac{1}{2}ay)K_{\frac{1}{4}-\frac{1}{2}\nu}(\frac{1}{2}ay)$
2.17	$x^{-\nu-\frac{1}{2}}r^{-1}(r+a)^\nu$ $\mathrm{Re}\,\nu<\frac{1}{2}$	$\frac{1}{2}\pi^{\frac{1}{2}}\sec(\frac{1}{4}\pi+\frac{1}{2}\pi\nu)D_{\nu-\frac{1}{2}}(z)\left(D_{-\nu-\frac{1}{2}}(z)+D_{-\nu-\frac{1}{2}}(-z)\right)$ $\cdot(\frac{1}{2}\pi/a)^{\frac{1}{2}} \qquad z=(2ay)^{\frac{1}{2}}$
2.18	$(u+x)^{-\nu}$ $\mathrm{Re}\,\nu>0$	$\pi\nu\,\mathrm{cosec}(\pi\nu)a^{-\nu}y^{-1}\left(\sin(\frac{1}{2}\pi\nu)I_\nu(ay)\right.$ $\left.+\frac{1}{2}i\,\mathbf{J}_\nu(iay)-\frac{1}{2}i\,\mathbf{J}_\nu(-iay)\right)$
2.19	$u^{-1}(u+x)^{-\nu}$ $\mathrm{Re}\,\nu>-1$	$\pi\,\mathrm{cosec}(\pi\nu)a^{-\nu}\left(\frac{1}{2}\,\mathbf{J}_\nu(iay)+\frac{1}{2}\mathbf{J}_\nu(-iay)\right.$ $\left.-\cos(\frac{1}{2}\pi\nu)I_\nu(ay)\right)$
2.20	$x^{2n}u^{-2\nu-1}$ $0\le n<\frac{1}{2}+\mathrm{Re}\,\nu$	$(-1)^n\pi^{\frac{1}{2}}(2a)^{-\nu}\left(\Gamma(\frac{1}{2}+\nu)\right)^{-1}d^{2n}/dy^{2n}\left(y^\nu K_\nu(ay)\right)$
2.21	$x^\nu u^{-2\mu-2}$ $-1<\mathrm{Re}\,\nu<2+2\mathrm{Re}\,\mu$	$\frac{1}{2}a^{\nu-2\mu-1}B(\frac{1}{2}+\frac{1}{2}\nu,\frac{1}{2}-\frac{1}{2}\nu+\mu)$ ${}_1F_2(\frac{1}{2}+\frac{1}{2}\nu;\frac{1}{2}+\frac{1}{2}\nu-\mu,\frac{1}{2};\frac{1}{4}a^2y^2)+\frac{1}{2}\pi^{\frac{1}{2}}(\frac{1}{2}y)^{1+2\mu-\nu}$ $+\dfrac{\Gamma(\frac{1}{2}\nu-\mu-\frac{1}{2})}{\Gamma(1+\mu-\frac{1}{2}\nu)}{}_1F_2(1+\mu;1+\mu-\frac{1}{2}\nu,\frac{3}{2}+\mu-\frac{1}{2}\nu;\frac{1}{4}a^2y^2)$

$r=(a^2+x^2)^{\frac{1}{2}}$

$u=(a^2+x^2)^{\frac{1}{2}}, v=(a^2-x^2)^{\frac{1}{2}}, w=(x^2-a^2)^{\frac{1}{2}}$

	$f(x)$	$g_c(y)$
2.22	$x^{-\nu-\frac{1}{2}}v^{-1}$ $\cdot \left((a+v)^{\nu} + (a-v)^{\nu} \right)$ $x<a$ 0 $x>a$	$(\tfrac{1}{2}\pi/a)^{\frac{1}{2}} \sec(\pi\nu) \left(D_{-\nu-\frac{1}{2}}(z) + D_{-\nu-\frac{1}{2}}(-z) \right)$ $\cdot \left(D_{\nu-\frac{1}{2}}(z) + D_{\nu-\frac{1}{2}}(-z) \right); z=(2iay)^{\frac{1}{2}}$ $-\tfrac{1}{2}<\mathrm{Re}\,\nu<\tfrac{1}{2}$
2.23	$x^{\nu}v^{2\mu}$ $x<a$ 0 $x>a$ $\mathrm{Re}(\nu,\mu)>-1$	$\tfrac{1}{2}a^{\nu+2\mu+1}B(1+\mu,\tfrac{1}{2}+\tfrac{1}{2}\nu)\,_1F_2(\tfrac{1}{2}+\tfrac{1}{2}\nu;\tfrac{1}{2},\mu+\tfrac{1}{2}\nu$ $+\tfrac{3}{2};-\tfrac{1}{4}a^2y^2)$
2.24	0 $x<a$ $w^{-1}\left((x+w)^{\nu} + (x-w)^{\nu} \right)$ $-1<\mathrm{Re}\,\nu<1$ $x>a$	$-\pi a^{\nu}\left(\sin(\tfrac{1}{2}\pi\nu)J_{\nu}(ay) + \cos(\tfrac{1}{2}\pi\nu)Y_{\nu}(ay) \right)$
2.25	0 $x<a$ $x^{-\frac{1}{2}}w^{-1}\left((x+w)^{\nu} \right.$ $\left. + (x-w)^{\nu} \right)$ $x>a$ $-\tfrac{1}{2}<\mathrm{Re}\,\nu<\tfrac{1}{2}$	$-\tfrac{1}{2}\pi a\nu(\tfrac{1}{2}\pi y)^{\frac{1}{2}}\left(J_{\frac{1}{2}\nu-\frac{1}{4}}(z)Y_{-\frac{1}{2}\nu-\frac{1}{4}}(z) \right.$ $\left. + J_{-\frac{1}{2}\nu-\frac{1}{4}}(z)Y_{\frac{1}{2}\nu-\frac{1}{4}}(z) \right)$, $z=\tfrac{1}{2}ay$ $w=(x^2-a^2)^{\frac{1}{2}}$
2.26	$x^{\nu}(a^2-x^2)^{-1}$ $-1<\mathrm{Re}\,\nu<2$ Cauchy principal value	$\tfrac{1}{2}\pi a^{\nu-1}\sin(ay) + (\pi/a)^{\frac{1}{2}}\dfrac{(2a)^{\nu}}{\Gamma(-\frac{1}{2}\nu)}\Gamma(\tfrac{1}{2}+\tfrac{1}{2}\nu)y^{\frac{1}{2}}$ $\cdot S_{-\frac{1}{2}-\nu,\frac{1}{2}}(ay)$
2.27	$x^{\nu}(1+x)^{-1}$ $-1<\mathrm{Re}\,\nu<1$	$\tfrac{1}{2}a^{\nu}\Gamma(1+\nu)\{e^{-iay}\Gamma(-\nu,-iay)$ $+ e^{iay}\Gamma(-\nu,iay)\}$

$$u=(a^2+x^2)^{\frac{1}{2}}, \quad v=(a^2-x^2)^{\frac{1}{2}}, \quad w=(x^2-a^2)^{\frac{1}{2}}$$

1.3   Exponential Functions

	$f(x)$	$g_c(y)$
3.1	e^{-ax}	$a(a^2+y^2)^{-1}$
3.2	$x^{-1}(e^{-ax}-e^{-bx})$	$\frac{1}{2}\log(\frac{b^2+y^2}{a^2+y^2})$
3.3	$x^{\frac{1}{2}}e^{-ax}$	$\frac{1}{2}\pi^{\frac{1}{2}}(a^2+y^2)^{-3/4}\cos(\frac{3}{2}\text{arc tan}(y/a))$
3.4	$x^{-\frac{1}{2}}e^{-ax}$	$(\frac{1}{2}\pi)^{\frac{1}{2}}(a^2+y^2)^{-\frac{1}{2}}(a+(a^2+y^2)^{\frac{1}{2}})^{\frac{1}{2}}$
3.5	$x^n e^{-ax}$ $n=0,1,2,\ldots$	$n!\,a^{n+1}(a^2+y^2)^{-n-1}\sum_m (-1)^m\binom{n+1}{2m}(y/a)^{2m}$ Summation over m such that $0\leq 2m\leq n+1$
3.6	$x^{n-\frac{1}{2}}e^{-ax}$ $n=1,2,3,\ldots$	$(-1)^n(\frac{1}{2}\pi)^{\frac{1}{2}}\frac{d^n}{da^n}\ (a^2+y^2)^{-\frac{1}{2}}((a^2+y^2)^{\frac{1}{2}}-a)^{-\frac{1}{2}}$
3.7	$x^{\nu-1}e^{-ax}$ $\text{Re}\,\nu>0$	$\Gamma(\nu)(a^2+y^2)^{-\frac{1}{2}\nu}\cos(\nu\,\text{arc tan}(y/a))$
3.8	$\begin{matrix}0 & x<b \\ (x-b)^\nu e^{-ax} & x>b\end{matrix}$ $\text{Re}\,\nu>-1$	$\Gamma(1+\nu)(a^2+y^2)^{-\frac{1}{2}-\frac{1}{2}\nu}e^{-ab}$ $\cdot\cos(by+(\nu+1)\text{arc tan})y/a))$
3.9	$(e^{ax}+1)^{-1}$	$\frac{1}{4}a^{-1}(\Psi(\frac{1}{2}iy/a)+\Psi(-\frac{1}{2}iy/a)-\Psi(\frac{1}{2}+\frac{1}{2}iy/a)$ $-\Psi(\frac{1}{2}-\frac{1}{2}iy/a))$
3.10	$(e^x-1)^{-1}-x^{-1}$	$\log y-\frac{1}{2}(\Psi(iy)+\Psi(-iy))$
3.11	$x^{-1}(\frac{1}{2}-x^{-1}+(e^x-1)^{-1})$	$-\frac{1}{2}\log(1-e^{-2\pi y})$
3.12	$x(e^{ax}-1)^{-1}$	$\frac{1}{2}y^{-2}-\frac{1}{2}(\frac{\pi}{a}\text{cosech}(\pi y/a))^2$
3.13	$x^{\nu-1}(e^{ax}-1)^{-1}$ $\text{Re}\,\nu>-1$	$\frac{1}{2}a^{-\nu}\Gamma(\nu)(\zeta(\nu,1+iy/a)+\zeta(\nu,1-iy/a))$
3.14	$x^{\nu-1}(e^{ax}+1)^{-1}$ $\text{Re}\,\nu>0$	$\Gamma(\nu)\{y^{-\nu}\cos(\frac{1}{2}\pi\nu)+\frac{1}{2}(2a)^{-\nu}(\zeta(\nu,\frac{1}{2}+\frac{1}{2}iy/a)$ $+\zeta(\nu,\frac{1}{2}-\frac{1}{2}iy/a)-\zeta(\nu,\frac{1}{2}iy/a)-\zeta(\nu,-\frac{1}{2}iy/a))\}$
3.15	$x^{-2}(1-e^{-ax})^2$	$a\,\log(\frac{y^2+4a^2}{y^2+a^2})-y\,\text{arc cot}(\frac{1}{2}(y/a)^3+\frac{3}{2}y/a)$
3.16	$e^{-ax}(1-e^{-bx})^{\nu-1}$ $\text{Re}\,\nu>0$	$\frac{1}{2}b^{-1}(B(\nu,\frac{a-iy}{b})+B(\nu,\frac{a+iy}{b}))$
3.17	e^{-ax^2}	$\frac{1}{2}(\pi/a)^{\frac{1}{2}}\exp(-\frac{1}{4}y^2/a)$

	$f(x)$	$g_c(y)$
3.18	$x^{\frac{1}{2}}e^{-ax^2}$	$\frac{1}{4}\pi(\frac{1}{2}y/a)^{3/2}e^{-z}\left(I_{-\frac{3}{4}}(z)-I_{\frac{1}{4}}(z)\right),\ z=\frac{1}{8}y^2/a$
3.19	$x^{-\frac{1}{2}}e^{-ax^2}$	$\frac{1}{2}\pi(\frac{1}{2}y/a)^{\frac{1}{2}}e^{-z}I_{-\frac{1}{4}}(z)\qquad\qquad,\ z=\frac{1}{8}y^2/a$
3.20	$x^{2n}e^{-a^2x^2}=$ $n=1,2,3,\ldots$	$(-1)^n\pi^{\frac{1}{2}}2^{-n-1}a^{-2n-1}\exp(-\frac{1}{4}y^2/a^2)He_{2n}(\frac{2^{-\frac{1}{2}}}{a}y)$
3.21	$x^{\nu}e^{-ax^2}$ $\mathrm{Re}\,\nu>-1$	$\frac{1}{2}(\frac{1}{2}\pi)^{\frac{1}{2}}(2a)^{-\frac{1}{2}-\frac{1}{2}\nu}\sec(\frac{1}{2}\pi\nu)\exp(-\frac{1}{8}y^2/a)$ $\cdot\left(D_{\nu}(z)+D_{\nu}(-z)\right)\qquad\quad,\ z=(2a)^{-\frac{1}{2}}y$
3.22	$(b^2+x^2)^{-1}e^{-a^2x^2}$	$\frac{1}{4}\pi b^{-1}\exp(a^2b^2)\left(e^{-by}\mathrm{Erfc}(u)+e^{by}\mathrm{Erfc}(v)\right)$ $\genfrac{}{}{0pt}{}{v}{u}=(ab\pm\frac{1}{2}y/a)$
3.23	$(b^2+x^2)^{-m-1}e^{-ax^2}$ $m=1,2,3,\ldots$	$\frac{1}{2}\pi^{\frac{1}{2}}2^{-\frac{1}{2}m-\frac{1}{2}}a^{\frac{1}{2}m}b^{-m-1}(m!)^{-1}\exp(-\frac{1}{8}y^2/a+\frac{1}{2}ab^2)$ $\cdot\sum_{k=0}^{m}(m+k)!\,2^{-\frac{3}{2}k}a^{-\frac{1}{2}k}n^{-k}[e^{\frac{1}{2}by}D_{-\nu}(u)$ $+e^{-\frac{1}{2}by}D_{-\nu}(v)]$ $\genfrac{}{}{0pt}{}{u}{v}=(2a)^{\frac{1}{2}}(1\pm\frac{1}{2}y/a),\ \nu=m+1-k$
3.24	$((b+ix)^{\nu}+(b-ix)^{\nu})$ $\cdot e^{-ax^2}$	$(\frac{1}{2}\pi)^{\frac{1}{2}}(2a)^{-\frac{1}{2}\nu-\frac{1}{2}}\exp(\frac{1}{2}ab^2-\frac{1}{8}y^2/a)$ $\cdot(e^{\frac{1}{2}by}D_{\nu}(u)+e^{-\frac{1}{2}by}D_{\nu}(v)),\genfrac{}{}{0pt}{}{u}{v}=(2a)^{\frac{1}{2}}(b\pm\frac{1}{2}y/a)$
3.25	$\exp(-ax-bx^2)$	$\frac{1}{4}(\pi/b)^{\frac{1}{2}}(e^{u^2}\mathrm{Erfc}(u)+e^{v^2}\mathrm{Erfc}(v)),$ $\genfrac{}{}{0pt}{}{u}{v}=\frac{1}{2}(a\pm iy)b^{-\frac{1}{2}}$
3.26	$x^2\exp(-ax^2)$	$\frac{1}{4}\pi^{\frac{1}{2}}a^{-3/2}(1-\frac{1}{2}y^2/a)\exp(-\frac{1}{4}y^2/a)$
3.27	$x^{\nu-1}\exp(ax-bx^2)$ $\mathrm{Re}\,\nu>0$	$\frac{1}{2}\Gamma(\nu)(2b)^{-\frac{1}{2}\nu}\exp(\frac{1}{8}(a^2-y^2)/b)$ $\cdot\left(\exp(\frac{1}{4}iay/b)D_{-\nu}(u)+\exp(-\frac{1}{4}iay/b)D_{-\nu}(v)\right)$ $\genfrac{}{}{0pt}{}{u}{v}=(2b)^{-\frac{1}{2}}(a\pm iy)$
3.28	$x^{-\frac{1}{2}}e^{-a/x}$	$(\frac{1}{2}\pi/y)^{\frac{1}{2}}e^{-z}(\cos z-\sin z),\ z=(2ay)^{-\frac{1}{2}}$
3.29	$x^{-3/2}e^{-a/x}$	$(\pi/a)^{\frac{1}{2}}e^{-z}\cos z,\qquad\qquad z=(2ay)^{\frac{1}{2}}$

	$f(x)$	$g_c(y)$
3.30	$x^{-\nu-1}e^{-a/x}$ $\mathrm{Re}\,\nu>-1$	$(y/a)^{\frac{1}{2}}\left(\exp(\tfrac{1}{4}i\pi\nu)K_\nu(u)+\exp(-\tfrac{1}{4}i\pi\nu)K_\nu(v)\right)$ $\genfrac{}{}{0pt}{}{u}{v}=2(\pm iay)^{\frac{1}{2}}$
3.31	$x^{-\frac{1}{2}}\exp(-ax-b^2/x)$	$\pi^{\frac{1}{2}}(a^2+y^2)^{-\frac{1}{2}}e^{-2bu}\left(u\cos(2bv)-v\sin(2bu)\right)$ $\genfrac{}{}{0pt}{}{u}{v}=2^{-\frac{1}{2}}\left((a^2+y^2)^{\frac{1}{2}}\pm a\right)^{\frac{1}{2}}$
3.32	$x^{-3/2}\exp(-ax-b^2/x)$	$\pi^{\frac{1}{2}}b^{-1}e^{-2bu}\cos(2bv),\qquad u,v\ \text{as before}$
3.33	$x^{\nu-1}\exp(-ax-b^2/x)$	$b^\nu\left(u^{-\frac{1}{2}\nu}K_\nu(2bu)+v^{-\frac{1}{2}\nu}K_\nu(2bv)\right)\genfrac{}{}{0pt}{}{u}{v}=(a\pm iy)^{\frac{1}{2}}$
3.34	$x^{-2}\exp(-a^2/x^2)$	$\tfrac{1}{2}\pi a^{-1}\sum_0^\infty \dfrac{(-ay)^n}{n!\,\Gamma(\tfrac{1}{2}+\tfrac{1}{2}n)}$
3.35	$e^{-ax^{\frac{1}{2}}}$	$(\tfrac{1}{2}\pi)^{\frac{1}{2}}ay^{-3/2}\left\{\cos z\left(\tfrac{1}{2}-C(z)\right)+\sin z\left(\tfrac{1}{2}-S(z)\right)\right\},$ $z=\tfrac{1}{4}a^2/y$
3.36	$x^{-\frac{1}{2}}e^{-ax^{\frac{1}{2}}}$	$(2\pi/y)^{\frac{1}{2}}\left\{\cos z\left(\tfrac{1}{2}-S(z)\right)-\sin z\left(\tfrac{1}{2}-C(z)\right)\right\},$ $z=\tfrac{1}{4}a^2/y$
3.37	$x^{-3/4}e^{-ax^{\frac{1}{2}}}$	$\tfrac{1}{4}\,(a/y)^{\frac{1}{2}}\,J_{\frac{1}{4}}(z)\sin(z+\tfrac{\pi}{8})-Y_{1/4}(z)\cos(z+\tfrac{\pi}{8}))$ $z=\tfrac{1}{8}a^2/y$
3.38	$x^{\nu-1}e^{-ax^{\frac{1}{2}}}$ $\mathrm{Re}\,\nu>0$	$(2y)^{-\nu}\Gamma(2\nu)\left\{e^{-\tau-z}D_{-2\nu}\left(\tfrac{1}{2}ay^{-\frac{1}{2}}(1-i)\right)\right\}$ $+\,e^{\tau+z}D_{-2\nu}\left(\tfrac{1}{2}ay^{-\frac{1}{2}}(1+i)\right),\ z=\tfrac{1}{8}ia^2/y,\tau=\tfrac{1}{2}i\pi\nu$
3.39	e^{-br}	$\mathrm{abs}^{-1}K_1(as)$
3.40	$r^{-2}e^{-br}$	$\displaystyle\int_b^\infty K_0\{a(t^2+y^2)^{\frac{1}{2}}\}dt$
3.41	$r^{-1}e^{-br}$	$K_0(as)$
3.42	$x^{-\frac{1}{2}}r^{-1}e^{-br}$	$(\tfrac{1}{2}\pi y)^{\frac{1}{2}}I_{-\frac{1}{4}}(\tfrac{1}{2}as-\tfrac{1}{2}ab)K_{\frac{1}{4}}(\tfrac{1}{2}as+\tfrac{1}{2}ab)$
3.43	$r^{-3/2}e^{-br}$	$(\tfrac{1}{2}b/\pi)^{\frac{1}{2}}K_{\frac{1}{4}}(\tfrac{1}{2}as-\tfrac{1}{2}ay)K_{\frac{1}{4}}(\tfrac{1}{2}as+\tfrac{1}{2}ay)$

$$r=(a^2+x^2)^{\frac{1}{2}}\qquad\qquad s=(b^2+y^2)^{\frac{1}{2}}$$

	$f(x)$	$g_c(y)$
3.44	$r^{-1}(r+a)^{-\frac{1}{2}}e^{-br}$	$\pi(2a)^{-\frac{1}{2}}e^{ab}\mathrm{Erfc}\{(sa^{\frac{1}{2}}+ba^{\frac{1}{2}})\}$
3.45	$r^{-1}(r+a)^{\frac{1}{2}}e^{-br}$	$(\tfrac{1}{2}\pi)^{\frac{1}{2}}s^{-1}(b+s)^{\frac{1}{2}}e^{-as}$
3.46	$r^{-1}\{(r+x)^{\nu}+(r-x)^{\nu}\}$ $\cdot\exp(-br)$	$2a^{\nu}\cos\{\nu(\arctan(y/b)\}K_{\nu}(as)$
3.47	$x^{\nu-\frac{1}{2}}(r+a)^{-\nu-1}r^{-1}e^{-br}$ $\mathrm{Re}\,\nu>-\frac{1}{2}$	$\tfrac{1}{2}(\tfrac{\pi}{a})^{\frac{1}{2}}\mathrm{cosec}(\tfrac{1}{2}\pi\nu+\tfrac{\pi}{4})D_{-\nu-\frac{1}{2}}(u)\{D_{\nu-\frac{1}{2}}(v)$ $+D_{\nu-\frac{1}{2}}(-v)\}$ $= (2a)^{-\frac{1}{2}}\Gamma(\tfrac{1}{2}+\nu)D_{-\nu-\frac{1}{2}}(u)\{D_{-\nu-\frac{1}{2}}(iv)$ $+ D_{-\nu-\frac{1}{2}}(-iv)\}$ $\genfrac{}{}{0pt}{}{u}{v} = (2a)^{\frac{1}{2}}(s\pm b)^{\frac{1}{2}}$
3.48	$r^{-1}(r+a)^{\frac{1}{2}}\exp(-bx^2)$	$\tfrac{1}{2}\exp(\tfrac{1}{4}a^2b-\tfrac{1}{8}y^2/b)(e^{\frac{1}{2}ay}u^{\frac{1}{2}}K_{\frac{1}{4}}(\tfrac{1}{2}bu^2)+e^{-\frac{1}{2}ay}v^{\frac{1}{2}}$ $\cdot K_{\frac{1}{4}}(\tfrac{1}{2}bv^2))$; $\genfrac{}{}{0pt}{}{u}{v} = a\pm\tfrac{1}{2}y/b$
3.49	$(r+x)^{\nu}r^{-1}e^{-br}$	$a^{\nu}\mathrm{cosec}(\pi\nu)\left(\pi\cos\{\nu\arctan(y/b)\}I_{-\nu}(as)\right.$ $\left.- \int_{0}^{\pi}\exp(ab\cos t)\cosh(ay\sin t)\cos(\nu t)dt\right)$
3.50	$u^{-3/2}e^{-bu}$ $x<a$ 0 $x>a$	$\tfrac{1}{2}\pi(\tfrac{1}{2}b)^{\frac{1}{2}}\left(J_{-\frac{1}{4}}(\tfrac{1}{2}ay-\tfrac{1}{2}aV)J_{-\frac{1}{4}}(\tfrac{1}{2}ay+\tfrac{1}{2}aV)\right.$ $\left.-J_{\frac{1}{4}}(\tfrac{1}{2}ay-\tfrac{1}{2}aV)J_{\frac{1}{4}}(\tfrac{1}{2}ay+\tfrac{1}{2}aV)\right)$
3.51	$u^{-3/2}e^{bu}$ $x<a$ 0 $x>a$	$\tfrac{1}{2}\pi(\tfrac{1}{2}b)^{\frac{1}{2}}\left(J_{-\frac{1}{4}}(\tfrac{1}{2}ay-\tfrac{1}{2}aV)J_{-\frac{1}{4}}(\tfrac{1}{2}ay+\tfrac{1}{2}aV)\right.$ $\left.+J_{\frac{1}{4}}(\tfrac{1}{2}ay-\tfrac{1}{2}aV)J_{\frac{1}{4}}(\tfrac{1}{2}ay+\tfrac{1}{2}aV)\right)$
3.52	$w^{-3/2}e^{-bw}$ $x<2a$ 0 $x>2a$	$\pi(\tfrac{1}{2}b)^{\frac{1}{2}}\cos(ay)\left(J_{-\frac{1}{4}}(\tfrac{1}{2}ay-\tfrac{1}{2}aV)J_{-\frac{1}{4}}(\tfrac{1}{2}ay+\tfrac{1}{2}aV)\right.$ $\left.+ J_{\frac{1}{4}}(\tfrac{1}{2}ay-\tfrac{1}{2}aV)J_{\frac{1}{4}}(\tfrac{1}{2}ay+\tfrac{1}{2}aV)\right)$
3.53	$w^{-3/2}e^{bw}$ $x<2a$ 0 $x>2a$	$\pi(\tfrac{1}{2}b)^{\frac{1}{2}}\cos(ay)\left(J_{-\frac{1}{4}}(\tfrac{1}{2}ay-\tfrac{1}{2}aV)J_{-\frac{1}{4}}(\tfrac{1}{2}ay+\tfrac{1}{2}aV)\right.$ $\left.+ J_{\frac{1}{4}}(\tfrac{1}{2}ay-\tfrac{1}{2}aV)J_{\frac{1}{4}}(\tfrac{1}{2}ay+\tfrac{1}{2}aV)\right)$

$r = (a^2+x^2)^{\frac{1}{2}}$

$u = (a^2-x^2)^{\frac{1}{2}}$, $w = (2ax-x^2)^{\frac{1}{2}}$ $v = (b^2-y^2)^{\frac{1}{2}}$, $V = (y^2-b^2)^{\frac{1}{2}}$

	$f(x)$	$g_c(y)$
3.54	$x^{-2\nu-\frac{1}{2}}\big((a-u)^{2\nu}e^{bu}$ $+(a+u)^{2\nu}e^{-bu}\big)u^{-1}$ $\qquad\qquad x<a$ $\qquad 0 \qquad x>a$ $-\frac{1}{4}<\mathrm{Re}\,\nu<\frac{1}{4}$	$\frac{1}{2}(\pi/a)^{\frac{1}{2}}\sec(2\pi\nu)\big(D_{2\nu-\frac{1}{2}}(z_1)+D_{2\nu-\frac{1}{2}}(-z_1)\big)$ $\qquad\qquad \cdot\,\big(D_{-2\nu-\frac{1}{2}}(z_2)+D_{-2\nu-\frac{1}{2}}(-z_2)\big)$ $z_1 \atop 2 = (2a)^{\frac{1}{2}}(b\pm v)^{\frac{1}{2}}$
3.55	$x^{\nu}\exp(-ax^c)$ $\mathrm{Re}\,\nu>-1,\; 0<c<1$	$-y^{-\nu-1}\sum_{0}^{\infty}(-a)^n(n!)^{-1}\Gamma(1+\nu+nc)\sin\{\tfrac{\pi}{2}(\nu+nc)\}y^{-nc}$
3.56	$x^{\nu}\exp(-ax^c)$ $\mathrm{Re}\,\nu>-1,\qquad c>1$	$c^{-1}\sum_{0}^{\infty}(-1)^n a^{-(1+2n+\nu)/c}\Gamma\{(1+2n+\nu)/c\}y^{2n}/(2n)!$
3.57	$\exp(-ax^3)$	$\frac{1}{2}sy^{-1}(e^{\frac{1}{4}i\pi}S_{0,\nu}(p)+e^{-\frac{1}{4}i\pi}S_{0,\nu}(q)),\;\nu=1/3$ $s=2(\tfrac{1}{3}y/a)^{3/2},\qquad {p\atop q}=s\,\exp(\pm\tfrac{3}{4}\pi i)$

$u=(a^2-x^2)^{\frac{1}{2}},\,w=(2ax-x^2)^{\frac{1}{2}}\qquad\qquad v=(b^2-y^2)^{\frac{1}{2}},\qquad V=(y^2-b^2)^{\frac{1}{2}}$

1.4   Logarithmic Functions

	$f(x)$	$g_c(y)$
4.1	$\log x \quad x<a$ $0 \qquad x>a$	$y^{-1}(\sin(ay)\log a - \mathrm{Si}(ay))$
4.2	$x^{-\frac{1}{2}}\log x$	$-(2y/\pi)^{-\frac{1}{2}}(\gamma+\tfrac{1}{2}\pi+\log(4y))$
4.3	$(x^2-a^2)^{-1}\log(x/a)$	$\tfrac{1}{2}\pi a^{-1}(\sin(ay)\mathrm{Ci}(ay)-\cos(ay)\,\mathrm{si}(ay))$
4.4	$(x^2-a^2)^{-1}\log(bx)$ Cauchy principal value	$\tfrac{1}{2}\pi a^{-1}\left\{\sin(ay)(\mathrm{Ci}(ay)-\log(ab))-\cos(ay)\,\mathrm{si}(ay)\right\}$
4.5	$(a^2+x^2)^{-1}\log(bx)$	$\tfrac{1}{2}\pi a^{-1}(2e^{-ay}\log(ab)+e^{ay}\mathrm{Ei}(-ay)-e^{-ay}\overline{\mathrm{Ei}}(ay))$
4.6	$\quad 0 \qquad\quad x<1$ $x^{-1}\log(2x-1), \quad x>1$	$\tfrac{1}{2}\left(\mathrm{Ci}(\tfrac{1}{2}y)\right)^2-\tfrac{1}{2}\left(\mathrm{si}(\tfrac{1}{2}y)\right)^2$
4.7	$x^{-1}\log(1+x)$	$\tfrac{1}{2}\left(\mathrm{Ci}(\tfrac{1}{2}y)\right)^2+\tfrac{1}{2}\left(\mathrm{si}(\tfrac{1}{2}y)\right)^2$
4.8	$\log\lvert(a+x)/(b-x)\rvert$	$y^{-1}\big\{\tfrac{1}{2}\pi\cos(by)-\tfrac{1}{2}\pi\cos(ay)+\cos(by)\mathrm{Si}(by)$ $+\cos(ay)\mathrm{Si}(ay)-\sin(ay)\mathrm{Ci}(ay)-\sin(by)\mathrm{Ci}(by)\big\}$
4.9	$\log(a-x) \quad x<a$ $0 \qquad\quad x>a$	$y^{-1}\big\{\sin(ay)(\mathrm{Ci}(ay)-\gamma-\log y)-\cos(ay)\mathrm{Si}(ay)\big\}$
4.10	$\log(a-x) \quad x<b$ $0 \qquad\quad x>b$ $\qquad\qquad\quad a>b$	$y^{-1}\big\{\sin(by)\log(a-b)$ $+\sin(ay)(\mathrm{Ci}(ay)-\mathrm{Ci}(ay-by))$ $-\cos(ay)(\mathrm{Si}(ay)-\mathrm{Si}(ay-by))\big\}$
4.11	$\log(a+x) \quad x<b$ $0 \qquad\quad x>b$	$y^{-1}\big\{\sin(by)\log(a+b)-\cos(ay)(\mathrm{si}(ay+by)-\mathrm{si}(ay))$ $+\sin(ay)(\mathrm{Ci}(ay+by)-\mathrm{Ci}(ay))\big\}$
4.12	$x^{\nu-1}\log x$ $0<\mathrm{Re}\nu<1$	$\Gamma(\nu)y^{-\nu}\cos(\tfrac{1}{2}\pi\nu)\left(\Psi(\nu)-\tfrac{1}{2}\pi\tan(\tfrac{1}{2}\pi\nu)-\log y\right)$
4.13	$(a+ix)^{-1}\log(a+ix)$ $+(a-ix)^{-1}\log(a-ix)$	$\pi e^{-ay}(\gamma+\log y)$
4.14	$(a^2+x^2)^{-1}\log(a^2+x^2)$	$-\tfrac{1}{2}\pi a^{-1}(\{\gamma-\log(2a/y)\}e^{-ay}-e^{ay}\mathrm{Ei}(-2ay))$

	$f(x)$	$g_c(y)$
4.15	$(a^2-x^2)^{-\frac{1}{2}}\log\left(\frac{x}{a}\right)$, $x<a$ 0 $\qquad\qquad x>a$	$-\frac{1}{2}\pi J_0(ay)\log(2)-\frac{1}{2}\pi\sum_{n\equiv 0}^{\infty}n^{-1}J_{2n}(ay)$
4.16	$x^{-1}\log\left(1+x+(1+x)^{\frac{1}{2}}\right)$	$\int_y^{\infty}t^{-1}K_0(t)\,dt$
4.17	$\log(a^2-x^2)$ $\qquad x<b$ 0 $\qquad\qquad x>b$ $\qquad\qquad\qquad a>b$	$y^{-1}\Big\{\sin(by)\log(a^2-b^2)+\cos(ay)\big(\mathrm{si}(ay+by)$ $-\mathrm{si}(ay-by)\big)+\sin(ay)\big(\mathrm{Ci}(ay+by)-\mathrm{Ci}(ay-by)\big)\Big\}$
4.18	$\log\left((a^2+x^2)/(b^2+x^2)\right)$	$\pi y^{-1}(e^{-by}-e^{-ay})$
4.19	$\log\left|(a^2+x^2)/(b^2-x^2)\right|$	$\pi y^{-1}\left(\cos(by)-e^{-ay}\right)$
4.20	$x^{-1}\log(1+x^2/a^2)$	$\mathrm{Ei}(-ay)\overline{\mathrm{Ei}}(ay)$
4.21	$x^{-1}\log\left|\frac{a+x}{a-x}\right|$	$-\pi\,\mathrm{si}(ay)$
4.22	$(a^2+x^2)^{-\frac{1}{2}}\log(a^2+x^2)$	$-\left(\gamma+\log(2y/a)\right)K_0(ay)$
4.23	$\log(1+a^2/x^2)$	$\pi y^{-1}(1-e^{-ay})$
4.24	$\log(1+a/x)$	$y^{-1}\left(\frac{1}{2}\pi+\cos(ay)\,\mathrm{si}(ay)-\sin(ay)\,\mathrm{Ci}(ay)\right)$
4.25	$\log\left|1-a^2/x^2\right|$	$2\pi y^{-1}\sin^2(\frac{1}{2}ay)$
4.26	$\log(a^2-x^2)$ $\qquad x<a$ 0 $\qquad\qquad x<a$	$y^{-1}\sin(ay)\left(\mathrm{Ci}(2ay)-\gamma-\log(\frac{1}{2}y/a)\right)$ $-y^{-1}\cos(ay)\,\mathrm{Si}(2ay)$
4.27	$(a^2-x^2)^{-\frac{1}{2}}\log(a^2-x^2)$ $\qquad\qquad\qquad x<a$ 0 $\qquad\qquad x>a$	$\frac{1}{4}\pi^2 Y_0(ay)-\frac{1}{2}\pi J_0(ay)\{\gamma+\log(2y/a)\}$
4.28	0 $\qquad\qquad x<a$ $(x^2-a^2)^{-\frac{1}{2}}\log(x^2-a^2)$ $\qquad\qquad\qquad x>a$	$-\frac{1}{4}\pi^2 J_0(ay+\frac{1}{2}\pi Y_0(ay)\{\gamma+\log(2y/a)\}$
4.29	$(2ax-x^2)^{-\frac{1}{2}}$ $\cdot\log(2ax-x^2)$ $\quad x<2a$ 0 $\qquad\qquad x>2a$	$\frac{1}{2}\pi\cos(ay)\left(\pi Y_0(ay)-2\{\gamma+\log(2y/a)\}J_0(ay)\right)$

	$f(x)$	$g_c(y)$
4.30	$(x^2+2ax)^{-\frac{1}{2}}$ $\cdot \log(x^2+2ax)$	$\frac{1}{2}\pi\left\{\{\gamma+\log(2\frac{y}{a})\}\left(Y_o(ay)\cos(ay)+J_o(ay)\sin(ay)\right)\right\}$ $-\frac{1}{2}\pi\left(J_o(ay)\cos(ay)-Y_o(ay)\sin(ay)\right)$
4.31	$\log\left(\frac{1}{2}+\frac{1}{2}(1+a^2/x^2)^{\frac{1}{2}}\right)$	$\frac{1}{2}\pi y^{-1}\left(1+\mathbf{L}_o(ay)-I_o(ay)\right)$
4.32	$(a^2+x^2)^{-\frac{1}{2}}$ $\cdot\log\left((1+a^2/x^2)^{\frac{1}{2}}+a/x\right)$	$\frac{1}{4}\pi^2\{I_o(ay)-\mathbf{L}_o(ay)\}$
4.33	$(1+x^2)^{-\frac{1}{2}}\left((1+x^2)^{\frac{1}{2}}-1\right)^{\frac{1}{2}}$ $\cdot\log\{x+(1+x^2)^{\frac{1}{2}}\}$	$-(2y/\pi)^{-\frac{1}{2}}\left(e^y\mathrm{Ei}(-2y)+\frac{1}{2}\pi e^{-y}\right)$
4.34	$(a^2+x^2)^{-\frac{1}{2}}$ $\cdot\log\{x+(a^2+x^2)^{\frac{1}{2}}\}$	$\frac{1}{2}\left(S_{-1,o}(iay)+S_{-1,o}(-iay)+\log a\ K_o(ay)\right)$
4.35	$\log(\frac{1}{2}u+\frac{1}{2}v)\quad x>a$ $\begin{matrix}u\\v\end{matrix}=1\pm a/x,\quad 0\quad x<a$	$\frac{1}{2}y^{-1}\left(\frac{1}{2}\pi J_o(ay)+\mathrm{si}(ay)\right)$
4.36	$\log\left(\frac{1}{2}+\frac{1}{2}(1+a/x)^{\frac{1}{2}}\right)$	$\frac{1}{4}\pi y^{-1}\left(1-\cos(\frac{1}{2}ay)J_o(\frac{1}{2}ay)-\sin(\frac{1}{2}ay)Y_o(\frac{1}{2}ay)\right)$
4.37	$\begin{matrix}0\quad x<1\\(1+x)^{-\frac{1}{2}}\\\cdot\log\left(x+(x^2-1)^{\frac{1}{2}}\right)\ x>1\end{matrix}$	$-(\pi/y)^{\frac{1}{2}}\left(\cos(y-\frac{1}{4}\pi)\mathrm{Ci}(2y)\sin(y-\frac{1}{4}\pi)\mathrm{si}(2y)\right)$
4.38	$\begin{matrix}0\quad x<1\\x^{-1}\log\left(x+(x^2-1)^{\frac{1}{2}}\right)\\ \qquad\qquad x>1\end{matrix}$	$-\frac{1}{2}\pi\int_y^\infty t^{-1}Y_o(t)\,dt$
4.39	$e^{-ax}\log x$	$-(a^2+y^2)^{-1}\left(a\gamma+\frac{1}{2}a\log(a^2+y^2)+y\arctan(y/a)\right)$
4.40	$x^{\nu-1}e^{-ax}\log x$ $\mathrm{Re}\,\nu>0$	$(a^2+y^2)^{-\frac{1}{2}\nu}\Gamma(\nu)\left\{\cos(\nu z)\left(\Psi(\nu)-\frac{1}{2}\log(a^2+y^2)\right)-z\sin(\nu z)\right\};\ z=\arctan(y/a)$
4.41	$x^{-1}\exp(-\frac{1}{4}a^2/x)\log x$	$\frac{1}{4}i\pi\left(K_o(v)-K_o(u)\right)-\log(2y^{\frac{1}{2}}/a)\left(K_o(v)+K_o(u)\right)$ $\begin{matrix}u\\v\end{matrix}=a(\pm iy)^{\frac{1}{2}}$

	$f(x)$	$g_c(y)$
4.42	$e^{-ax}(\log x)^2$	$(a^2+y^2)^{-1}\left\{\frac{1}{6}\pi^2 a + \left(\gamma+\frac{1}{2}\log(a^2+y^2)\right) \cdot \left(\gamma a+\frac{1}{2}a \log(a^2+y^2)\right)+2yz-az^2\right\}$ $z=\arctan(y/a)$
4.43	$\log(1+e^{-ax})$	$\frac{1}{2}ay^{-2}-\frac{1}{2}\pi y^{-1}\mathrm{cosech}\,(\pi y/a)$
4.44	$\log(1-e^{-ax})$	$\frac{1}{2}ay^{-2}-\frac{1}{2}\pi y^{-1}\coth(\pi y/a)$
4.45	$(a^2+x^2)^{-n-\frac{1}{2}}$ $\cdot\log(a^2+x^2)$ $n=1,2,3,\ldots$	$\dfrac{-n!}{(2n)!}(2y/a)^n\left\{K_n(ay)\left(\gamma+\log(2y/a)-2\sum_{k=1}^{n}\dfrac{1}{2k-1}\right)\right.$ $\left.+\frac{1}{2}n!\sum_{k=0}^{n-1}(\frac{1}{2}ay)^{k-n}\dfrac{K_k(ay)}{(n-k)k!}\right\}$
4.46	$r^{-1}\left((r+x)^n-(r-x)^n\right)$ $\cdot e^{-br}\log(\frac{x}{a}+\frac{r}{a})$ $n=1,2,3,;r=(a^2+x^2)^{\frac{1}{2}}$	$n!\cos(nz)\sum_{k=0}^{n-1}a^k\dfrac{(\frac{1}{2}s)^{k-n}}{(n-k)k!}K_k(as)$ $-2a^n z\,\sin(nz)K_n(as)\,;\,z=\arctan(y/b)\,;$ $s=(b^2+y^2)^{\frac{1}{2}}$
4.47	$(a^2-x^2)^{n-\frac{1}{2}}$ $\cdot\log(a^2-x^2)\quad x<a$ $\qquad\quad 0 \qquad\qquad x>a$ $n=1,2,3,\ldots$	$\frac{1}{2}\pi(2y/a)^{-n}\dfrac{(2n)!}{n!}\left(\frac{1}{2}\pi Y_n(ay)\right.$ $+2J_n(ay)\sum_{k=1}^{n}\dfrac{1}{2k-1}$ $-\{\gamma+\log(2y/a)\}J_n(ay)+\frac{1}{2}n!\sum_{k=0}^{n-1}\dfrac{(\frac{1}{2}ay)^{k-n}}{(n-k)k!}J_k(ay)\Big)$
4.48	$(2ax-x^2)^{n-\frac{1}{2}}$ $\cdot\log(2ax-x^2)\quad x<2a$ $\qquad\quad 0 \qquad\qquad x>2a$ $n=1,2,3,\ldots$	$2\cos(ay)A(y)\quad$ with $\quad A(y)=g_c(y)$ as given in the previous result 4.47

1.5   Trigonometric Functions

	$f(x)$	$g_c(y)$
5.1	$x^{-1}\sin(ax)$	$\tfrac{1}{2}\pi$ $y<a$ $\tfrac{1}{2}\pi$ $y=a$ 0 $y>a$
5.2	$\dfrac{\sin(a_1 x)}{x}\ \ldots\ldots \ldots\ \dfrac{\sin(a_n x)}{x}$	0 $y>a_1+a_2+\ldots a_n$
5.3	$x^{\nu-1}\sin(ax)$ $-1<\mathrm{Re}\,\nu<1$	$\tfrac{1}{4}\pi\dfrac{\sec(\tfrac{1}{2}\pi\nu)}{(1-\nu)}\left((y+a)^{-\nu}-\operatorname{sgn}(y-a)\,|y-a|^{-\nu}\right)$
5.4	$x(b^2+x^2)^{-1}\sin(ax)$	$\tfrac{1}{2}\pi e^{-ab}\cosh(by)$ $y<a$ $-\tfrac{1}{2}\pi e^{-by}\sinh(ab)$ $y>a$
5.5	$x^{-1}(b^2+x^2)^{-1}\sin(ax)$	$\tfrac{1}{2}\pi b^{-2}\left(1-e^{-ab}\cosh(by)\right)$ $y<a$ $\tfrac{1}{2}\pi b^{-2}e^{-by}\sinh(ab)$ $y>a$
5.6	$e^{-bx}\sin(ax)$	$\tfrac{1}{2}(a+y)\left(b^2+(a+y)^2\right)^{-1}+\tfrac{1}{2}(a-y)\left(b^2+(a-y)^2\right)^{-1}$
5.7	$x^{-1}e^{-bx}\sin(ax)$	$\tfrac{1}{2}\arctan\left(\dfrac{2ab}{b^2-a^2+y^2}\right)$
5.8	$x^{-1}\sin(ax)$	$\tfrac{1}{4}\log\left|1-(2a/y)^2\right|$
5.9	$x^{-1}\sin(ax)\sin(bx)$	$\tfrac{1}{4}\log\left|\dfrac{(a+b)^2-y^2}{(a-b)^2-y^2}\right|$
5.10	$x^{-2}\sin^2(ax)$	$\tfrac{1}{2}\pi(a-\tfrac{1}{2}y)$ $y<2a$ 0 $y>2a$
5.11	$x^{-3}\sin^3(ax)$	$\tfrac{1}{8}\pi(3a^2-y^2)$ $y<a$ $\tfrac{1}{4}\pi y^2$ $y=a$ $\pi(3a-y)^2/16$ $a<y<3a$ 0 $y>3a$
5.12	$\left(x^{-1}\sin(ax)\right)^{2m}$ $m=1,2,3,\ldots$	$(-1)^m 2^{-2m}m\pi\left((m!)^{-2}y^{2m-1}+\displaystyle\sum_{n=1}^{m}(-1)^n A_n\right),\ y\le 2am$ 0 $y\ge 2am$ $A_n=\dfrac{(2an+y)^{2m-1}+|2an-y|^{2m-1}}{(m+n)!\,(m-n)!}$

	$f(x)$	$g_c(y)$
5.13	$\left(x^{-1}\sin(ax)\right)^{2m+1}$ $m=1,2,3,\ldots$	$(-1)^m 2^{-2m-2}(2m+1)\,\pi F(y)$ $F(y) = \displaystyle\sum_{n=0}^{m} A_n \qquad\qquad y\leq a$ $F(y) = \displaystyle\sum_{n=0}^{k-1} B_n + \sum_{n=k}^{m} A_n \quad (2k-1)a\leq y\leq(2k+1)a$ $F(y) = \qquad\qquad\qquad 0 \qquad\qquad y\geq(2m+1)a$ $\begin{matrix}A_n\\B_n\end{matrix} = (-1)^n \dfrac{\left((2n+1)a+y\right)^{2m} \pm \left((2n+1)a-y\right)^{2m}}{(n+1+m)!\,(m-n)!}$ $k=1,2,3,\ldots m$
5.14	$e^{-ax}(\sin x)^{2n}$ $n=0,1,2,\ldots$	$(-1)^n\dfrac{2^{-2n-2}}{2n+1}(A_n^{-1}-B_n^{-1}) ; \begin{matrix}A_n\\B_n\end{matrix} = \left(\begin{matrix}n+\frac12 y\pm\frac12 ia\\2n+1\end{matrix}\right)i$
5.15	$e^{-ax}(\sin x)^{2n-1}$ $n=1,2,3,\ldots$	$(-1)^n n^{-1} 2^{-2n-2}(A_n^{-1}+B_n^{-1}) ; \begin{matrix}A_n\\B_n\end{matrix}=\left(\begin{matrix}n-\frac12+\frac12 y\pm\frac12 ia\\2n\end{matrix}\right)$
5.16	$(\sin\pi x)^{\nu-1} \quad x<1$ $\qquad 0 \qquad x>1$ $\mathrm{Re}\,\nu>0$	$2^{1-\nu}\cos\left(\tfrac12\tfrac{y}{\pi}\right)\Gamma(\nu)\left(\Gamma(\tfrac12+\tfrac12\nu+\tfrac12 y/\pi)\,\Gamma(\tfrac12+\tfrac12\nu-\tfrac12 y/\pi)\right)^{-1}$
5.17	$x^{-2}\sin^3(ax)$	$\tfrac18\big((y+3a)\log(y+3a)+(y-3a)\log|y-3a|$ $\quad -(y+a)\log(y+a)-(y-a)\log|y-a|\big)$
5.18	$x^{-1}(1-\cos ax)$	$\tfrac12\log|1-a^2/y^2)|$
5.19	$x^2(1-\cos ax)$	$\tfrac12\pi(a-y) \qquad\qquad y<a$ $\qquad 0 \qquad\qquad\qquad y>a$
5.20	$x^{-1}-x^{-2}\sin x$	$-1+\tfrac12 y\left(\log(1+y)-\log|1-y|+\tfrac12\log|1-y^{-2}|\right)$
5.21	$x^{\nu-1}\cos(ax)$ $0<\mathrm{Re}\,\nu<1$	$\tfrac12\Gamma(\nu)\cos(\tfrac12\pi\nu)\left(|y-a|^{-\nu}+(y+a)^{-\nu}\right)$
5.22	$(b^2+x^2)^{-1}\cos(ax)$	$\tfrac12\pi b^{-1}e^{-ab}\cosh(by) \qquad y<a$ $\tfrac12\pi b^{-1}e^{-by}\cosh(ab) \qquad y>a$
5.23	$e^{-bx}\cos(ax)$	$\tfrac12 b\left\{\left(b^2+(a-y)^2\right)^{-1}+\left(b^2+(a+y)^2\right)^{-1}\right\}$
5.24	$e^{-bx^2}\cos(ax)$	$\tfrac12(\pi/b)^{\frac12}\exp\left(-\tfrac14(a^2+y^2)/b\right)\cosh(\tfrac12 ay/b)$

	$f(x)$	$g_c(y)$
5.25	$(\cos(\tfrac{1}{2}\pi x))^{\nu-1}$ $x<1$ 0 $\quad x>1$ $\mathrm{Re}\,\nu>0$	$2^{1-\nu}\Gamma(\nu)\left(\Gamma(\tfrac{1}{2}+\tfrac{1}{2}\nu+y/\pi)\,\Gamma(\tfrac{1}{2}+\tfrac{1}{2}\nu-y/\pi)\right)^{-1}$
5.26	$(\cosh a-\cos x)^{-1}, x<\pi$ 0 $\quad\quad, x>\pi$	$-y\sin(\pi y)\operatorname{cosech} a\sum\limits_{n=0}^{\infty}(-1)^n\varepsilon_n(n^2-y^2)^{-1}e^{-na}$
5.27	$(\cosh a-\cos x)^{-\nu}, x<\pi$ 0 $\quad\quad, x>\pi$	$-y\sin(\pi y)(\sinh a)^{-\nu}\sum\limits_{n=0}^{\infty}(-1)^n\varepsilon_n(\nu)_n(n^2-y^2)^{-1}$ $P_{-\nu}^{-n}(\coth a)$
5.28	$(\cos z-\cos a)^{\nu-\frac{1}{2}}, x<a$ 0 $\quad\quad, x>a$	$(\tfrac{1}{2}\pi)^{\frac{1}{2}}\Gamma(\tfrac{1}{2}+\nu)(\sin a)^{\nu}P_{-\frac{1}{2}+y}^{-\nu}(\cos a)$
5.29	$(a^2+x^2)^{-1}$ $\cdot(1-2b\cos x+b^2)^{-1}$ $\|b\|<1$	$\tfrac{1}{2}\pi a^{-1}(1-b^2)^{-1}(e^a-b)^{-1}(e^{a-ay}+be^{ay})$ $\quad y<1$
5.30	$(a^2+x^2)^{-1}$ $\cdot(1-2b\cos x+b^2)^{-1}$ $\|b\|<1$	$\tfrac{1}{2}\pi a^{-1}(1-b^2)^{-1}(e^{-ay}$ $+(e^{-a}-b)^{-1}(be^{-ay}-b^{n+1}e^{-a\lambda})$ $+(e^a-b)^{-1}(be^{-an-a\lambda}+b^{n+1}e^{a\lambda}))$ $y=n+\lambda,\ 0\le\lambda<1,\ n=1,2,3,\ldots$
5.31	$(a^2+x^2)^{-1}(\cos x-b)$ $\cdot(1-2b\cos x+b^2)^{-1}$ $\|b\|<1$	$\tfrac{1}{2}\pi a^{-1}(e^a-b)^{-1}\cos h(ay)$ $\quad\quad y<1$
5.32	$\sin(ax^2)$	$\tfrac{1}{4}(2\pi/a)^{\frac{1}{2}}(\cos z-\sin z);\ z=\tfrac{1}{4}y^2/a$
5.33	$\cos(ax^2)$	$\tfrac{1}{4}(2\pi/a)^{\frac{1}{2}}(\cos z+\sin z);\ z=\tfrac{1}{4}y^2/a$
5.34	$x^{-1}\sin(ax^2)$	$\tfrac{1}{2}\pi\left(\tfrac{1}{2}-C^2(z)-S^2(z)\right)$ $\quad;\ z=\tfrac{1}{4}y^2/a$
5.35	$x^{-2}\sin(ax^2)$	$\tfrac{1}{2}\pi y\left(S(z)-C(z)\right)+(\pi a)^{\frac{1}{2}}\sin(z+\tfrac{1}{4}\pi)\ ;\ z=\tfrac{1}{4}y^2/a$
5.36	$e^{-ax^2}\sin(bx^2)$	$\tfrac{1}{2}\pi^{\frac{1}{2}}(a^2+b^2)^{-\frac{1}{4}}\exp\left(-\tfrac{1}{4}ay^2(a^2+b^2)^{-1}\right)$ $\cdot\sin\left(\tfrac{1}{2}\arctan(b/a)-\tfrac{1}{4}by^2(a^2+b^2)^{-1}\right)$
5.37	$e^{-ax^2}\cos(bx^2)$	$\tfrac{1}{2}\pi^{\frac{1}{2}}(a^2+b^2)^{-\frac{1}{4}}\exp\left(-\tfrac{1}{4}ay^2(a^2+b^2)^{-1}\right)$ $\cdot\cos\left(\tfrac{1}{2}\arctan(b/a)-\tfrac{1}{4}by^2(a^2+b^2)^{-1}\right)$

	$f(x)$	$g_c(y)$
5.38	$x^{\frac{1}{2}}\sin(ax^2)$	$(\tfrac{1}{2}\pi)^{\frac{1}{2}}(y/a)^{3/2}\left(\cos(z+\tfrac{1}{8}\pi)J_{-\frac{3}{4}}(z)-\sin(z+\tfrac{1}{8}\pi)J_{\frac{1}{4}}(z)\right)$
5.39	$x^{\frac{1}{2}}\cos(ax^2)$	$\tfrac{1}{4}\pi(\tfrac{1}{2}y/a)^{3/2}\left(\cos(z+\tfrac{1}{8}\pi)J_{\frac{1}{4}}(z)+\sin(z+\tfrac{1}{8}\pi)J_{-\frac{3}{4}}(z)\right)$ $z=\tfrac{1}{8}y^2/a$
5.40	$x^{-\frac{1}{2}}\sin(ax^2)$	$-\tfrac{1}{2}\pi(\tfrac{1}{2}y/a)^{\frac{1}{2}}\sin(z-\tfrac{1}{8}\pi)J_{-\frac{3}{4}}(z)$, $\quad z=\tfrac{1}{8}y^2/a$
5.41	$x^{-\frac{1}{2}}\cos(ax^2)$	$\tfrac{1}{2}\pi(\tfrac{1}{2}y/a)^{\frac{1}{2}}\cos(z-\tfrac{1}{8}\pi)J_{-\frac{1}{4}}(z)$, $\quad z=\tfrac{1}{8}y^2/a$
5.42	$x^{\nu-1}\sin(ax^2)$ $-2<\mathrm{Re}\,\nu<2$	$\tfrac{1}{4}(2a)^{-\frac{1}{2}\nu}\Gamma(\nu)i\left(\exp(\tfrac{1}{4}u-\tfrac{1}{4}i\pi\nu)\left(D_{-\nu}(u^{\frac{1}{2}})+D_{-\nu}(-u^{\frac{1}{2}})\right)\right.$ $\left.-\exp(-\tfrac{1}{4}u+\tfrac{1}{4}i\pi\nu)\,D_{-\nu}(v^{\frac{1}{2}})+D_{-\nu}(-v^{\frac{1}{2}})\right)$, $u=-v=\tfrac{1}{2}iy^2/a$
5.43	$x^{\nu-1}\cos(ax^2)$ $0<\mathrm{Re}\,\nu<2$	$\tfrac{1}{4}(2a)^{-\frac{1}{2}\nu}\Gamma(\nu)\left\{\exp(\tfrac{1}{4}u-\tfrac{1}{4}i\pi\nu)\left(D_{-\nu}(u^{\frac{1}{2}})+D_{-\nu}(-u^{\frac{1}{2}})\right)\right.$ $\left.+\exp(-\tfrac{1}{2}u+\tfrac{1}{4}i\pi\nu)\left(D_{-\nu}v^{\frac{1}{2}}+D_{-\nu}(-v^{\frac{1}{2}})\right)\right\}$ $u=-v\tfrac{1}{2}iy^2/a$
5.44	$\sin(a^3x^3)$	$\dfrac{\pi}{6a}(y/a)^{\frac{1}{2}}\left(I_{-\frac{1}{3}}(z)+I_{\frac{1}{3}}(z)+J_{-\frac{1}{3}}(z)-J_{\frac{1}{3}}(z)\right.$ $\left.-2J_{\frac{1}{3}}(z)(z)+2J_{-\frac{1}{3}}(z)+2iJ_{\frac{1}{3}}(iz)-2iJ_{-\frac{1}{3}}(iz)\right)$ $z=2(3a/y)^{-\frac{3}{2}}$
5.45	$\cos(a^3x^3)$	$\dfrac{1}{6a}\left(\dfrac{y}{a}\right)^{\frac{1}{2}}3^{-\frac{1}{2}}\left(3^{\frac{1}{2}}K_{\frac{1}{3}}(u)+J_{\frac{1}{3}}(u)+J_{-\frac{1}{3}}(u)\right)$; $u=2(3a/y)^{-\frac{3}{2}}$
5.46	$\sin(a/x)$	$-\tfrac{1}{2}(y/a)^{-\frac{1}{2}}\left(\pi Y_1(z)+2K_1(z)\right)$
5.47	$x^{-1}\sin(a/x)$	$\tfrac{1}{2}\pi J_0(z)$
5.48	$x^{-\frac{1}{2}}\sin(a/x)$	$\tfrac{1}{2}(\tfrac{1}{2}\pi/y)^{\frac{1}{2}}\left(\sin(z)+\cos(z)-e^{-z}\right)$
5.49	$x^{-3/2}\sin(a/x)$	$\tfrac{1}{2}(\tfrac{1}{2}\pi/a)^{\frac{1}{2}}\left(\sin(z)+\cos(z)+e^{-z}\right)$
5.50	$x^{-1}\cos(a/x)$	$\tfrac{1}{2}\pi\left(2\pi^{-1}K_0(z)-Y_0(z)\right)$
5.51	$x^{-\frac{1}{2}}\cos(a/x)$	$\tfrac{1}{2}(\tfrac{1}{2}\pi/y)^{\frac{1}{2}}\left(\cos(z)-\sin(z)-e^{-z}\right)$

	$f(x)$	$g_c(y)$
5.52	$x^{-3/2}\cos(a/x)$	$\frac{1}{2}(\frac{1}{2}\pi/a)^{\frac{1}{2}}(\cos(z)-\sin(z)+e^{-z})$
5.53	$x^{\nu-1}\sin(a/x)$ $-1<\text{Re}\nu<2$	$\frac{1}{4}\pi(y/a)^{-\frac{1}{2}\nu}\sec(\frac{1}{2}\pi\nu)\left(J_\nu(z)+J_{-\nu}(z)+I_\nu(z)-I_{-\nu}(z)\right)$
5.54	$x^{\nu-1}\cos(a/x)$ $-1<\text{Re}\nu<1$	$\frac{1}{4}\pi(a/y)^{\frac{1}{2}\nu}\mathrm{cosec}(\frac{1}{2}\pi\nu)\left(J_{-\nu}(z)-J_\nu(z)+I_{-\nu}(z)-I_\nu(z)\right)$ $=\frac{1}{2}\pi(\frac{a}{y})^{\frac{1}{2}\nu}\left(\cos(\frac{1}{2}\pi\nu)\{2\pi^{-1}K_\nu(z)-Y_\nu(z)\}\right.$ $\left.-\sin(\frac{1}{2}\pi\nu)J_\nu(z)\right)$
5.55	$x^{-1}\log(bx)\sin(a/x)$	$\frac{1}{2}\pi\left(J_0(z)\log b+\frac{1}{2}J_0(z)\log(a/y)-K_0(z)\right)$
5.56	$x^{-1}\log(bx)\cos(a/x)$	$\log\left(b(a/y)^{\frac{1}{2}}\right)\left(K_0(z)-\frac{1}{2}\pi Y_0(z)\right)$; $z=2(ay)^{\frac{1}{2}}$
5.57	$x^{-1}\sin(ax^{\frac{1}{2}})$	$\pi\left(S(\frac{1}{4}a^2/y)+C(\frac{1}{4}a^2/y)\right)$
5.58	$x^{-\frac{1}{2}}\sin(ax^{\frac{1}{2}})$	$(2\pi/y)^{\frac{1}{2}}\left(C(z)\sin z-S(z)\cos z\right)$; $z=\frac{1}{4}a^2/y$
5.59	$x^{-\frac{1}{4}}\sin(ax^{\frac{1}{2}})$	$-\frac{1}{2}\pi(\frac{1}{2}a/y)^{3/2}\left(\sin(z-\frac{1}{8}\pi)J_{-\frac{1}{4}}(z)+\cos(z-\frac{1}{8}\pi)J_{\frac{3}{4}}(z)\right)$ $z=\frac{1}{8}a^2/y$
5.60	$x^{-3/4}\sin(ax^{\frac{1}{2}})$	$\pi(\frac{1}{2}a/y)^{\frac{1}{2}}\cos(z+\frac{1}{8}\pi)J_{\frac{1}{4}}(z)$; $z=\frac{1}{8}a^2/y$
5.61	$e^{-bx}\sin(ax^{\frac{1}{2}})$	$\frac{1}{2}\pi^{\frac{1}{2}}a(b^2+y^2)^{-3/4}\exp\left(-\frac{1}{4}a^2b(b^2+y^2)^{-1}\right)$ $\cdot\cos\left(\frac{3}{2}\arctan(y/b)-\frac{1}{4}a^2y(b^2+y^2)^{-1}\right)$
5.62	$x^{-1}e^{-ax^{\frac{1}{2}}}\sin(ax^{\frac{1}{2}})$	$\frac{1}{2}\pi\mathrm{Erf}(z)$; $z=a(2y)^{-\frac{1}{2}}$
5.63	$x^{\nu-1}\sin(ax^{\frac{1}{2}}-\frac{1}{2}\pi\nu)$ $\cdot e^{-ax^{\frac{1}{2}}}$; $\text{Re}\nu>0$	$-(\frac{1}{2}\pi)^{\frac{1}{2}}(2y)^{-\nu}\exp(-\frac{1}{4}a^2/y)D_{2\nu-1}(ay^{-\frac{1}{2}})$
5.64	$x^{\nu-1}\sin(ax^{\frac{1}{2}})$ $-\frac{1}{2}<\text{Re}\nu<1$	$-\frac{1}{2}(\frac{1}{2}\pi)^{\frac{1}{2}}(2y)^{-\nu}\sec(\pi\nu)$ $\cdot\left\{\exp(\frac{1}{4}iz^2-\frac{1}{2}i\pi\nu)\left(D_{2\nu-1}(u)-D_{2\nu-1}(-u)\right)\right.$ $\left.+\exp(-\frac{1}{4}iz^2+\frac{1}{2}i\pi\nu)D_{2\nu-1}(v)-D_{2\nu-1}(-v)\right)\}$ $u=a(2y)^{-\frac{1}{2}}(-i)^{\frac{1}{2}}$, $v=a(2y)^{-\frac{1}{2}}(i)^{\frac{1}{2}}$

$$z=2(ay)^{\frac{1}{2}}$$

	$f(x)$	$g_c(y)$
5.65	$e^{-ax^{\frac{1}{2}}}\cos(ax^{\frac{1}{2}})$	$a\pi^{\frac{1}{2}}(2y)^{-3/2}e^{-\frac{1}{2}a^2/y}$
5.66	$x^{-\frac{1}{2}}\cos(ax^{\frac{1}{2}})$	$(\pi/y)^{\frac{1}{2}}\sin(\tfrac{1}{4}\pi+\tfrac{1}{4}a^2/y)$
5.67	$x^{-\frac{1}{4}}\cos(ax^{\frac{1}{2}})$	$-\tfrac{1}{2}\pi(\tfrac{1}{2}a/y)^{3/2}\left(\sin(z+\tfrac{1}{8}\pi)J_{-\frac{3}{4}}(z)+\cos(z+\tfrac{1}{8}\pi)J_{-\frac{1}{4}}(z)\right)$ $z=\tfrac{1}{8}a^2/y$
5.68	$x^{-3/4}\cos(ax^{\frac{1}{2}})$	$\pi(\tfrac{1}{2}ay)^{\frac{1}{2}}\cos(z-\tfrac{1}{8}\pi)J_{-\frac{1}{4}}(z);\;\; z=\tfrac{1}{8}a^2/y$
5.69	$x^{-\frac{1}{2}}e^{-bx}\cos(ax^{\frac{1}{2}})$	$\pi^{\frac{1}{2}}(b^2+y^2)^{-\frac{1}{4}}\exp\left(-\tfrac{1}{4}a^2b(b^2+y^2)^{-1}\right)$ $\cdot\cos\left(\tfrac{1}{2}\arctan(y/b)-\tfrac{1}{4}a^2y(b^2+y^2)^{-1}\right)$
5.70	$x^{-\frac{1}{2}}e^{-ax^{\frac{1}{2}}}$ $\cdot\cos(ax^{\frac{1}{2}})-\sin(ax^{\frac{1}{2}})$	$(\tfrac{1}{2}\pi/y)^{\frac{1}{2}}e^{-\frac{1}{2}a^2/y}$
5.71	$x^{\nu-1}\cos(ax^{\frac{1}{2}})$ $0<\mathrm{Re}\,\nu<1$	$\tfrac{1}{2}(\tfrac{1}{2}\pi)^{\frac{1}{2}}(2y)^{-\nu}\mathrm{cosec}(\pi\nu)$ $\cdot\Big\{\exp(\tfrac{1}{4}iz^2-\tfrac{1}{2}i\pi\nu)\left(D_{2\nu-1}(u)+D_{2\nu-1}(-u)\right)$ $+\exp(-\tfrac{1}{4}iz^2+\tfrac{1}{2}i\pi\nu)\left(D_{2\nu-1}(v)+D_{2\nu-1}(-v)\right)\Big\}$ $u=a(2y)^{-\frac{1}{2}}(-i)^{\frac{1}{2}},\;\; v=a(2y)^{-\frac{1}{2}}(i)^{\frac{1}{2}}$
5.72	$x(b^2+x^2)^{-1}\tan(ax)$	$\pi\cosh(by)(1+e^{2ab})^{-1},$ (Cauchy principal value)
5.73	$x(b^2+x^2)^{-1}\cot(ax)$	$\pi\cosh(by)(e^{2ab}-1)^{-1},$ (Cauchy principal value)
5.74	$(a^2+x^2)^{-1}\sec(bx)$	$\tfrac{1}{2}\pi a^{-1}\cosh(ay)\mathrm{sech}(ab),$ (Cauchy principal value) $y<b$
5.75	$x/(a^2+x^2)\csc(bx)$	$\tfrac{1}{2}\pi\cosh(ay)\mathrm{csch}(ab),$ (Cauchy principal value), $y<b$
5.76	$x^{-2/3}\sin(ax^{1/3})$ $z=2(a/3)^{3/2}y^{-\frac{1}{2}}$	$\tfrac{1}{2}\pi(a/y)^{\frac{1}{2}}\big(I_{-1/3}(z)+I_{1/3}(z)-J_{-1/3}(z)+J_{1/3}(z)$ $+2iJ_{1/3}(iz)-2iJ_{-1/3}(iz)+2J_{1/3}(z)-2J_{-1/3}(z)\big)$

	$f(x)$	$g_c(y)$
5.77	$x^{-2/3}\cos(ax^{1/3})$	$\frac{1}{2}(3y/a)^{-\frac{1}{2}}\left(3^{\frac{1}{2}}K_{1/3}(z)+\pi J_{1/3}(z)+\pi J_{-1/3}(z)\right),z,\text{same as before}$
5.78	$x^{\nu}\sin(ax^c),\ c<1$ $-c-1<\mathrm{Re}\,\nu<0$	$-y^{-\nu-c-1}\sum_{n=0}^{\infty}(-1)^n\sin\left(\frac{1}{2}\pi\{\nu+(2n+1)c\}\right)\Gamma\{1+\nu+(2n+1)c\}$ $\cdot y^{-2nc}a^{2n+1}/(2n+1)!,\qquad |a|<\infty,\ \ y>0$
	$x^{\nu}\sin(ax^c),\ c>1$ $-c-1<\mathrm{Re}\,\nu<c-1$	$c^{-1}a^{-(1+\nu)/c}\sum_{n=0}^{\infty}(-1^n\sin\{\frac{1}{2}\pi(1+\nu+2n)/c\}\Gamma\{(1+\nu+2n)/c\}$ $\cdot a^{-2n/c}y^{2n}/(2n)!,\qquad a>0\ ,\ |y|<\infty$
5.79	$x^{\nu}\cos(ax^c),\ c<1$ $-1<\mathrm{Re}\,\nu<0$	$-y^{-\nu-1}\sum_{n=0}^{\infty}(-1)^n\sin\{\frac{1}{2}\pi(\nu+2nc)\Gamma(1+\nu+2nc)$ $\cdot y^{-2nc}a^{2n}/(2n)!,\qquad |a|<\infty,\ y>0$
	$x^{\nu}\cos(ax^c),\ c>1$ $-1<\mathrm{Re}\,\nu<c-1$	$c^{-1}a^{-(1+\nu)/c}\sum_{n=0}^{\infty}(-1)^n\cos\{\frac{1}{2}\pi(1+\nu+2n)/c\}\Gamma\{(1+\nu+2n)/c\}$ $\cdot a^{-2n/c}y^{2n}/(2n)!\qquad a>0\quad |y|<\infty$
5.80	$u^{-1}\cos(bu)$	$\cos(ay)K_0(as)\qquad\qquad y<b$ $\frac{1}{2}\pi\left(\sin(ay)J_0(aS)-\cos(ay)Y_0(aS)\right)\qquad y>b$
5.81	$u^{-3/2}\sin(bu)$	$\frac{1}{2}\pi(\frac{1}{2}\pi b)^{\frac{1}{2}}J_{\frac{1}{4}}(\frac{1}{2}ay-\frac{1}{2}aS)$ $\cdot\left(\sin(ay)J_{-\frac{1}{4}}(\frac{1}{2}ay+\frac{1}{2}aS)-\cos(ay)Y_{\frac{1}{4}}(\frac{1}{2}ay+\frac{1}{2}aS)\right)\ y>b$
5.82	$u^{-3/2}\cos(bu)$ $u=(2ax+x^2)^{\frac{1}{2}}$	$\frac{1}{2}\pi(\frac{1}{2}\pi b)^{\frac{1}{2}}J_{-\frac{1}{4}}(\frac{1}{2}ay-\frac{1}{2}aS)\left(\sin(ay)J_{\frac{1}{4}}(\frac{1}{2}ay+\frac{1}{2}aS)\right.$ $\left.-\cos(ay)Y_{\frac{1}{4}}(\frac{1}{2}ay+\frac{1}{2}aS)\right)\qquad y>b$
5.83	$r^{-1}\sin(br)$	$\frac{1}{2}\pi J_0(as)\qquad\qquad y<b$ $0\qquad\qquad y>b$
5.84	$x^{-\frac{1}{2}}r^{-1}\sin(br)$	$(\frac{1}{2}\pi)^{3/2}y^{\frac{1}{2}}J_{-\frac{1}{4}}(\frac{1}{2}ab-\frac{1}{2}as)J_{\frac{1}{4}}(\frac{1}{2}ab+\frac{1}{2}as)$
5.85	$x^{-\frac{1}{2}}r^{-1}\cos(br)$	$-(\frac{1}{2}\pi y)^{\frac{1}{2}}\frac{1}{2}\pi J_{-\frac{1}{4}}(\frac{1}{2}ab-\frac{1}{2}as)Y_{\frac{1}{4}}(\frac{1}{2}ab+\frac{1}{2}as)\qquad y<b$
5.86	$r^{-3/2}\sin(br)$	$\frac{1}{2}(\pi b)^{\frac{1}{2}}I_{\frac{1}{4}}(\frac{1}{2}ay-\frac{1}{2}aS)K_{\frac{1}{4}}(\frac{1}{2}ay+\frac{1}{2}aS)\qquad y>b$
5.87	$r^{-3/2}\cos(br)$	$\frac{1}{2}(\pi b)^{\frac{1}{2}}I_{-\frac{1}{4}}(\frac{1}{2}ay-\frac{1}{2}aS)K_{\frac{1}{4}}ay+\frac{1}{2}aS)\qquad y>b$

$u=(x^2+2ax)^{\frac{1}{2}}$ $s=(b^2-y^2)^{\frac{1}{2}};\quad S=(y^2-b^2)^{\frac{1}{2}}$

	$f(x)$	$g_c(y)$	
5.88	$r^{-1}\cos(br)$	$-\tfrac{1}{2}\pi Y_0(aS)$	$y<b$
		$K_0(aS)$	$y>b$
5.89	$r^{-5/2}\sin(br)$	$b(2\pi b)^{\frac{1}{2}}\left(I_{\frac{3}{4}}(\tfrac{1}{2}ay-\tfrac{1}{2}aS)K_{\frac{3}{4}}(\tfrac{1}{2}ay+\tfrac{1}{2}aS)\right.$	
		$\left.-I_{-\frac{1}{4}}(\tfrac{1}{2}ay-\tfrac{1}{2}aS)K_{\frac{1}{4}}(\tfrac{1}{2}ay+\tfrac{1}{2}aS)\right)$	$y>b$
5.90	$r^{-5/4}\cos(br)$	$b(2\pi b)^{\frac{1}{2}}\left(I_{-\frac{3}{4}}(\tfrac{1}{2}ay-\tfrac{1}{2}aS)K_{\frac{3}{4}}(\tfrac{1}{2}ay+\tfrac{1}{2}aS)\right.$	
		$\left.-I_{\frac{1}{4}}(\tfrac{1}{2}ay=\tfrac{1}{2}aS)K_{\frac{1}{4}}(\tfrac{1}{2}ay+\tfrac{1}{2}aS)\right)$	$y>b$
5.91	$xr^{-1}(r-a)^{-\frac{1}{2}}\sin(br)$	$(\tfrac{1}{2}\pi)^{\frac{1}{2}}s^{-1}(b+s)^{\frac{1}{2}}\cos(as-\tfrac{1}{4}\pi)$	$y<b$
		$-(\tfrac{1}{2}\pi/y)^{\frac{1}{2}}S^{-1}e^{-aS}\sin(\tfrac{1}{2}\arcsin(b/y)$	$y>b$
5.92	$xr^{-1}(r-a)^{-\frac{1}{2}}\cos(br)$	$-(\tfrac{1}{2}\pi)^{\frac{1}{2}}s^{-1}(b+s)^{\frac{1}{2}}\sin(as-\tfrac{1}{4}\pi)$	$y<b$
		$\tfrac{1}{2}(\pi/y)^{\frac{1}{2}}S^{-1}e^{-aS}\cos(\tfrac{1}{2}\arcsin(b/y)$	$y>b$
5.93	$\left((r+x)^{\nu}+(r-x)^{\nu}\right)$ $\cdot r^{-1}\sin(br)$ $-1<\mathrm{Re}\,\nu<1$	$\tfrac{1}{2}\pi a^{\nu}\left(\{(b+y)/(b-y)\}^{\frac{1}{2}\nu}+\{(b-y)/(b+y)\}^{\frac{1}{2}\nu}\right)$ $\cdot\left(\cos(\tfrac{1}{2}\pi\nu)J_{\nu}(as)-\sin(\tfrac{1}{2}\pi\nu)Y_{\nu}(as)\right)$ $-a^{\nu}\sin(\tfrac{1}{2}\pi\nu)\left(\{(y+b)/(y-b)\}^{\frac{1}{2}\nu}-\{(y-b)/(y+b)\}^{\frac{1}{2}\nu}\right)$ $K_{\nu}(aS)$	$y<b$ $y>b$
5.94	$\left((r+x)^{\nu}+(r-x)^{\nu}\right)$ $\cdot r^{-1}\cos(br)$ $-1<\mathrm{Re}\,\nu<1$	$-\tfrac{1}{2}\pi a^{\nu}\left(\{(b+y)/(b-y)\}^{\frac{1}{2}\nu}+\{(b-y)/(b+y)\}^{\frac{1}{2}\nu}\right)$ $\cdot\left(\sin(\tfrac{1}{2}\pi\nu)J_{\nu}(as)+\cos(\tfrac{1}{2}\pi\nu)Y_{\nu}(as)\right)$ $a^{\nu}\cos(\tfrac{1}{2}\pi\nu)\left(\{(y+b)/(y-b)\}^{\frac{1}{2}\nu}+\{(y-b)/(y+b)\}^{\frac{1}{2}\nu}\right)$ $\cdot K_{\nu}(as)$	$y<b$ $y>b$
5.95	$(c^2+x^2)^{-1}r^{-1}\sin(br)$	$\tfrac{1}{2}\pi c^{-1}(a^2-c^2)^{-\frac{1}{2}}e^{-cy}\sin\{b(a^2-c^2)^{\frac{1}{2}}\}$	$y>b$
5.96	$r^{-2}\sin(br)$	$\tfrac{1}{2}a^{-1}\left(e^{-ay}\mathrm{Ei}(ay)-e^{ay}\mathrm{Ei}(-ay)\right)$	
		$-\tfrac{1}{2}\pi\displaystyle\int_{y}^{b}Y_0\{a(t^2-y^2)\}dt$	$y<b$
		$\displaystyle\int_{0}^{b}K_0\{a(y^2-t^2)^{\frac{1}{2}}\}dt$	$y>b$

	$r=(a^2+x^2)^{\frac{1}{2}}$	$s=(b^2-y^2)^{\frac{1}{2}}$; $S=(y^2-b^2)^{\frac{1}{2}}$

	$f(x)$	$g_c(y)$
5.97	$r^{-2}\cos(br)$ $r=(a^2+x^2)^{-\frac{1}{2}}$	$\frac{1}{2}\pi a^{-1}e^{-ay}$ $y>b$ $\frac{1}{2}\pi a^{-1}e^{-ay}-\frac{1}{2}\pi\int_y^b J_0\{a(t^2-y^2)^{\frac{1}{2}}\}dt$ $y<b$
5.98	$\sin(bu)$ $x<a$ 0 $x>a$	$\frac{1}{2}\pi abv^{-1}J_1(av)$
5.99	$u^{-1}\cos(bu)$ $x<a$ 0 $x>a$	$\frac{1}{2}\pi J_0(av)$
5.100	$x^{-\frac{1}{2}}u^{-1}\cos(bu)$ $x<a$ 0 $x>a$	$(\frac{1}{2}\pi)^{3/2}y^{\frac{1}{2}}J_{-\frac{1}{4}}(\frac{1}{2}av-\frac{1}{2}ab)J_{-\frac{1}{4}}(\frac{1}{2}av+\frac{1}{2}ab)$
5.101	$u^{-3/2}\sin(bu)$ $x<a$ 0 $x>a$	$(\frac{1}{2}\pi)^{3/2}b^{\frac{1}{2}}J_{\frac{1}{4}}(\frac{1}{2}av-\frac{1}{2}ay)J_{\frac{1}{4}}(\frac{1}{2}av+\frac{1}{2}ay)$
5.102	$u^{-3/2}\cos(bu)$ $x<a$ 0 $x>a$	$(\frac{1}{2}\pi)^{3/2}b^{\frac{1}{2}}J_{-\frac{1}{4}}(\frac{1}{2}av-\frac{1}{2}ay)J_{-\frac{1}{4}}(\frac{1}{2}av+\frac{1}{2}ay)$
5.103	$w^{-1}\cos(bw)$ $x<2a$ 0 $x>2a$	$\pi\cos(ay)J_0(av)$
5.104	$w^{-3/2}\cos(bw)$ $x<2a$ 0 $x>2a$	$\pi(\frac{1}{2}\pi b)^{\frac{1}{2}}\cos(ay)J_{-\frac{1}{4}}(\frac{1}{2}av-\frac{1}{2}ay)J_{-\frac{1}{4}}(\frac{1}{2}av+\frac{1}{2}ay)$
5.105	$w^{-3/2}\sin(bw)$ $x<2a$ 0 $x>2a$	$\pi(\frac{1}{2}\pi b)^{\frac{1}{2}}\cos(ay)J_{\frac{1}{4}}(\frac{1}{2}av-\frac{1}{2}ay)J_{\frac{1}{4}}(\frac{1}{2}av+\frac{1}{2}ay)$
5.106	$\sin(bw)$ $x<2a$ 0 $x>2a$	$\pi ab\cos(ay)v^{-1}J_1(av)$
5.107	0 $x<a$ $U^{-1}\cos(bU)$ $x>a$	$K_0(as)$ $y<b$ $-\frac{1}{2}\pi Y_0(aS)$ $y>b$
5.108	0 $x<a$ $U^{-3/2}\cos(bU)$ $x>a$	$-(\frac{1}{2}\pi)^{3/2}b^{\frac{1}{2}}J_{-\frac{1}{4}}(\frac{1}{2}ay-\frac{1}{2}aS)Y_{\frac{1}{4}}(\frac{1}{2}ay+\frac{1}{2}aS)$ $y>b$

$u=(a^2-x^2)^{\frac{1}{2}}$; $w=(2ax-x-x^2)^{\frac{1}{2}}$

$U=(x^2-a^2)^{\frac{1}{2}}$

$v=(b^2+y^2)^{\frac{1}{2}}$, $s=(b^2-y^2)^{\frac{1}{2}}$, $S=(y^2-b^2)^{\frac{1}{2}}$

	$f(x)$	$g_c(y)$
5.109	$\begin{array}{ll} 0 & x<a \\ U^{-3/2}\sin(bU) & x>a \end{array}$	$-(\tfrac{1}{2}\pi)^{3/2}b^{\frac{1}{2}}J_{\frac{1}{4}}(\tfrac{1}{2}ay-\tfrac{1}{2}aS)Y_{\frac{1}{4}}(\tfrac{1}{2}ay+\tfrac{1}{2}aS) \qquad y>b$
5.110	$\begin{array}{ll} 0 & x<a \\ x^{-\frac{1}{2}}U^{-1}\cos(bU) & x>a \end{array}$	$\tfrac{1}{2}(\pi y)^{\frac{1}{2}}I_{-\frac{1}{4}}(\tfrac{1}{2}ab-\tfrac{1}{2}as)K_{\frac{1}{4}}(\tfrac{1}{2}ay+\tfrac{1}{2}as) \qquad y<b$
5.111	$\begin{array}{ll} 0 & x<a \\ x(k^2+x^2)^{-1}U^{-1} & \\ \cdot\cos(bU) & x>a \end{array}$	$(\tfrac{1}{2}\pi)^{\frac{1}{2}}(a^2+k^2)^{-\frac{1}{2}}\exp\left(-b(a^2+k^2)^{\frac{1}{2}}\right)\cosh(ky) \qquad y<b$
5.112	$\begin{array}{ll} u^{-1}\sin(bu) & x<a \\ -U^{-1}e^{-bU} & x>a \end{array}$	$\tfrac{1}{2}\pi Y_o(aw)$
5.113	$\begin{array}{ll} u^{-1}e^{-bu} & x<a \\ -U^{-1}\sin(bU) & x>a \end{array}$	$\begin{array}{ll} 0 & y<b \\ \tfrac{1}{2}\pi J_o(aS) & y>b \end{array}$
5.114	$\begin{array}{ll} x^{-\frac{1}{2}}u^{-1}\sin(bu) & x<a \\ -x^{-\frac{1}{2}}U^{-1}e^{-bU} & x>a \end{array}$	$(\tfrac{1}{2}\pi)^{3/2}y^{\frac{1}{2}}J_{-\frac{1}{4}}(\tfrac{1}{2}aw-\tfrac{1}{2}ab)Y_{-\frac{1}{4}}(\tfrac{1}{2}aw+\tfrac{1}{2}ab)$
5.115	$\begin{array}{ll} \exp(a\cos x) & x<\pi \\ 0 & x>\pi \end{array}$	$\pi I_y(a)+\sin(\pi y)\int\limits_{o}^{\infty}\exp(-a\cosh t-yt)dt$
5.116	$\begin{array}{ll} (\sin x)^{-\frac{1}{2}}e^{-2a\sin x}, & \\ & x<\pi \\ 0 & x>\pi \end{array}$	$\pi(\pi a)^{\frac{1}{2}}\cos(\tfrac{1}{2}\pi y)\left(I_{-\frac{1}{4}-\frac{1}{2}y}(a)I_{-\frac{1}{4}+\frac{1}{2}y}(a) \right. $ $\left. -I_{\frac{1}{4}-\frac{1}{2}y}(a)I_{\frac{1}{4}+\frac{1}{2}y}(a)\right)$
5.117	$\begin{array}{ll} (\sin x)^{-\frac{1}{2}}e^{2a\sin x}, & \\ & x<\pi \\ 0 & x>\pi \end{array}$	$\pi(\pi a)^{\frac{1}{2}}\cos(\tfrac{1}{2}\pi y)\left(I_{-\frac{1}{4}-\frac{1}{2}y}(a)I_{-\frac{1}{4}-\frac{1}{2}y}(a) \right. $ $\left. +I_{\frac{1}{4}+\frac{1}{2}y}(a)I_{\frac{1}{4}-\frac{1}{2}y}(a)\right)$
5.118	$\begin{array}{ll} (\cos x)^{-\frac{1}{2}}e^{-2a\cos x}, & \\ & x<\tfrac{1}{2}\pi \\ 0 & x>\tfrac{1}{2}\pi \end{array}$	$\tfrac{1}{2}\pi(\pi a)^{\frac{1}{2}}\left(I_{-\frac{1}{4}-\frac{1}{2}y}(a)I_{-\frac{1}{4}+\frac{1}{2}y}(a) \right. $ $\left. -I_{\frac{1}{4}+\frac{1}{2}y}(a)I_{\frac{1}{4}-\frac{1}{2}y}(a)\right)$
5.119	$\begin{array}{ll} (\cos x)^{-\frac{1}{2}}e^{2a\cos x}, & \\ & x<\tfrac{1}{2}\pi \\ 0 & x>\tfrac{1}{2}\pi \end{array}$	$\tfrac{1}{2}\pi(\pi a)^{\frac{1}{2}}\left(I_{-\frac{1}{4}-\frac{1}{2}y}(a)I_{-\frac{1}{4}+\frac{1}{2}y}(a) \right. $ $\left. +I_{\frac{1}{4}+\frac{1}{2}y}(a)I_{\frac{1}{4}-\frac{1}{2}y}(a)\right)$

$u=(a^2-x^2)^{\frac{1}{2}}; U=(x^2-a^2)^{\frac{1}{2}}$ $\qquad\qquad$ $s=(b^2-y^2)^{\frac{1}{2}}; S=(y^2-b^2)^{\frac{1}{2}}; w=(b^2+y^2)^{\frac{1}{2}}$

	$f(x)$	$g_c(y)$
5.120	$(\cos x)^{-\frac{1}{2}}e^{-a \sec x}$, $\qquad$ $x<\frac{1}{2}\pi$ $\qquad\qquad$ 0 $\qquad$ $x>\frac{1}{2}\pi$	$\pi^{\frac{1}{2}}D_{y-\frac{1}{2}}(z)D_{-y-\frac{1}{2}}(z)$ $\quad;\quad$ $z=(2a)^{\frac{1}{2}}$
5.121	$(\cos x)^{-\frac{1}{2}}$ $\cdot\sin(2a\cos x)$ $\quad x<\frac{1}{2}\pi$ $\qquad\qquad$ 0 $\qquad$ $x>\frac{1}{2}\pi$	$\frac{1}{2}\pi(\pi a)^{\frac{1}{2}}J_{\frac{1}{4}+\frac{1}{2}y}(a)J_{\frac{1}{4}-\frac{1}{2}y}(a)$
5.122	$(\cos x)^{-\frac{1}{2}}$ $\cos(2a\cos x)$ $\quad x<\frac{1}{2}\pi$ $\qquad\qquad$ 0 $\qquad$ $x>\frac{1}{2}\pi$	$\frac{1}{2}\pi(\pi a)^{\frac{1}{2}}J_{-\frac{1}{4}+\frac{1}{2}y}(a)J_{-\frac{1}{4}-\frac{1}{2}y}(a)$
5.123	$(\sin x)^{-\frac{1}{2}}$ $\cdot\sin(2a\sin x)$ $\quad x<\pi$ $\qquad\qquad$ 0 $\qquad$ $x>\pi$	$\pi(a\pi)^{\frac{1}{2}}\cos(\frac{1}{2}\pi y)J_{\frac{1}{4}+\frac{1}{2}y}(a)J_{\frac{1}{4}-\frac{1}{2}y}(a)$
5.124	$(\sin x)^{-\frac{1}{2}}$ $\cos(2a\sin x)$ $\quad x<\pi$ $\qquad\qquad$ 0 $\qquad$ $x>\pi$	$\pi(a\pi)^{\frac{1}{2}}\cos(\frac{1}{2}\pi y)J_{-\frac{1}{4}+\frac{1}{2}y}(a)J_{-\frac{1}{4}-\frac{1}{2}y}(a)$
5.125	$\sin(a\sin x)$ $\quad x<\pi$ $\qquad\qquad$ 0 $\qquad$ $x>\pi$	$\frac{1}{2}\pi\cot(\frac{1}{2}\pi y)\{\mathbf{J}_y(a)-\mathbf{J}_{-y}(a)\}=-\frac{1}{2}\pi\{\mathbf{E}_y(a)+\mathbf{E}_{-y}(a)\}$
5.126	$\cos(a\sin x)$ $\quad x<\pi$ $\qquad\qquad$ 0 $\qquad$ $x>\pi$	$\frac{1}{2}\pi\{\mathbf{J}_y(a)+\mathbf{J}_{-y}(a)\}=\frac{1}{2}\pi\cot(\frac{1}{2}\pi y)\{\mathbf{E}_y(a)-\mathbf{E}_{-y}(a)\}$
5.127	$\sin(a\cos x)$ $\quad x<\frac{1}{2}\pi$ $\qquad\qquad$ 0 $\qquad$ $x<\frac{1}{2}$	$(\frac{1}{2}\pi y)\{\mathbf{J}_y(a)-\mathbf{J}_{-y}(a)=-\frac{1}{4}\pi\sec(\frac{1}{2}\pi y)$ $\cdot\{\mathbf{E}_y(a)+\mathbf{E}_{-y}(a)\}$
5.128	$\cos(a\cos x)$ $\quad x<\frac{1}{2}\pi$ $\qquad\qquad$ 0 $\qquad$ $x>\frac{1}{2}\pi$	$\frac{1}{4}\pi\sec(\frac{1}{2}\pi y)\{\mathbf{J}_y(a)+\mathbf{J}_{-y}(a)\}=\frac{1}{4}\pi\cos\mathrm{ec}(\frac{1}{2}\pi y)$ $\cdot\{\mathbf{E}_y(a)-\mathbf{E}_{-y}(a)\}$
5.129	$(\sin x)^{-3/2}$ $\quad x<\pi$ $\cdot\sin(2a\sin x)$ $\qquad\qquad$ 0 $\qquad$ $x>\pi$	$2(\pi a)^{3/2}\cos(\frac{1}{2}\pi y)\left(J_{-\frac{1}{4}-\frac{1}{2}y}(a)J_{-\frac{1}{4}+\frac{1}{2}y}(a)\right.$ $\left.+J_{3/4-\frac{1}{2}y}(a)J_{3/4+\frac{1}{2}y}(a)\right)$
5.130	$(\cos x)^{-3/2}$ $\quad x<\pi$ $\cdot\sin(2a\cos(\frac{1}{2}x)$ $\qquad\qquad$ 0 $\qquad$ $x>\pi$	$2(\pi a)^{3/2}\{J_{-\frac{1}{4}-y}(a)J_{-\frac{1}{4}+y}(a)$ $+J_{3/4+y}(a)J_{3/4-y}(a)\}$

	f(x)	$g_c(y)$
5.131	$\log\{\sin(\pi x)\}$ $x<1$ 0 $x>1$	$-y^{-1}\sin\, y\left[\gamma+\log 2+\tfrac{1}{2}\Psi(1+\tfrac{1}{2}y/\pi)+\tfrac{1}{2}\Psi(1-\tfrac{1}{2}y/\pi)\right]$
5.132	$\log\{\cos(\tfrac{1}{2}\pi x)\}$ $x<1$ 0 $x>1$	$-y^{-1}\sin\, y\left[\gamma+\log 2+\tfrac{1}{2}\Psi(1+y/\pi)+\tfrac{1}{2}\Psi(1-y/\pi)\right]$
5.133	$\{\sin(\pi x)\}^{\nu-1}$ $x<1$ $\cdot\log\{\sin(\pi x)\}$ 0 $x>1$ Re $\nu>0$	$2^{1-\nu}\Gamma(\nu)\cos(\tfrac{1}{2}y)\left[\Gamma(\tfrac{1}{2}+\tfrac{1}{2}\nu+\tfrac{1}{2}y/\pi)\,\Gamma(\tfrac{1}{2}+\tfrac{1}{2}\nu-\tfrac{1}{2}y/\pi)\right]^{-1}$ $\cdot\left[\Psi(\nu)-\tfrac{1}{2}\Psi(\tfrac{1}{2}+\tfrac{1}{2}\nu+\tfrac{1}{2}y/\pi)-\tfrac{1}{2}\Psi(\tfrac{1}{2}+\tfrac{1}{2}\nu-\tfrac{1}{2}y/\pi)-\log 2\right]$
5.134	$\{\cos(\tfrac{1}{2}\pi x)\}^{\nu-1}$ $x<1$ $\cdot\log\{\cos(\tfrac{1}{2}\pi x)\}$ 0 $x<1$ Re $\nu>0$	$2^{1-\nu}\Gamma(\nu)\left[\Gamma(\tfrac{1}{2}+\tfrac{1}{2}\nu+y/\pi)\,\Gamma(\tfrac{1}{2}+\tfrac{1}{2}\nu-y/\pi)\right]^{-1}$ $\cdot\left[\Psi(\nu)-\log 2-\tfrac{1}{2}\Psi(\tfrac{1}{2}+\tfrac{1}{2}\nu+y/\pi)-\tfrac{1}{2}\Psi(\tfrac{1}{2}+\tfrac{1}{2}\nu-y/\pi)\right]$
5.135	$(\cos x-\cos a)^{-\frac{1}{2}}$ $\log(\cos x-\cos a), x<a$ 0 $,x>a$	$2^{-\frac{1}{2}}\pi\Big(P_{-\frac{1}{2}+y}(\cos a)\{\log(\sin a)-\gamma-\log 4$ $-\Psi(\tfrac{1}{2}-y)\}-Q_{-\frac{1}{2}-y}(\cos a)\Big]$
5.136	$(a^2+x^2)^{-1}$ $\cdot\log(r\cosh u\pm r\cos(bx)$	$\pi a^{-1}\Big\{\{\tfrac{1}{2}u+\tfrac{1}{2}\log(\tfrac{1}{2}r)\}e^{-ay}$ $+\cosh(ay)\log(1\pm e^{-u-ab})$ $+\sum_{n=1}^{m}(\mp 1)^{n}n^{-1}e^{-au}\sinh(ay-abn)\Big\};$ $m<\tfrac{1}{2}y/b<m+1;\ u>0$
5.137	$(a^2+x^2)^{-1}$ $\cdot\log\lvert r\cos(bx)\rvert$	$\tfrac{1}{2}\pi a^{-1}\Big(\cosh(ay)\log(1+e^{-2ab})+\log(\tfrac{1}{2}r)e^{-ay}$ $+\sum_{n=1}^{m}(-1)^{n}n^{-1}\sinh(ay-2abn)\Big)$; $m<\tfrac{1}{2}y/b<m+1$
5.138	$x^{-2}(a^2+x^2)^{-1}$ $\log\lvert\cos(bx)\rvert$	$\tfrac{1}{2}\pi a^{-3}\Big((ay+e^{-ay})\log 2-\cosh(ay)\log(1+e^{-2ab})$ $-ab-\sum_{n=1}^{m}(-1)^{n}n^{-1}\{\sinh(ay-2abn)+\sinh(2abn)$ $-ay\cosh(2abn)\}\Big]$ $m\leq\tfrac{1}{2}y/b<m+1$

1.6   Inverse Trigonometric Functions

	$f(x)$	$g_c(y)$
6.1	arcsin x $\quad$ x<1 0 $\qquad$ x>1	$\frac{1}{2}\pi y^{-1}\{\sin y - \mathbf{H}_o(y)\}$
6.2	arccos x $\quad$ x<1 0 $\qquad$ x>1	$\frac{1}{2}\pi y^{-1}\mathbf{H}_o(y)$
6.3	$(1-x^2)^{-\frac{1}{2}}$ $\cdot\cos(\nu\arccos x)$ x<1 0 $\qquad$ x>1	$-\nu \sin(\frac{1}{2}\pi\nu)s_{-1,\nu}(y)$ $=\frac{1}{4}\pi\sec(\frac{1}{2}\pi\nu)\{\mathbf{J}_\nu(y)+\mathbf{J}_{-\nu}(y)\}$
6.4	$x^{-\frac{1}{2}}(1-x^2)^{-\frac{1}{2}}$ $\cdot\cos(\nu\arccos x)$ x<1 0 $\qquad$ x>1	$\frac{1}{2}\pi(\frac{1}{2}\pi)^{\frac{1}{2}}J_{\frac{1}{2}\nu-\frac{1}{4}}(\frac{1}{2}y)J_{-\frac{1}{2}\nu-\frac{1}{4}}(\frac{1}{2}y)$
6.5	$x^{-1}\arctan(x/a)$	$-\frac{1}{2}\pi\mathrm{Ei}(-ay)$
6.6	$(a^2+x^2)^{-\frac{1}{2}\nu}$ $\cdot\cos(\nu\arctan(x/a)$ Re$\nu>0$	$\frac{1}{2}\pi y^{\nu-1}\left[\Gamma(\nu)\right]^{-1}e^{-ay}$
6.7	$x^\nu(1+x^2)^{\frac{1}{2}\nu}$ $\cdot\sin(\nu\mathrm{arccot}\, x)$ $-1<$Re$\nu<0$	$\frac{1}{2}\pi^{\frac{1}{2}}\Gamma(1+\nu)\left(I_{-\frac{1}{2}-\nu}(\frac{1}{2}y)\sinh(\frac{1}{2}y)\right.$ $\left.-I_{\frac{1}{2}+\nu}(\frac{1}{2}y)\cosh(\frac{1}{2}y)\right)y^{-\nu-\frac{1}{2}}$
6.8	$x^\nu(1+x^2)^{\frac{1}{2}\nu}$ $\cos(\nu\mathrm{arccot}\, x)$ $-1<$Re$\nu<0$	$-\pi^{-\frac{1}{2}}\Gamma(1+\nu)y^{-\nu-\frac{1}{2}}\sin(\pi\nu)\cosh(\frac{1}{2}y)K_{\frac{1}{2}+\nu}(\frac{1}{2}y)$
6.9	$\arctan(a/x)$	$\frac{1}{2}y^{-1}\left(e^{-ay}\overline{\mathrm{Ei}}(ay)-e^{ay}\mathrm{Ei}(-ay)\right)$
6.10	$\arctan(a/x^2)$	$\pi y^{-1}e^{-yz}\sin(yz), \qquad z=(\frac{1}{2}a)^{\frac{1}{2}}$
6.11	$\arctan\{(a/x)^n\}$	$-\frac{1}{2}\pi y^{-1}\sum_{m=1}^{n}(-1)^m\exp\left(-ay\,\sin\{(m-\frac{1}{2})\pi/n\}\right)$ $\cdot\sin\left(ay\,\cos\{(m-\frac{1}{2})\pi/n\}\right)$

	$f(x)$	$g_c(y)$
6.12	$\arctan(ax^{-\frac{1}{2}}), z=a^2y$	$\frac{1}{2}\pi y^{-1}$ $\cos z$ $C(z)-S(z)$ $-\sin z$ $1-C(z)-S(z)$
6.13	$x^{-1}\arcsin x \qquad x<1$ $0 \qquad\qquad x>1$	$\frac{1}{2}\pi\left(\mathrm{Ci}(y) + \int_y^\infty t^{-1}J_o(t)\,dt\right)$
6.14	$(a^2-x^2)^{-\frac{1}{2}}$ $\cdot\log\left(\frac{1}{x}\{a+(a^2-x^2)^{\frac{1}{2}}\}\right) \qquad x<a$ $(x^2-a^2)^{-\frac{1}{2}}$ $\cdot\arctan\left(\frac{1}{a}(x^2-a^2)^{\frac{1}{2}}\right) \qquad x>a$	$\frac{1}{4}\pi^2\left(\mathbf{H}_o(ay)-Y_o(ay)\right)$

1.7 Hyperbolic Functions

	$f(x)$	$g_c(y)$
7.1	$\mathrm{sech}(ax)$	$\frac{1}{2}\pi a^{-1}\mathrm{sech}(\frac{1}{2}\pi y/a)$
7.2	$\left(\mathrm{sech}(ax)\right)^2$	$\frac{1}{2}\pi a^{-2}y\,\mathrm{cosech}(\frac{1}{2}\pi y/a)$
7.3	$\left(\mathrm{sech}(ax)\right)^{2n}$ $n=2,3,\ldots$	$\frac{2^{2n-3}\pi y a^{-2}}{(2n-1)!}\mathrm{cosech}(\frac{1}{2}\pi y/a)\prod_{m=1}^{n-1}\left((\frac{1}{2}y/a)^2+m^2\right)$
7.4	$\left(\mathrm{sech}(ax)\right)^{2n+1}$ $n=1,2,3\ldots$	$\frac{2^{2n-1}\pi a^{-1}}{(2n)!}\,\mathrm{sech}(\frac{1}{2}\pi y/a)\prod_{m=1}^{n}\left((\frac{1}{2}y/a)^2+(m-\frac{1}{2})^2\right)$
7.5	$\left(\mathrm{sech}(ax)\right)^\nu$ $\mathrm{Re}\,\nu>0$	$2^{\nu-2}a^{-1}\left(\Gamma(\nu)\right)^{-1}\Gamma(\frac{1}{2}\nu+i\frac{1}{2}y/a)\,\Gamma(\frac{1}{2}\nu-i\frac{1}{2}y/a)$
7.6	$\left(\mathrm{cosech}(ax)\right)^\nu$ $0<\mathrm{Re}\,\nu<1$	$2^\nu\pi a^{-1}\sin(\frac{1}{2}\pi\nu)\,\Gamma(1-\nu)\cosh(\frac{1}{2}\pi y/a)A^{-1}$ $A = \left(\cosh(\pi y/a)\right.$ $\left.-\cos(\pi\nu)\right)\Gamma(1-\frac{1}{2}\nu+\frac{1}{2}iy/a)\,\Gamma(1-\frac{1}{2}\nu-\frac{1}{2}iy/a)$
7.7	$\left(\cosh(ax)+\cos b\right)^{-1}$	$\pi a^{-1}\mathrm{cosec}\,b\,\mathrm{cosech}(\pi y/a)\sinh(by/a) \qquad b<\pi$
7.8	$\left(\cosh(ax)+\cosh b\right)^{-1}$	$a^{-1}\mathrm{cosech}\,b\,\mathrm{cosech}(\pi y/a)\sin(by/a)$

	$f(x)$	$g_c(y)$
7.9	$\left(\cosh(ax)-\cosh b\right)^{-1}$	$-\pi a^1 \text{cosech}\, b\, \cosh(\pi y/a)\sin(by/a)$ (Cauchy principal value)
7.10	$\left(\cosh(ax)+\cos b\right)^{-\frac{1}{2}}$	$2^{-\frac{1}{2}}\pi a^{-1}\text{sech}(\pi y/a)P_{-\frac{1}{2}+iy/a}(\cos b)$
7.11	$\left(\cosh(ax)+\cosh b\right)^{-\frac{1}{2}}$	$2^{-\frac{1}{2}}\pi a^{-1}\text{sech}(\pi y/a)P_{-\frac{1}{2}+iy/a}(\cosh b)$
7.12	$\left(\cosh(ax-\cos b\right)^{-\frac{1}{2}}$ $0<b<\pi$	$2^{-\frac{1}{2}}a^{-1}\left[Q_{-\frac{1}{2}+iy/a}(\cos b)+Q_{-\frac{1}{2}-iy/a}(\cos b)\right]$
7.13	$(a+\cosh x)^{\frac{1}{2}}$ $-(b+\cosh x)^{\frac{1}{2}};a,b>-1$	$\frac{1}{4}i2^{-\frac{1}{2}}\pi y^{-1}\text{sech}(\pi y)\left[P_{\frac{1}{2}+iy}(b)-P_{\frac{1}{2}-iy}(b)\right.$ $\left. -P_{\frac{1}{2}+iy}(a)+P_{\frac{1}{2}-iy}(a)\right]$
7.14	$(1+2\cosh s)^{-1}$ $s=x(2\pi/3)^{\frac{1}{2}}$	$(\tfrac{1}{2}\pi)^{\frac{1}{2}}(1+\cosh t)^{-1}\;;\quad t=y(2\pi/3)^{\frac{1}{2}}$
7.15	$(\cosh a+\cosh x)^{-\nu}$ $\text{Re}\,\nu>0$	$(\tfrac{1}{2}\pi)^{\frac{1}{2}}\left[\Gamma(\nu)\right]^{-1}(\sinh a)^{\frac{1}{2}-\nu}\Gamma(\nu+iy)\,\Gamma(\nu-iy)$ $\cdot\,P_{-\frac{1}{2}+iy}^{\frac{1}{2}-\nu}(\cosh a)$
7.16	$(\cos a+\cosh x)^{-\nu}$ $0<a<\pi,\quad \text{Re}\,\nu>0$	$(\tfrac{1}{2}\pi)^{\frac{1}{2}}\left[\Gamma(\nu)\right]^{-1}(\sin a)^{\frac{1}{2}-\nu}\Gamma(\nu+iy)\,\Gamma(\nu-iy)$ $\cdot P_{-\frac{1}{2}+iy}^{\frac{1}{2}-\nu}(\cos a)$
7.17	$(\cosh a-\cosh x)^{-\nu}$ $\qquad\qquad ,x<a$ $\qquad 0\qquad ,x>a$ $\text{Re}\,\nu<1$	$(\tfrac{1}{2}\pi)^{\frac{1}{2}}\Gamma(1-\nu)\sinh a)^{\frac{1}{2}-\nu}p_{-\frac{1}{2}+iy}^{-\frac{1}{2}+\nu}(\cosh a)$
7.18	$\qquad 0\qquad x<a$ $(\cosh x-\cosh a)^{\nu-\frac{1}{2}}$ $\qquad\qquad ,x>a$ $-\frac{1}{2}<\text{Re}\,\nu<\frac{1}{2}$	$(2\pi)^{-\frac{1}{2}}\Gamma(\tfrac{1}{2}+\nu)(\sinh a)^{\nu}e^{i\pi\nu}$ $\cdot\left(q_{-\frac{1}{2}-iy}^{-\nu}(\cosh a)+q_{-\frac{1}{2}+iy}^{-\nu}(\cosh a)\right]$
7.19	$\dfrac{\cosh(ax)}{\cosh(\pi bx)},a<\pi b$	$b^{-1}\dfrac{\cos(\frac{1}{2}a/b)\cosh(\frac{1}{2}y/b)}{\cos(a/b)+\cosh(y/b)}$

	$f(x)$	$g_c(y)$
7.20	$\dfrac{\sinh(ax)}{\sinh(\pi bx)}$, $a<\pi b$	$\frac{1}{2}b^{-1}\dfrac{\sin(a/b)}{\cos(a/b)+\cosh(y/b)}$
7.21	$x\,\mathrm{cosech}(\pi ax)$	$\frac{1}{4}a^{-2}\left(\mathrm{sech}(\frac{1}{2}y/a)\right)^2$
7.22	$x^{-1}(\mathrm{cosech}\,x-x^{-1})$	$-\log(1+e^{-\pi y})$
7.23	$x^{-1}(1-\mathrm{sech}\,x)$	$\log\left[\Gamma(\frac{3}{4}+\frac{1}{2}iy)\,\Gamma(\frac{3}{4}-\frac{1}{2}iy)\right]$ $-\log\left[\Gamma(\frac{1}{4}+\frac{1}{2}iy)\,\Gamma(\frac{1}{4}-\frac{1}{2}iy)\right]-\log(\frac{1}{4}y)$
7.24	$\dfrac{\sinh(ax)}{\cosh(bx)}$ $\quad a<b$	$\frac{1}{2}\pi b^{-1}\sin(\pi a/b)\left[\cos(\pi a/b)+\cosh(\pi y/b)\right]$ $+\frac{1}{4}b^{-1}\left[\Psi(\frac{3}{4}-\frac{1}{4}\frac{a}{b}+\frac{1}{4}i\frac{y}{b})+\Psi(\frac{3}{4}-\frac{1}{4}\frac{a}{b}-\frac{1}{4}i\frac{y}{b})\right.$ $\left.-\Psi(\frac{3}{4}+\frac{1}{4}\frac{a}{b}-\frac{1}{4}i\frac{y}{b})-\Psi(\frac{3}{4}+\frac{1}{4}\frac{a}{b}+\frac{1}{4}i\frac{y}{b})\right]$
7.25	$x^{-1}\dfrac{\sinh(ax)}{\cosh(\pi bx)}$, $a\leq\pi b$	$\frac{1}{2}\log\left[\dfrac{\cosh(\frac{1}{2}y/b)+\sin(\frac{1}{2}a/b)}{\cosh(\frac{1}{2}y/b)-\sin(\frac{1}{2}a/b)}\right]$
7.26	$x^{-1}\dfrac{\sinh^2(ax)}{\sinh(\pi bx)}$, $a\leq\frac{1}{2}\pi b$	$\frac{1}{4}\log\left[\dfrac{1+\cosh(y/b)}{\cos(2a/b)+\cosh(y/b)}\right]$
7.27	$x^{\nu-1}(\mathrm{cosech}\,x-x^{-1})$ $-1<\mathrm{Re}\,\nu<2$	$2^{-\nu}\Gamma(\nu)\left[\zeta(\nu,\frac{1}{2}+\frac{1}{2}iy)+\zeta(\nu,\frac{1}{2}-\frac{1}{2}iy)\right]$ $-\sin(\frac{1}{2}\pi\nu)\,\Gamma(\nu-1)\,y^{1-\nu}$
7.28	$\dfrac{\cosh(\frac{1}{2}ax)}{\cosh b+\cosh(ax)}$	$\frac{1}{2}\pi a^{-1}\cos(by/a)\,\mathrm{sech}(\frac{1}{2}b)\,\mathrm{sech}(\pi y/a)$
7.29	$\dfrac{\cosh(\frac{1}{2}ax)}{\cos b+\cosh(ax)}$ $0<b<\pi$	$\frac{1}{2}\pi a^{-1}\cosh(by/a)\,\sec(\frac{1}{2}b)\,\mathrm{sech}(\pi y/a)$
7.30	$\dfrac{\cosh(ax)-\cosh(bx)}{x\sinh(\pi cx)}$ $a,b<\pi c$	$\frac{1}{2}\log\left[\dfrac{\cos(b/c)+\cosh(y/c)}{\cos(a/c)+\cosh(y/c)}\right]$
7.31	$(a^2+x^2)\,\mathrm{sech}(\frac{1}{2}\pi x/a)$	$2a^3\left(\mathrm{sech}(ay)\right)^3$
7.32	$x(a^2+x^2)\,\mathrm{cosech}(\pi x/a)$	$\frac{3}{8}a^4\left(\mathrm{sech}(\frac{1}{2}ay)\right)^4$

	$f(x)$	$g_c(y)$
7.33	$(1+x^2)^{-1}\text{sech}(\pi x)$	$2\cosh(\tfrac{1}{2}y)-e^y\arctan(e^{-\tfrac{1}{2}y})-e^{-y}\arctan(e^{\tfrac{1}{2}y})$
7.34	$(1+x^2)^{-1}\text{sech}(\tfrac{1}{2}\pi x)$	$ye^{-y}+\cosh y \log(1+e^{-2y})$
7.35	$x^{-1}e^{-ax}\sinh(bx)$ $a\leq b$	$\tfrac{1}{4}\log\left(\dfrac{y^2+(a+b)^2}{y^2+(a-b)^2}\right)$
7.36	$1-\tanh(ax)$	$\tfrac{1}{4}a^{-1}\Big(\Psi(\tfrac{1}{4}iy/a)+\Psi(-\tfrac{1}{4}iy/a)$ $\qquad -\Psi(\tfrac{1}{2}+\tfrac{1}{4}iy/a)-\Psi(\tfrac{1}{2}-\tfrac{1}{4}iy/a)\Big)$
7.37	$x^{-1}\tanh(ax)$	$\log\{\coth(\tfrac{1}{4}\pi y/a)\}$
7.38	$x(1+x^2)^{-1}\tanh(\tfrac{1}{2}\pi x)$	$-ye^{-y}-\cosh y \log(1-e^{-2y})$
7.39	$x(1+x^2)^{-1}\tanh(\tfrac{1}{4}\pi x)$	$\cosh y \log\coth(\tfrac{1}{2}y)-\tfrac{1}{2}\pi e^{-y}$ $\qquad\qquad -2\sin y \arctan(e^{-y})$
7.40	$(1+e^{\pi bx})^{-1}\sinh(ax)$ $a<b\pi$	$-\tfrac{1}{2}a(a^2+y^2)^{-1}+b^{-1}\dfrac{\sin(a/b)\cos(y/b)}{\cosh(2y/b)-\cos(2a/b)}$
7.41	$(e^{\pi bx}-1)^{-1}\sinh(ax)$ $a<b\pi$	$\tfrac{1}{2}a(a^2+y^2)^{-1}+\tfrac{1}{2}b^{-1}\dfrac{\sin(2a/b)}{\cosh(2y/b)-\cos(2a/b)}$
7.42	$ur^{-2}\text{cosech}(bu)$	$\tfrac{1}{2}\pi c^{-1}e^{-cy}v\,\text{cosec}(bv)$ $-\pi b^{-1}\sum\limits_{n=1}^{\infty}(-1)^n c_n^2(v^2-c_n^2)^{-1}v_n^{-1}e^{-yv_n}; c_n=n\pi/b$
7.43	$r^{-2}\text{sech}(bu)$	$\tfrac{1}{2}\pi c^{-1}e^{-cy}\sec(bv)+\pi b^{-1}\sum\limits_{n=0}^{\infty}(-1)^n c_n(v^2-c_n^2)^{-1}$ $\quad\cdot v_n^{-1}e^{-yv_n}; c_n=(n+\tfrac{1}{2})\pi/b$
7.44	$r^{-2}\dfrac{\sinh(au)}{\sinh(bu)}$; $a\leq b$	$\tfrac{1}{2}\pi c^{-1}e^{-cy}\sin(av)\text{cosec}(bv)-\pi b^{-1}\sum\limits_{n=1}^{\infty}(-1)^n c_n$ $\quad\cdot\sin(ac_n)(v^2-c_n^2)^{-1}v_n^{-1}e^{-yv_n}$; $c_n=n\pi/b$

$r=(c^2+x^2)^{\tfrac{1}{2}}; u=(k^2+x^2)^{\tfrac{1}{2}}$ $\qquad$ $v=(c^2-k^2)^{\tfrac{1}{2}}; v_n=(k^2+c_n^2)^{\tfrac{1}{2}}$

	$f(x)$	$g_c(y)$
7.45	$r^{-2}\dfrac{\cosh(au)}{\cosh(bu)}$; $\underline{a \leq b}$	$\tfrac{1}{2}\pi c^{-1}e^{-cy}\cos(av)\sec(bv)+\pi b^{-1}\sum\limits_{n=0}^{\infty}(-1)^{n}c_{n}$ $\cdot\cos(ac_{n})(v^{2}-c_{n}^{2})^{-1}v_{n}^{-1}\,e^{-yv_{n}}$; $c_{n}=(n+\tfrac{1}{2})\pi/b$
7.46	$r^{-2}\dfrac{\sinh(ax)}{\sinh(bx)}$; $\underline{a \leq b}$	$\tfrac{1}{2}\pi c^{-1}e^{-cy}\sin(ac)\operatorname{cosec}(bc)-b^{-1}\sum\limits_{n=1}^{\infty}(-1)^{n}$ $\cdot\sin(ac_{n})(c^{2}-c_{n}^{2})^{-1}e^{-yc_{n}}$; $c_{n}=n\pi/b$
7.47	$r^{-2}\dfrac{\cosh(ax)}{\cosh(bx)}$; $\underline{a \leq b}$	$\tfrac{1}{2}\pi c^{-1}e^{-cy}\cos(ac)\sec(bc)+\pi b^{-1}\sum\limits_{n=0}^{\infty}(-1)^{n}$ $\cdot\cos(ac_{n})(c^{2}-c_{n}^{2})^{-1}e^{-yc_{n}}$; $c_{n}=(n+\tfrac{1}{2})\pi/b$
7.48	$\sinh(au)\operatorname{cosech}(bu)$ $a<b$	$-\pi b^{-1}\sum\limits_{n=1}^{\infty}(-1)^{n}c_{n}\sin(ac_{n})v_{n}^{-1}e^{-yv_{n}}$ $c_{n}=n\pi/b$
7.49	$\cosh(au)\operatorname{sech}(bu)$ $\underline{a \leq b}$	$\pi b^{-1}\sum\limits_{n=0}^{\infty}(-1)^{n}c_{n}\cos(ac_{n})v_{n}^{-1}e^{-yv_{n}}$ $c_{n}=(n+\tfrac{1}{2})\pi/b$
7.50	$u^{-1}\sinh(au)\operatorname{sech}(bu)$ $\underline{a \leq b}$	$\pi b^{-1}\sum\limits_{n=0}^{\infty}(-1)^{n}\sin(ac_{n})v_{n}^{-1}e^{-yv_{n}}$ $c_{n}=(n+\tfrac{1}{2})\pi/b$
7.51	$\dfrac{1}{u}\cosh(au)\operatorname{cosech}(bu)$ $\underline{a \leq b}$	$\tfrac{1}{2}\pi b^{-1}\sum\limits_{n=0}^{\infty}(-1)^{n}\,\varepsilon_{n}v_{n}^{-1}e^{-yv_{n}}\cos(ac_{n})$ $c_{n}=n\pi/b$
7.52	$u^{-2}\sinh(au)\operatorname{cosech}(bu)$ $\underline{a \leq b}$	$\tfrac{1}{2}\pi a(bk)^{-1}e^{-yk}+\pi b^{-1}\sum\limits_{n=1}^{\infty}(-1)^{n}c_{n}^{-1}\sin(ac_{n})v_{n}^{-1}$ $c_{n}=n\pi/b \qquad \cdot e^{-yv_{n}}$
7.53	$u^{-2}\cosh(au)\operatorname{sech}(bu)$ $\underline{a \leq b}$	$\tfrac{1}{2}\pi k^{-1}e^{-yk}-\pi b^{-1}\sum\limits_{n=0}^{\infty}(-1)^{n}c_{n}^{-1}\cos(ac_{n})v_{n}^{-1}e^{-yv_{n}}$ $c_{n}=(n+1)\pi/b$

$$u=(k^{2}+x^{2})\ ,\quad r=(c^{2}+x^{2})^{\frac{1}{2}}\qquad v_{n}=(k^{2}+c_{n}^{2})^{\frac{1}{2}}\qquad v=(c^{2}-k^{2})^{\frac{1}{2}}$$

	$f(x)$	$g_c(y)$
7.54	$x^{-\frac{1}{2}}\mathrm{sech}(ax^{\frac{1}{2}})$	L.G.Mordell;Acta Mathematica,Vol.61,323-360
7.55	$\mathrm{sech}^2(ax^{\frac{1}{2}})$	as before
7.56	$e^{-bx}\sinh(ax^{\frac{1}{2}})$	$\frac{1}{2}a\pi^{\frac{1}{2}}(b^2+y^2)^{-3/2}\exp\left\{\frac{1}{4}a^2b(b^2+y^2)^{-1}\right\}$ $\cos\left\{\frac{3}{2}\arctan(y/b)-\frac{1}{4}a^2y(b^2+y^2)^{-1}\right\}$
7.57	$u^{-1}\cosh(bu) \quad x<a$ $\qquad 0 \qquad\quad x>a$	$\frac{1}{2}\pi J_0(av) \qquad y<b$ $\frac{1}{2}\pi J_0(aV) \qquad y>b$
7.58	$\sinh(bu) \qquad x<a$ $\qquad 0 \qquad\quad x>a$	$\frac{1}{2}\pi abv^{-1}I_1(av) \qquad y<b$ $\frac{1}{2}\pi abv^{-1}J_1(aV) \qquad y>b$
7.59	$u^{3/2}\sinh(bu) \quad x<a$ $\qquad 0 \qquad\qquad x>a$	$\frac{1}{2}\pi(\frac{1}{2}\pi b)^{\frac{1}{2}}J_{\frac{1}{4}}(\frac{1}{2}ay+\frac{1}{2}av)J_{\frac{1}{4}}(\frac{1}{2}ay-\frac{1}{2}av)$
7.60	$u^{-3/2}\cosh(bu) \quad x<a$ $\qquad 0 \qquad\qquad x>a$	$\frac{1}{2}\pi(\frac{1}{2}\pi b)^{\frac{1}{2}}J_{-\frac{1}{4}}(\frac{1}{2}ay+\frac{1}{2}av)J_{-\frac{1}{4}}(\frac{1}{2}ay-\frac{1}{2}av)$
7.61	$u^{-1}\sinh(bu) \quad x<a$ $U^{-1}\sin(bU) \quad x>a$	$\frac{1}{2}\pi I_0(av) \qquad y<b$ $\qquad 0 \qquad\qquad y>b$
7.62	$x^{-\frac{1}{2}}u^{-1}\sinh(bu) \quad x<a$ $x^{-\frac{1}{2}}U^{-1}\sin(bU) \quad x>a$	$\frac{1}{2}\pi(\frac{1}{2}\pi y)^{\frac{1}{2}}I_{-\frac{1}{4}}(\frac{1}{2}ab-\frac{1}{2}av)$ $\left\{I_{-\frac{1}{4}}(\frac{1}{2}ab-\frac{1}{2}av)-2^{\frac{1}{2}}\pi^{-1}K_{\frac{1}{4}}(\frac{1}{2}ab+\frac{1}{2}av)\right\} \qquad y<b$
7.63	$x^{-\frac{1}{2}}u^{-1}\cosh(bu) \quad x<a$ $\qquad 0 \qquad\qquad x>a$	$\frac{1}{2}\pi(\frac{1}{2}\pi y)^{\frac{1}{2}}I_{-\frac{1}{4}}\frac{1}{2}ab+\frac{1}{2}av)I_{-\frac{1}{4}}(\frac{1}{2}ab+\frac{1}{2}av)$
7.64	$\qquad 0 \qquad\qquad x>2a$ $(2ax-x^2)^{-3/4}$ $\cdot\cosh\{b(2ax-x^2)^{\frac{1}{2}}\}$ $\qquad\qquad x<2a$	$\pi(\frac{1}{2}\pi b)^{\frac{1}{2}}\cos(ay)J_{-\frac{1}{4}}(\frac{1}{2}ay+\frac{1}{2}aV)J_{-\frac{1}{4}}(\frac{1}{2}ay-\frac{1}{2}aV)$

$u=(a^2-x^2)^{\frac{1}{2}}, U=(x^2-a^2)^{\frac{1}{2}}$ $\qquad\qquad$ $v=(b^2-y^2)^{\frac{1}{2}}, V=(y^2-b^2)^{\frac{1}{2}}$

	$f(x)$	$g_c(y)$
7.65	$(2ax-x^2)^{-3/4}$ $\sinh\{b(2ax-x^2)^{\frac12}\}$ $\quad x<2a$ $\qquad 0 \qquad x>2a$	$\pi(\tfrac12\pi b)^{\frac12}\cos(ay)J_{\frac14}(\tfrac12 ay+\tfrac12 az)J_{\frac14}(\tfrac12 ay-\tfrac12 az)$ $z=(y^2-b^2)^{\frac12}$
7.66	$u^{-1}\sinh(bu) \quad x<a$ $v^{-1}\sin(bv) \quad x>a$ $u=(a^2-x^2)^{\frac12}$ $v=(x^2-b^2)^{\frac12}$	$\tfrac12\pi I_0\{a(b^2-y^2)^{\frac12}\} \qquad\qquad y<b$ $0 \qquad\qquad\qquad\qquad\qquad y>b$
7.67	$e^{-ax}\left(\sinh(bx)\right)^{\nu}$ $\mathrm{Re}\nu>-1, b<\mathrm{Re}\nu<a$	$2^{-\nu-2}b^{-1}\Gamma(1+\nu)\left\{\dfrac{\Gamma\{\tfrac12(a-\nu b+iy)/b\}}{\Gamma\{1+\tfrac12(a+\nu b+iy)/b\}}\right.$ $\left.+\dfrac{\Gamma\{\tfrac12(a-\nu b-iy)/b\}}{\Gamma\{1+\tfrac12(a+\nu b-iy)/b\}}\right\}$
7.68	$e^{-bx^2}\cosh(ax)$	$\tfrac12\pi^{\frac12}b^{-1}\exp\{\tfrac14(a^2-y^2)/b\}\cos(\tfrac12 ay/b)$
7.69	$x^{-2}\sinh(x^2)$	$(\tfrac12\pi)^{\frac12}e^{-x^{2/8}}-\tfrac14\pi y\mathrm{Erfc}(y2^{-3/2})$
7.70	$e^{-ax}\coth(bx^{\frac12})$	Mordell,L.J.:Mess.Math.Vol.49,65-72,(1920)
7.71	$e^{-ax}\tanh(bx^{\frac12})$	as before
7.72	$\sin(x^2/\pi)\,\mathrm{sech}\,x$	$\tfrac12\pi\,\mathrm{sech}(\tfrac12\pi y)\{\cos(\tfrac14\pi y^2)-2^{-\frac12}\}$
7.73	$\cos(x^2/\pi)\,\mathrm{sech}(x)$	$\tfrac12\pi\,\mathrm{sech}(\tfrac12\pi y)\{\sin(\tfrac14\pi y^2)+2^{-\frac12}\}$
7.74	$\sin(ax^2)$	Erdelyi, A. et.al.: Tables of Integral
7.75	$\cos(ax^2)$ $\mathrm{sech}(bx$	Transforms Vol.1,p.10, New York, 1953
7.76	$\exp(-a\cosh x)$	$K_{iy}(a)$
7.77	$\exp(-a\sinh x)$	$S_{0,iy}(a)=\tfrac12\pi i\,\mathrm{cosech}(\pi y)\left[J_{iy}(a)-J_{-iy}(a)\right.$ $\left.+\ \mathbf{J}_{-iy}(a)-\mathbf{J}_{iy}(a)\right]$

	$f(x)$	$g_c(y)$
7.78	$\exp(-a\cosh x)$ $\cdot\cos(b\sinh x)$	$\cosh\!\left(y\arctan(b/a)\right)K_{iy}\!\left((a^2+b^2)^{\frac{1}{2}}\right)$
7.79	$(\cosh x)^{-\frac{1}{2}}$ $\cdot\exp(-2a\cosh x)$	$(a/\pi)^{\frac{1}{2}}K_{\frac{1}{4}+\frac{1}{2}iy}(a)\,K_{\frac{1}{4}-\frac{1}{2}iy}(a)$
7.80	$(\sinh x)^{-\frac{1}{2}}$ $\cdot\exp(-2a\sinh x)$	$\frac{1}{4}\pi(a\pi)^{\frac{1}{2}}\Big(J_{\frac{1}{4}-\frac{1}{2}iy}(a)Y_{-\frac{1}{4}-\frac{1}{2}iy}(a)$ $-J_{-\frac{1}{4}-\frac{1}{2}iy}(a)Y_{\frac{1}{4}-\frac{1}{2}iy}(a)+J_{\frac{1}{4}+\frac{1}{2}iy}(a)Y_{-\frac{1}{4}+\frac{1}{2}iy}(a)$ $-J_{\frac{1}{4}+\frac{1}{2}iy}(a)Y_{\frac{1}{4}+\frac{1}{2}iy}(a)\Big)$
7.81	$(\sinh x)^{-\frac{1}{2}}$ $\cdot\exp(-a\,\mathrm{cosech}\,x)$	$2^{-\frac{1}{2}}\Big(\Gamma(\tfrac{1}{2}+iy)D_{-\frac{1}{2}-iy}(u)D_{-\frac{1}{2}-iy}(v)$ $+\Gamma(\tfrac{1}{2}-iy)D_{-\frac{1}{2}+iy}(u)D_{-\frac{1}{2}+iy}(v)\Big);\ \begin{smallmatrix}u\\v\end{smallmatrix}\ (\pm 2ia)^{\frac{1}{2}}$
7.82	$(\sin x)^{-3/2}$ $\cdot\sinh(2a\sin x)$ $\qquad,\ x<\pi$ $\qquad 0\qquad,\ x>\pi$	$2(\pi a)^{3/2}\cos(\tfrac{1}{2}\pi y)\Big(I_{-\frac{1}{4}-\frac{1}{2}y}(a)I_{-\frac{1}{4}+\frac{1}{2}y}(a)$ $-I_{3/4+\frac{1}{2}y}(a)I_{3/4-\frac{1}{2}y}(a)\Big)$
7.83	$(\cos x)^{-3/2}$ $\cdot\sinh(2a\cos x)$ $\qquad,\ x<\tfrac{1}{2}\pi$ $\qquad 0\qquad x>\tfrac{1}{2}\pi$	$(\pi a)^{3/2}\Big(I_{\frac{1}{4}-\frac{1}{2}y}(a)I_{-\frac{1}{4}+\frac{1}{2}y}(a)$ $-I_{3/4-\frac{1}{2}y}(a)I_{3/4+\frac{1}{2}y}(a)\Big)$
7.84	$(\sin x)^{-\frac{1}{2}}$ $\cdot\sinh(2a\sin x),\ x<\pi$ $\qquad 0\qquad,\ x>\pi$	$\pi(a\pi)^{\frac{1}{2}}\cos(\tfrac{1}{2}\pi y)I_{\frac{1}{4}-\frac{1}{2}y}(a)I_{\frac{1}{4}+\frac{1}{2}y}(a)$
7.85	$(\sin x)^{-\frac{1}{2}}$ $\cdot\cosh(2a\sin x),\ x<\pi$ $\qquad 0\qquad,\ x>\pi$	$\pi(a\pi)^{\frac{1}{2}}\cos(\tfrac{1}{2}\pi y)I_{-\frac{1}{4}+\frac{1}{2}y}(a)I_{-\frac{1}{4}-\frac{1}{2}y}(a)$
7.86	$(\cos x)^{-\frac{1}{2}}$ $\cdot\sinh(2a\cos x),\ x<\tfrac{\pi}{2}$ $\qquad 0\qquad,\ x>\tfrac{\pi}{2}$	$\tfrac{1}{2}\pi(\pi a)^{\frac{1}{2}}I_{\frac{1}{4}+\frac{1}{2}y}(a)I_{\frac{1}{4}-\frac{1}{2}y}(a)$

	$f(x)$	$g_c(y)$
7.87	$(\cos x)^{-\frac{1}{2}}$ $\cdot \cosh(2a\cos x), x < \frac{\pi}{2}$ $\qquad 0 \quad , x > \frac{\pi}{2}$	$\frac{1}{2}\pi(\pi a)^{\frac{1}{2}} I_{-\frac{1}{4}+\frac{1}{2}y}(a) I_{-\frac{1}{4}-\frac{1}{2}y}(a)$
7.88	$\sinh(a\sin x) \quad x<\pi$ $\qquad 0 \qquad x>\pi$	$-\frac{1}{2}\pi\{\mathbf{E}_y(ia)+\mathbf{E}_{-y}(ia)\}$
7.89	$\cosh(a\sin x) \quad x<\pi$ $\qquad 0 \qquad x>\pi$	$\frac{1}{2}\pi\{\mathbf{J}_y(ia)+\mathbf{J}_{-y}(ia)\}$
7.90	$\sinh(a\cos x) \quad x<\frac{1}{2}\pi$ $\qquad 0 \qquad x>\frac{1}{2}$	$\frac{1}{4}i\pi\sec(\frac{1}{2}\pi y)\{\mathbf{E}_y(ia)+\mathbf{E}_{-y}(ia)\}$
7.91	$\cosh(a\cos x) \quad x<\frac{1}{2}\pi$ $\qquad 0 \qquad x>\frac{1}{2}\pi$	$\frac{1}{4}\pi\sec(\frac{1}{2}\pi y)\{\mathbf{J}_y(ia)+\mathbf{J}_{-y}(ia)\}$
7.92	$\sin(a\cosh x)$	$\frac{1}{4}\pi\,\mathrm{sech}(\frac{1}{2}\pi y)\{J_{iy}(a)+J_{-iy}(a)\}$
7.93	$\cos(a\cosh x)$	$\frac{1}{4}i\pi\,\mathrm{cosech}(\frac{1}{2}\pi y)\{J_{iy}(a)-J_{-iy}(a)\}$
7.94	$\cos(a\sinh x)$	$\cosh(\frac{1}{2}\pi y) K_{iy}(a)$
7.95	$\sin(a\sinh x)$	$\frac{1}{4}\pi\,\mathrm{sech}(\frac{1}{2}\pi y)\left[I_{iy}(a)+I_{-iy}(a) \right.$ $\left. -2\,\mathrm{sech}(\frac{1}{2}\pi y)\{\mathbf{E}_{iy}(ia)+\mathbf{E}_{-iy}(ia)\}\right]$
7.96	$\cos(a\cosh x)$ $\cdot \cos(b\sinh x)$	$\cosh(\frac{1}{2}\pi y)\cos\left[\frac{1}{2}y\,\log(\frac{b+a}{b-a})\right]K_{iy}\{(b^2-a^2)^{\frac{1}{2}}\}$ $\quad , \ a<b$ $\frac{1}{4}i\pi\,\mathrm{cosech}(\frac{1}{2}\pi y)\cos\left[\frac{1}{2}y\,\log(\frac{a+b}{a-b})\right]$ $\cdot\left[J_{iy}\{(a^2-b^2)^{\frac{1}{2}}\}-J_{-iy}\{(a^2-b^2)^{\frac{1}{2}}\}\right] \quad , \ a>b$
7.97	$\sin(a\cosh x)$ $\cdot \cos(b\sinh x)$	$\sinh(\frac{1}{2}\pi y)\sin\left[\frac{1}{2}y\,\log(\frac{b+a}{b-a})\right]K_{iy}\{(b^2-a^2)^{\frac{1}{2}}\}, a<b$ $\frac{1}{4}\pi\,\mathrm{sech}(\frac{1}{2}\pi y)\cos\left[\frac{1}{2}y\,\log(\frac{a+b}{a-b})\right]$ $\cdot\left[J_{iy}\{(a^2-b^2)^{\frac{1}{2}}\}+J_{-iy}\{(a^2-b^2)^{\frac{1}{2}}\}\right] \quad , \ a>b$
7.98	$(\cosh x)^{-\frac{1}{2}}$ $\cdot \sin(2a\cosh x)$	$-\frac{1}{4}\pi(a\pi)^{\frac{1}{2}}\{J_{\frac{1}{4}+\frac{1}{2}iy}(a)Y_{\frac{1}{4}-\frac{1}{2}iy}(a)$ $+J_{\frac{1}{4}-\frac{1}{2}iy}(a)Y_{\frac{1}{4}+\frac{1}{2}iy}(a)\}$

	$f(x)$	$g_c(y)$
7.99	$(\cosh x)^{-\frac{1}{2}}$ $\cdot \cos(2a\cosh x)$	$-\frac{1}{4}\pi(a\pi)^{\frac{1}{2}}\{J_{-\frac{1}{4}+\frac{1}{2}iy}(a)Y_{-\frac{1}{4}-\frac{1}{2}iy}(a)$ $+J_{-\frac{1}{4}-\frac{1}{2}iy}(a)Y_{-\frac{1}{4}+\frac{1}{2}iy}(a)\}$
7.100	$(\sinh x)^{-\frac{1}{2}}$ $\cdot \sin(2a\sinh x)$	$\frac{1}{2}(a\pi)^{\frac{1}{2}}\{I_{\frac{1}{4}-\frac{1}{2}iy}(a)K_{\frac{1}{4}+\frac{1}{2}iy}(a)$ $+I_{\frac{1}{4}+\frac{1}{2}iy}(a)K_{\frac{1}{4}-\frac{1}{2}iy}(a)\}$
7.101	$(\sinh x)^{-\frac{1}{2}}$ $\cdot \cos(2a\sinh x)$	$\frac{1}{2}(a\pi)^{\frac{1}{2}}\{I_{-\frac{1}{4}-\frac{1}{2}iy}(a)K_{\frac{1}{4}-\frac{1}{2}iy}(a)$ $+I_{-\frac{1}{4}+\frac{1}{2}iy}(a)K_{\frac{1}{4}+\frac{1}{2}iy}(a)\}$
7.102	$\cosh x \, \log(1-e^{-2x})$	$-\frac{1}{2}\pi y(1+y^2)^{-\frac{1}{2}}\tanh(\frac{\pi}{2}y)+(y^2-1)(y^2+1)^{-2}$
7.103	$(\cosh x+a)^{-\frac{1}{2}}$ $\cdot \log(\cosh x+a), -1<a<1$	$\pi2^{-\frac{1}{2}}\text{sech}(\pi y)\Big(P_{-\frac{1}{2}+iy}(a)\{-\gamma+\frac{1}{2}\log(1-a^2)$ $-\log4-\frac{1}{2}\Psi(\frac{1}{2}+iy)-\frac{1}{2}\Psi(\frac{1}{2}-iy)\}+\frac{1}{2}Q_{-\frac{1}{2}+iy}(a)$ $+\frac{1}{2}Q_{-\frac{1}{2}-iy}(a)\Big)$
7.104	$(\cosh x+z)^{-\frac{1}{2}}$ $\cdot \log(\cosh x+z), \quad x>1$	$\pi2^{-\frac{1}{2}}\text{sech}(\pi y)\Big(P_{-\frac{1}{2}+iy}(z)\{-\gamma+\frac{1}{2}\log(z^2-1)$ $-\log4-\frac{1}{2}\Psi(\frac{1}{2}+iy)-\frac{1}{2}\Psi(\frac{1}{2}-iy)\}+\frac{1}{2}q_{-\frac{1}{2}+iy}(z)$ $+\frac{1}{2}q_{-\frac{1}{2}-iy}(z)\Big)$
7.105	$(z-\cosh x)^{-\frac{1}{2}}$ $\cdot \log(z-\cosh x)$ $\qquad , \cosh x<z$ $\quad 0 \quad , \cosh x>z$	$\pi2^{-\frac{1}{2}}\Big(P_{-\frac{1}{2}+iy}(z)\{-\gamma-\log4+\frac{1}{2}\log(z^2-1)$ $-\frac{1}{2}\Psi(\frac{1}{2}+iy)-\frac{1}{2}\Psi(\frac{1}{2}-iy)\}-\frac{1}{2}q_{-\frac{1}{2}+iy}(z)$ $-\frac{1}{2}q_{-\frac{1}{2}-iy}(z)\Big)$
7.106	$\quad 0 \quad , \cosh x<z$ $(\cosh x-z)^{-\frac{1}{2}}$ $\cdot \log(\cosh x-z)$	$2^{-\frac{1}{2}}\Big(\{q_{-\frac{1}{2}+iy}(z)+q_{-\frac{1}{2}-iy}(z)\}\{-\gamma+\frac{1}{2}\log(z^2-1)$ $-\log4\}-\Psi(\frac{1}{2}+iy)q_{-\frac{1}{2}+iy}(z)-\Psi(\frac{1}{2}-iy)q_{-\frac{1}{2}-iy}(z)\Big)$
7.107	$\log\left(\frac{1+\cosh(ax)}{\cos b+\cosh(ax)}\right)$	$2\pi y^{-1}\text{cosech}(\pi y/a)\sinh^2(\frac{1}{2}by/a) \qquad , \ b<\pi$
7.108	$\log\left(\frac{\cosh(ax)+\sin b}{\cosh(ax)-\sin b}\right)$	$\pi y^{-1}\sinh(by/a)\text{sech}(\frac{1}{2}\pi y/b) \qquad , \ b<\pi$

	$f(x)$	$g_c(y)$
7.109	$\log\left(\dfrac{\cosh(ax)+\cos b}{\cosh(ax)+\cos c}\right)$ $,b,c\leq\pi$	$\pi y^{-1}\mathrm{cosech}(\pi y/a)\left\{\cosh(cy/a)-\cosh(by/a)\right\}$
7.110	$\log(1+a^2\mathrm{sech}^2 x)$	$2\pi y^{-1}\mathrm{cosech}(\tfrac{1}{2}\pi y)\sin^2[\tfrac{1}{2}y\,\log\{a+(1+a^2)^{\frac{1}{2}}\}]$
7.111	$\log(1-a^2\mathrm{sech}^2 x),a<1$	$-2\pi y^{-1}\mathrm{cosech}(\tfrac{1}{2}\pi y)\sinh^2(\tfrac{1}{2}y\,\arcsin a)$
7.112	$\log\{\tanh(ax)\}$	$-\tfrac{1}{2}\pi y^{-1}\tanh(\tfrac{1}{4}y/a)$
7.113	$B^{-1}\log\{(A+B)/(A-B)\}$ $,-1<a<1$ $A=(\cosh x+1)^{\frac{1}{2}},$ $B=(\cosh x-a)^{\frac{1}{2}}$	$2^{-\frac{1}{2}}\pi^2\mathrm{sech}^2(\pi y)P_{-\frac{1}{2}+iy}(a)$
7.114	$(\cosh ax)^{-\nu}$ $\cdot\log(\cosh ax)$ $\mathrm{Re}\,\nu>0$	$2^{\nu-2}\{a\Gamma(\nu)\}^{-1}\Gamma(\tfrac{1}{2}\nu+\tfrac{1}{2}iy/a)\Gamma(\tfrac{1}{2}\nu-\tfrac{1}{2}iy/a)$ $\cdot\{\Psi(\nu)-\log 2-\tfrac{1}{2}\Psi(\tfrac{1}{2}\nu+\tfrac{1}{2}iy/a)-\tfrac{1}{2}\Psi(\tfrac{1}{2}\nu-\tfrac{1}{2}iy/a)\}$
7.115	$(1+x^2)^{-1}\mathrm{sech}(\tfrac{1}{4}\pi x)$	$2^{-\frac{1}{2}}\left(\pi e^{-y}+2\sinh y\,\arctan(2^{-\frac{1}{2}}\mathrm{cosech}\,y)\right.$ $\left.-\cosh y\,\log\{(\cosh y+2^{-\frac{1}{2}})(\cosh y-2^{-\frac{1}{2}})^{-1}\}\right]$
7.116	$x(1+x^2)^{-1}\mathrm{cosech}(\pi x)$	$\tfrac{1}{2}ye^{-y}-\tfrac{1}{2}+\cosh y\,\log(1+e^{-2y})$
7.117	$(1+x^2)^{-1}\sinh(ax)$ $\cdot\mathrm{cosech}(\pi x)$ $a\leq\pi$	$\tfrac{1}{2}e^{-y}(y\sin a-a\cos a)+\tfrac{1}{2}\sin a\cosh y$ $\cdot\log(1+2e^{-y}\cos a+e^{-2y})$ $-\cos a\sinh y\,\arctan\{\sin a(e^y+\cos a)^{-1}\}$
7.118	$(1+x^2)^{-1}\cosh(ax)$ $\mathrm{sech}(\tfrac{1}{2}\pi x)$ $a\leq\tfrac{1}{2}\pi$	$ye^{-y}\cos a+ae^{-y}\sin a+\tfrac{1}{2}\cos a\cosh y$ $\cdot\log\{1+2\cos(2a)e^{-2y}+e^{-4y})\}$ $+\sin a\sinh y\,\arctan\{\dfrac{e^{-2y}\sin(2a)}{1+e^{-2y}\cos(2a)}\}$
7.119	$(1+x^2)^{-1}\sinh(ax)$ $\cdot\mathrm{cosech}(\tfrac{1}{2}\pi x)$ $a\leq\tfrac{1}{2}\pi$	$\tfrac{1}{2}\pi e^{-y}\sin a+\sin a\sinh y\,\arctan(\dfrac{\cos a}{\sinh y})$ $-\tfrac{1}{2}\cos a\cosh y\,\log(\dfrac{\cosh y+\sin a}{\cosh y-\sin a})$
7.120	$\log(1+\cos a\,\mathrm{sech}\,x),a<\pi$	$\pi y^{-1}\mathrm{cosech}(\pi y)\{\cosh(\tfrac{1}{2}\pi y)-\cosh(ay)\}$

	$f(x)$	$g_c(y)$
7.121	$\log\{\coth(\tfrac{1}{2}x)\}\log(\sinh x)$	$\pi y^{-1}\tanh(\tfrac{1}{2}\pi y)\,\{-\gamma-\log 2-\tfrac{1}{4}\Psi(\tfrac{1}{2}+\tfrac{1}{2}iy)$ $-\tfrac{1}{4}\Psi(\tfrac{1}{2}-\tfrac{1}{2}iy)-\tfrac{1}{4}\Psi(1+\tfrac{1}{2}iy)-\tfrac{1}{4}\Psi(1-\tfrac{1}{2}iy)\}$
7.122	$\arctan\{\sinh a\ \operatorname{sech}(bx)\}$	$\tfrac{1}{2}\pi y^{-1}\sin(ay/b)\,\operatorname{sech}(\tfrac{1}{2}\pi y/b)$
7.123	$(1+z^2)^{-\frac{1}{2}\nu}\cos(\nu\arctan z)$ $\operatorname{Re}\nu>0\ ,\ z=c\sin hx$	$2(c/\pi)^{-\frac{1}{2}}\{\Gamma(\nu)\}^{-1}\cosh(\tfrac{1}{2}\pi y)(1-c^2)^{\frac{1}{4}-\frac{1}{2}\nu}$ $\cdot\Gamma(\nu-iy)\Gamma(\nu+iy)P^{\frac{1}{2}-\nu}_{-\frac{1}{2}+iy}(1/c)$

1.8 Orthogonal Polynomials

	$f(x)$		$g_c(y)$
8.1	$P_{2n}(x)$	$x<1$	$(-1)^n(\tfrac{1}{2}\pi/y)^{\frac{1}{2}}J_{2n+\frac{1}{2}}(y)$
	0	$x>1$	
8.2	$P_{2n+1}(x)$	$x<1$	$(-1)^{n+1}(2n+1)!\,(n!)^{-2}y^{-\frac{1}{2}}s_{-\frac{1}{2},2n+3/2}(y)$
	0	$x>1$	
8.3	$P_n(a-bx^2)$ $a-1<bx^2<a+1$ 0 otherwise $a\geq 1$		$\pi(2b)^{-\frac{1}{2}}J_{n+\frac{1}{2}}(u)Y_{n+\frac{1}{2}}(v)$ $\genfrac{}{}{0pt}{}{v}{u}=\tfrac{1}{2}yb^{-\frac{1}{2}}\{(a+1)^{\frac{1}{2}}\pm(a-1)^{\frac{1}{2}}\}$
8.4	$x^{\nu-1}P_n(x)$ 0 $\operatorname{Re}(\nu+n)>0$	$x<1$ $x>1$	$\pi^{\frac{1}{2}}2^{-\nu}\Gamma(\nu)\left(\Gamma(1+\tfrac{1}{2}\nu+\tfrac{1}{2}n)\Gamma(\tfrac{1}{2}+\tfrac{1}{2}\nu-\tfrac{1}{2}n)\right)^{-1}$ $_2F_3(\tfrac{1}{2}\nu,\tfrac{1}{2}+\tfrac{1}{2}\nu;\tfrac{1}{2},\tfrac{1}{2}+\tfrac{1}{2}\nu-\tfrac{1}{2}n,1+\tfrac{1}{2}\nu+\tfrac{1}{2}n;-\tfrac{1}{4}y^2)$
8.5	$(a^2+x^2)^{-n-\frac{1}{2}}$ $\cdot P_{2n}\{x(a^2+x^2)^{-\frac{1}{2}}\}$		$(-1)^n y^{2n}\{(2n)!\}^{-1}y^{2n}K_o(ay)$
8.6	$(\operatorname{sech} x)^{2n+1}$ $\cdot P_{2n}(\tanh x)$		$(-1)^n\pi^{-1}2^{2n-1}\{(2n)!\}^{-2}\cos h(\tfrac{1}{2}\pi y)$ $\cdot\left(\Gamma(\tfrac{1}{2}+n+\tfrac{1}{2}iy)\Gamma(\tfrac{1}{2}+n-\tfrac{1}{2}iy)\right)^2$
8.7	$u^{-n-\frac{1}{2}}P_{2n}(u^{-\frac{1}{2}}a\sinh x)$ $u=b^2+a^2\sinh^2 x$		$(-1)^n\tfrac{1}{4}\pi^{-1}(\tfrac{1}{2}a)^{-2n-1}\{(2n)!\}^{-2}\cos h(\tfrac{1}{2}\pi y)$ $\cdot\left(\Gamma(\tfrac{1}{2}+n+\tfrac{1}{2}iy)\Gamma(\tfrac{1}{2}+n-\tfrac{1}{2}iy)\right)^2$ $_2F_1(\tfrac{1}{2}+n+\tfrac{1}{2}iy,\tfrac{1}{2}+n-\tfrac{1}{2}iy;2n+1;1-b^2/a^2)$

	$f(x)$	$g_c(y)$
8.8	$(a^2-x^2)^{-\frac{1}{2}}T_{2n}\left(\frac{x}{a}\right)$, $x<a$ 0 , $x>a$	$(-1)^n \frac{1}{2}\pi J_{2n}(ay)$
8.9	$x^{-\frac{1}{2}}(a^2-x^2)^{-\frac{1}{2}}T_n\left(\frac{x}{a}\right)$, $x<a$ 0 , $x>a$	$\frac{1}{2}\pi(\frac{1}{2}\pi y)^{\frac{1}{2}}J_{\frac{1}{2}n-\frac{1}{4}}(\frac{1}{2}ay)J_{-\frac{1}{2}n-\frac{1}{4}}(\frac{1}{2}ay)$
8.10	$u^{-1}\cos(bu)T_{2n}(x/a)$, $x<a$ 0 , $x>a$ $u=(a^2-x^2)^{\frac{1}{2}}$	$(-1)^n \frac{1}{2}\pi T_{2n}(y/z)J_{2n}(az)$; $z=(b^2+y^2)^{\frac{1}{2}}$
8.11	$u^{-1}\cos(bu)T_{2n}(u/a)$, $x<a$ 0 , $x>a$ $u=(a^2-x^2)^{\frac{1}{2}}$	$(-1)^n \frac{1}{2}\pi T_{2n}(b/z)J_{2n}(az)$; $z=(b^2+y^2)^{\frac{1}{2}}$
8.12	$u^{-1}\sin(bu)T_{2n+1}(u/a)$, $x<a$ 0 , $x>a$ $u=(a^2-x^2)^{\frac{1}{2}}$	$(-1)^n \frac{1}{2}\pi T_{2n+1}(b/z)J_{2n+1}(az)$; $z=(b^2+y^2)^{\frac{1}{2}}$
8.13	$u^{-1}\cos(bu)T_{2n}(u/a)$, $x<2a$ 0 , $x>2a$ $u=(2ax-x^2)^{\frac{1}{2}}$	$(-1)^n \pi \cos(ay)T_{2n}(b/z)J_{2n}(az)$; $z=(b^2+y^2)^{\frac{1}{2}}$
8.14	$u^{-1}\sin(bu)T_{2n+1}\left(\frac{u}{a}\right)$, $x<2a$ 0 , $x>2a$ $u=(2ax-x^2)^{\frac{1}{2}}$	$(-1)^n \pi \cos(ay)T_{2n+1}(b/z)J_{2n+1}(az)$ $z=(b^2+y)^{\frac{1}{2}}$
8.15	$u^n T_n(au)$; $u=(a^2+x^2)^{-\frac{1}{2}}$	$\frac{1}{2}\pi\{(n-1)!\}^{-1}y^{n-1}e^{-ay}$
8.16	$u^n\exp(-ax^2)T_n(b/u)$ $u=(b^2+x^2)^{\frac{1}{2}}$	$\frac{1}{4}(2a)^{-\frac{1}{2}n}(\pi/a)^{\frac{1}{2}}\exp(-\frac{1}{4}y^2/a)$ $\cdot\left[He_n(z)+He_n(Z)\right]\;{}^z_Z=(2a)^{\frac{1}{2}}(b\pm\frac{1}{2}y/a)$
8.17	$\sin(bu)U_{2n}(x/a)$, $x<a$ 0 , $x>a$ $u=(a^2-x^2)^{\frac{1}{2}}$	$(-1)^n \frac{1}{2}\pi abz^{-1}U_{2n}(y/z)J_{2n+1}(bz)$; $z=(b^2+y^2)^{\frac{1}{2}}$

	$f(x)$	$g_c(y)$
8.18	$\exp(-bx^2) \mathrm{He}_{2n}(ax)$	$\frac{1}{2}\pi^{\frac{1}{2}} b^{-n-\frac{1}{2}} (-z)^n \exp(-\frac{1}{4}y^2/b) \mathrm{He}_{2n}\{\frac{1}{2}ay(bz)^{-\frac{1}{2}}\}$ $z = \frac{1}{4}a^2 - b$
8.19	$e^{-\frac{1}{2}x^2} \mathrm{He}_n(x) \mathrm{He}_{n+2m}(x)$	$(-1)^m n! \, (\frac{1}{2}\pi)^{\frac{1}{2}} y^{2m} \exp(-\frac{1}{2}y^2) L_n^{2m}(y^2)$
8.20	$u^{\frac{1}{2}+\frac{1}{2}n} \exp(-2a^2 u)$ $\cdot \mathrm{He}_n\{2a(1+u)^{\frac{1}{2}}\} \, x < \frac{1}{2}\pi$ $\qquad 0 \qquad x > \frac{1}{2}\pi$ $u = \sec x$	$\pi^{\frac{1}{2}} D_{\frac{1}{2}n-\frac{1}{2}+y}^{\frac{1}{2}n}(2a) D_{\frac{1}{2}n-\frac{1}{2}-y}^{\frac{1}{2}n}(2a)$
8.21	$e^{-\frac{1}{2}x^2} L_n(x^2)$	$(\frac{1}{2}\pi)^{\frac{1}{2}} (n!)^{-1} \exp(-\frac{1}{2}y^2) \{\mathrm{He}_n(y)\}^2$
8.22	$\exp(-\frac{1}{2}x^2) L_n(\frac{1}{2}x^2)$ $\cdot \mathrm{He}_{2n}(\frac{1}{2}x)$	$(\frac{1}{2}\pi)^{\frac{1}{2}} \exp(-\frac{1}{2}y^2) L_n(\frac{1}{2}y^2) \mathrm{He}_{2n}(\frac{1}{2}y)$
8.23	$e^{-ax} x^{\nu-2n} L_{2n-1}^{\nu-2n}(ax)$ $\mathrm{Re}\,\nu > 2n-1$	$\frac{1}{2}i(-1)^{n+1} \Gamma(\nu) \{(2n-1)!\}^{-1} y^{2n-1}$ $\cdot \left((a-iy)^{-\nu} - (a+iy)^{-\nu} \right)$
8.24	$e^{-ax} x^{\nu-2n-1}$ $\cdot L_{2n}^{\nu-2n-1}(ax)$ $\mathrm{Re}\,\nu > 2n$	$(-1)^n \frac{1}{2} \Gamma(\nu) \{(2n)!\}^{-1} y^{2n} \{(a+iy)^{-\nu} + (a-iy)^{-\nu}\}$
8,25	$x^{2m} e^{-\frac{1}{2}x^2} L_n^{2m}(x^2)$	$(-1)^m (\frac{1}{2}\pi)^{\frac{1}{2}} (n!)^{-1} \exp(-\frac{1}{2}y^2) \mathrm{He}_n(y) \mathrm{He}_{n+2m}(y)$
8.26	$x^{2n} e^{-\frac{1}{2}x^2} L_n^{n-\frac{1}{2}}(\frac{1}{2}x^2)$	$(\frac{1}{2}\pi)^{\frac{1}{2}} e^{-\frac{1}{2}y^2} y^{2n} L_n^{n+\frac{1}{2}}(\frac{1}{2}y^2)$
8.27	$e^{-\frac{1}{2}x^2} \{L_n^{-\frac{1}{4}}(\frac{1}{2}x^2)\}^2$	$(\frac{1}{2}\pi)^{\frac{1}{2}} e^{-\frac{1}{2}y^2} \{L_n^{-\frac{1}{4}}(\frac{1}{2}y^2)\}^2$
8.28	$e^{-\frac{1}{2}x^2} L_n^{\nu}(\frac{1}{2}x^2) L_n^{-\frac{1}{2}-\nu}(\frac{1}{2}x^2)$	$(\frac{1}{2}\pi)^{\frac{1}{2}} e^{-\frac{1}{2}y^2} L_n^{\nu}(\frac{1}{2}y^2) (L_n^{-\frac{1}{2}-\nu}(\frac{1}{2}y^2)$
8.29	$(a^2-x^2)^{\nu-\frac{1}{2}} C_{2n}^{\nu}(x/a), x < a$ $\qquad 0 \qquad , x > a$ $\mathrm{Re}\,\nu > -\frac{1}{2}$	$(-1)^n (2y/a)^{-\nu} \Gamma(2n+2\nu) \{(2n)! \Gamma(\nu)\}^{-1}$ $\cdot J_{\nu+2n}(ay)$

	$f(x)$	$g_c(y)$
8.30	$(a^2-x^2)^\nu P_{2n}^{(\nu,\nu)}(\frac{x}{a})$, $x<a$ 0 , $x>a$	$\frac{1}{2}(-1)^n(2a/y)^{\nu+\frac{1}{2}}\pi^{\frac{1}{2}}\{(2n)!\}^{-1}\Gamma(\nu+2n-1)$ $\cdot J_{\nu+2n+\frac{1}{2}}(ay)$
8.31	$\Big((1-x)^\nu(1+x)^\mu$ $+(1+x)^\nu(1-x)^\mu\Big)$ $P_{2n}^{(\nu,\mu)}(x)$, $x<1$ 0 , $x>1$	$(-1)^n 2^{2n+\nu+\mu}\{(2n)!\}^{-1}B(2n+\nu+1,2n+\mu+1)$ $\cdot y^{2n}\Big(e^{iy}\,_1F_1(2n+\nu+1;4n+\nu+\mu+2;-2iy)$ $+e^{-iy}\,_1F_1(2n+\nu+1;4n+\nu+\mu+2;;2iy)\Big)$
8.32	$\Big((1-x)^\nu(1+x)^\mu$ $-(1+x)^\nu(1-x)^\mu\Big)$ $P_{2n+1}^{(\nu,\mu)}(x)$; $x<1$ 0 ; $x>1$	$(-1)^{n+1}2^{2n+\nu+\mu+1}\{(2n+1)!\}^{-1}$ $\cdot B(2n+\nu+2,2n+\mu+2)y^{2n+1}$ $\cdot\Big(ie^{iy}\,_1F_1(2n+\nu+2;4n+\nu+\mu+4;-2iy)$ $-ie^{-iy}\,_1F_1(2n+\nu+2;4n+\nu+\mu+4;2iy)\Big)$

<u>1.9 Gamma- and Related Functions</u>

	$f(x)$	$g_c(y)$
9.1	$|\Gamma(a+ix)|^2$	$\pi 2^{-2a}\Gamma(2a)\{\operatorname{sech}(\frac{1}{2}y)\}^{2a}$
9.2	$\left|\dfrac{\Gamma(\frac{1}{4}+ibx)}{\Gamma(\frac{3}{4}+ibx)}\right|^2$	$b^{-1}\operatorname{sech}(\frac{1}{2}y/b)K\{\operatorname{sech}(\frac{1}{4}y/b)\}$
9.3	$|\Gamma(\frac{1}{4}+ibx)|^4$	$2\pi^{-2}b^{-1}\operatorname{sech}(\frac{1}{4}y/b)K\{\tanh(\frac{1}{4}y/b)\}$
9.4	$h(\pi x)|\Gamma\frac{3}{4}+ix)|\}^{-2}$	$\pi^{-1}(\frac{1}{2}\pi\cosh y)^{-\frac{1}{2}}\log\{(1+\cosh y)^{\frac{1}{2}}+\cosh y)^{\frac{1}{2}}\}$
9.5	$\{\Gamma(a+bx)\Gamma(a-bx)\}^{-1}$ $a>\frac{1}{2}$	$2^{2a-3}\{\Gamma(2a-1)\}^{-1}b^{-1}\{\cos(\frac{1}{2}y/b)\}^{2a-2}$, $y<\pi b$ 0 , $y>\pi b$
9.6	$\{\Gamma(a+bx)\Gamma(c-bx)\}^{-1}$ $+\{\Gamma(a-bx)\Gamma(c+bx)\}^{-1}$ $a+c>\frac{1}{2}$	$b^{-1}\{\Gamma(a+c-1)\}^{-1}\{2\cos(\frac{1}{2}y/b)\}^{a+c-2}$ $\cdot\cos\{\frac{1}{2}\pi y/b(c-a)\}$, $y<\pi b$ 0 , $y>\pi b$
9.7	$\Gamma(a+ibx)\Gamma(a-ibx)$ $\cdot\Gamma(\frac{1}{2}-a+ibx)\Gamma(\frac{1}{2}-a-ibx)$ $0<a<\frac{1}{2}$	$\pi^3 b^{-1}\operatorname{cosec}(2\pi a)P_{2a-1}\{\cosh(\frac{1}{2}y/b)\}$

	$f(x)$	$g_c(y)$
9.8	$\Gamma(b+iax)\Gamma(b-iax)$ $\cdot\Gamma(c+iax)\Gamma(c-iax)$ $a,b,c>0$	$2^{\frac{1}{2}-b-c}\pi^{3/2}a^{-1}\Gamma(2c)\Gamma(2b)\Gamma(b+c)$ $\cdot\left(\sinh(\frac{1}{2}y/a)\right)^{\frac{1}{2}-b-c}P_{b-c-\frac{1}{2}}^{\frac{1}{2}-c-b}\left(\cosh(\frac{1}{2}y/a)\right)$
9.9	$\Gamma(a+icx)\Gamma(a-icx)$ $\cdot\left[\Gamma(b+icx)\Gamma(b-icx)\right]^{-1}$ $b>a>0$	$\pi^{\frac{1}{2}}2^{b-a-\frac{1}{2}}c^{-1}e^{-i\pi(a-b+\frac{1}{2})}\{\Gamma(b-a)\}^{-1}$ $\left(\sinh(\frac{1}{2}y/c)\right)^{b-a-\frac{1}{2}}Q_{a+b-3/2}^{a-b+\frac{1}{2}}\left(\cosh(\frac{1}{2}y/c)\right)$
9.10	$\Psi(1+x)-\log x$	$\frac{1}{2}\Psi(1+\frac{1}{2}y/\pi)-\frac{1}{2}\log(\frac{1}{2}y/\pi)$
9.11	$2x^{-1}+\pi^{-1}\Psi(\frac{1}{2}x/\pi)$ $-\pi^{-1}\log(\frac{1}{2}x/\pi)$	$y^{-1}+\Psi(y)-\log y$
9.12	$\Psi(\frac{1}{2}+ix)+\Psi(\frac{1}{2}-ix)-\log 4$	$\pi\{y^{-1}-\frac{1}{2}\operatorname{cosech}(\frac{1}{2}y)\}$
9.13	$\Gamma(\nu+ix)\Gamma(\nu-ix)$ $\cdot\{\Psi(\nu+ix)+\Psi(\nu-ix)\}$ $\operatorname{Re}\nu>0$	$2\pi\Gamma(2\nu)\{2\cosh(\frac{1}{2}y)\}^{-2\nu}$ $\{\Psi(2\nu)-\log(2\cosh\frac{1}{2}y)\}$
9.14	$x^{-1}\sin x$ $\cdot\{\Psi(1+\frac{1}{2}x/\pi)+\Psi(1-\frac{1}{2}x/\pi)\}$	$-\pi\{\gamma+\log(2\sin\pi y)\}$ $\qquad y<1$ $0\qquad\qquad y>1$
9.15	$\zeta(\frac{1}{2}+ix)$	$2\pi^2\sum\limits_{n=o}^{\infty}(2\pi n^4 e^{-\frac{9}{2}y}-3n^2 e^{-\frac{5}{2}y})\exp(-n^2\pi e^{-2y})$
9.16	$(1+4x^2)^{-1}\zeta(\frac{1}{2}+ix)$	$\frac{1}{4}\pi\left(\cosh(\frac{1}{2}y)+\frac{1}{4}\theta_3(0,ie^{-2y})\right)$

1.10   The Error- and the Fresnel Integrals

	$f(x)$	$g_c(y)$
10.1	$\operatorname{Erfc}(x)$	$\pi^{-\frac{1}{2}}{}_1F_1(1;\frac{3}{2};-\frac{1}{4}y^2)=-iy^{-1}e^{-\frac{1}{4}y^2}\operatorname{Erf}(\frac{1}{2}iy)$
10.2	$x^{-1}\operatorname{Erf}(x)$	$-\frac{1}{2}\operatorname{Ei}(-\frac{1}{4}y^2)$
10.3	$x\operatorname{Erfc}(x)$	$\frac{1}{2}e^{-\frac{1}{4}y^2}-y^{-2}(1-e^{-\frac{1}{4}y^2})$
10.4	$x^{\nu-1}\operatorname{Erfc}(x)$ $\operatorname{Re}\nu>0$	$\pi^{-\frac{1}{2}}\nu^{-1}\Gamma(\frac{1}{2}+\frac{1}{2}\nu){}_2F_2(\frac{1}{2}\nu,\frac{1}{2}+\frac{1}{2}\nu;\frac{1}{2},1+\frac{1}{2}\nu;\frac{1}{4}y^2)$

	$f(x)$	$g_c(y)$
10.5	$x^{-1}\{\mathrm{Erfc}(ax) - \mathrm{Erfc}(bx)\}$	$\tfrac{1}{2}\mathrm{Ei}(\tfrac{1}{4}y^2/a^2) - \tfrac{1}{2}\mathrm{Ei}(-\tfrac{1}{4}y^2/b^2)$
10.6	$e^{x^2}\mathrm{Erfc}(x)$	$-\tfrac{1}{2}\pi^{-\frac{1}{2}} e^{\frac{1}{4}y^2} \mathrm{Ei}(-\tfrac{1}{4}y^2)$
10.7	$ie^{-x^2}\mathrm{Erf}(ix)$	$\tfrac{1}{2}\pi^{-\frac{1}{2}} e^{-\frac{1}{4}y^2} \overline{\mathrm{Ei}}(\tfrac{1}{4}y^2)$
10.8	$ix^{-1}e^{-x^2}\mathrm{Erf}(ix)$	$-\tfrac{1}{2}\pi\,\mathrm{Erfc}(\tfrac{1}{2}y)$
10.9	$e^{a^2x^2}\mathrm{Erfc}(ax+b)$	$-\tfrac{1}{2}\pi^{-\frac{1}{2}} a^{-1} e^{\frac{1}{4}y^2/a^2} \mathrm{Ei}(-b^2-\tfrac{1}{4}y^2/a^2)$
10.10	$\mathrm{Erfc}(x^{\frac{1}{2}})$	$2^{-\frac{1}{2}}(1+y^2)^{-\frac{1}{2}}\left[(1+y^2)^{\frac{1}{2}}+1\right]^{-\frac{1}{2}}$
10.11	$x^{-1}\mathrm{Erf}(x^{\frac{1}{2}})$	$\log y - \tfrac{1}{2}\log\left[1+(1+y^2)^{\frac{1}{2}}\right] - \log\left[\{1+(1+y^2)^{\frac{1}{2}}\}^{\frac{1}{2}}-2^{\frac{1}{2}}\right]$
10.12	$x^{-\frac{1}{2}}\mathrm{Erf}(x^{\frac{1}{2}})$	$(2\pi)^{-\frac{1}{2}}\arctan\left[(2y)^{\frac{1}{2}}(y-a)^{-1}\right]$ $-\log\left[(1+y^2)^{-\frac{1}{2}}\{y+1+(2y)^{\frac{1}{2}}\}\right]$
10.13	$x^{-1}\left[1-\mathrm{Erfc}(x^{\frac{1}{2}})\right]$	$\log y - \tfrac{1}{2}\log\{1+(1+y^2)^{\frac{1}{2}}\} - \log\left[\{1+(1+y^2)^{\frac{1}{2}}\}^{\frac{1}{2}}-2^{\frac{1}{2}}\right]$
10.14	$e^x\mathrm{Erfc}(x^{\frac{1}{2}})$	$(2y)^{-\frac{1}{2}}\{1+(2y)^{\frac{1}{2}}\}^{-1}$
10.15	$x^{-1}\mathrm{Erfc}(ax^{-\frac{1}{2}})$	$-\mathrm{Ei}\left[-a(2iy)^{\frac{1}{2}}\right] - \mathrm{Ei}\left[-a(-2iy)^{\frac{1}{2}}\right]$
10.16	$x^{-1}e^{-a/x}\mathrm{Erfc}\left[(\tfrac{a}{x})^{\frac{1}{2}}\right]$	$\tfrac{1}{2}\pi\left[\mathbf{H}_0(u)+\mathbf{H}_0(v)-Y_0(u)-Y_0(v)\right] \quad \begin{matrix}u\\v\end{matrix}=2(\pm iay)^{\frac{1}{2}}$
10.17	$\mathrm{Erfc}(a/x)$	$-2a\pi^{-\frac{1}{2}}\sum\limits_{n=0}^{\infty}(ay)^{2n}\{n!(2n+1)!\}$ $\cdot\left[\Psi(2n+2)+\tfrac{1}{2}\Psi(n+1)-\log(ay)\right]$
10.18	$\mathrm{Erfc}\left[a\{b+(b^2+x^2)^{\frac{1}{2}}\}^{\frac{1}{2}}\right]$	$2^{-\frac{1}{2}}a\,\exp(-ba^2)(a^4+y^2)^{-\frac{1}{2}}\left[(a^4+y^2)^{\frac{1}{2}}+a^2\right]^{-\frac{1}{2}}$ $\cdot\exp\{-b(a^4+y^2)^{\frac{1}{2}}\}$
10.19	$\mathrm{Erfc}\{a(1+\sec x)^{\frac{1}{2}}\}$ $x<\tfrac{1}{2}\pi$ $x>\tfrac{1}{2}\pi$	$e^{-a^2}D_{y-1}(2^{\frac{1}{2}}a)D_{-y-1}(2^{\frac{1}{2}}a)$

	$f(x)$	$g_c(y)$
10.20	$\text{Erf}\{b\,\text{sech}(ax)\}$	$\pi^{\frac{1}{2}}a^{-1}b\,\text{sech}(\tfrac{1}{2}\pi y/a)\,_2F_2(\tfrac{1}{2}+i\tfrac{y}{a},\tfrac{1}{2}-i\tfrac{y}{a};\tfrac{3}{2},1;-b^2)$
10.21	$\text{Erfc}(a\cosh x)$	$\tfrac{1}{2}a^{-1}W_{-\frac{1}{2},\frac{1}{2}iy}(a^2)\exp(-\tfrac{1}{2}a^2)$
10.22	$\exp\{\tfrac{1}{2}a^2\cosh(2x)\}$ $\cdot\text{Erfc}(a\cosh x)$	$\tfrac{1}{2}\,\text{sech}(\tfrac{1}{2}\pi y)K_{\frac{1}{2}iy}(\tfrac{1}{2}a^2)$
10.23	$\exp\{-\tfrac{1}{2}a^2\cosh(2x)\}$ $\cdot\text{Erf}(ia\cosh x)$	$\tfrac{1}{4}\pi i\,\text{sech}(\tfrac{1}{2}\pi y)\left[I_{iy}(\tfrac{1}{2}a^2)+I_{-iy}(\tfrac{1}{2}a^2)\right]$
10.24	$(\text{sech}\,x)^{\frac{1}{2}}e^{a^2\text{sech}\,x}$ $\cdot\text{Erfc}\,a(1+\text{sech}\,x)^{\frac{1}{2}}$	$\pi^{\frac{1}{2}}\text{sech}(\pi y)D_{-\frac{1}{2}+iy}(a2^{\frac{1}{2}})D_{-\frac{1}{2}-iy}(a2^{\frac{1}{2}})$
10.25	$x^{-\frac{1}{2}}S(ax)$	$(2\pi y)^{-\frac{1}{2}}\left\{\arctan\{(a/y)^{\frac{1}{2}}\}+\tfrac{1}{2}\log\lvert a^{\frac{1}{2}}-y^{\frac{1}{2}}\rvert\right.$ $\left.-\tfrac{1}{2}\log(a^{\frac{1}{2}}+y^{\frac{1}{2}})\right\}$
10.26	$\tfrac{1}{2}-S(ax)$	$\tfrac{1}{2}(\tfrac{1}{2}a)^{\frac{1}{2}}y^{-1}(a^2-y^2)^{-\frac{1}{2}}\{a-(a^2-y^2)^{\frac{1}{2}}\}^{\frac{1}{2}}$ $y<a$ $\tfrac{1}{2}(\tfrac{1}{2}a)^{\frac{1}{2}}y^{-1}(y^2-a^2)^{-\frac{1}{2}}\{y-(y^2-a^2)^{\frac{1}{2}}\}^{\frac{1}{2}}$ $y>a$
10.27	$\tfrac{1}{2}-C(x)$	$-\tfrac{1}{2}(\tfrac{1}{2}a)^{\frac{1}{2}}y^{-1}(a^2-y^2)^{-\frac{1}{2}}\{a-(a^2-y^2)^{\frac{1}{2}}\}^{\frac{1}{2}}$ $y<a$ $\tfrac{1}{2}(\tfrac{1}{2}a)^{\frac{1}{2}}y^{-1}(y^2-a^2)^{-\frac{1}{2}}\{y+(y^2-a^2)^{\frac{1}{2}}\}^{\frac{1}{2}}$ $y>a$
10.28	$x^{-\frac{1}{2}}C(ax)$	$\tfrac{1}{2}(\tfrac{1}{2}\pi/y)^{\frac{1}{2}}$ $y<a$ 0 $y>a$
10.29	$\tfrac{1}{2}-S(ax^2)$	$y^{-1}\left[C(\tfrac{1}{4}y^2/a)\cos(\tfrac{1}{4}y^2/a)+S(\tfrac{1}{4}y^2/a)\sin(\tfrac{1}{4}y^2/a)\right]$
10.30	$\tfrac{1}{2}-C(ax^2)$	$y^{-1}\left[C(\tfrac{1}{4}y^2/a)\sin(\tfrac{1}{4}y^2/a)-S(\tfrac{1}{4}y^2/a)\cos(\tfrac{1}{4}y^2/a)\right]$
10.31	$x^{-3/2}\{\sin(bx)C(bx)$ $-\cos(bx)S(bx)\}$	0 $y>b$
10.32	$\tfrac{1}{2}-\{C(ax)\}^2-\{S(ax)\}^2$	$y^{-1}\sin(\tfrac{1}{4}y^2/a)$
10.33	$1-C(ax)-S(ax)$	0 $y<a$ $\tfrac{1}{2}a^{\frac{1}{2}}y^{-1}(y-a)^{-\frac{1}{2}}$ $y>a$
10.34	$x^{-1}S(ax^2)$	$\tfrac{1}{4}\{\text{si}(\tfrac{1}{4}y^2/a)-\text{Ci}(\tfrac{1}{4}y^2/a)\}$

	$f(x)$	$g_c(y)$
10.35	$x^{-1}C(ax^2)$	$-\tfrac{1}{4}\{Ci(\tfrac{1}{4}y^2/a)+si(\tfrac{1}{4}y^2/a)\}$
10.36	$\cos(ax^2)S(ax^2)$ $-\sin(ax^2)C(ax^2)$	$\tfrac{1}{2}(2\pi a)^{-\tfrac{1}{2}}\Big(\cos(\tfrac{1}{4}y^2/a)Ci(\tfrac{1}{4}y^2/a)$ $+\sin(\tfrac{1}{4}y^2/a)\{\pi\ si(\tfrac{1}{4}y^2/a)\}\Big]$
10.37	$\sin(ax^2)S(ax^2)$ $+\cos(ax^2)C(ax^2)$	$\tfrac{1}{2}(2\pi a)^{-\tfrac{1}{2}}\Big(\cos(\tfrac{1}{4}y^2/a)\{\pi+si(\tfrac{1}{4}y^2/a)\}$ $-\sin(\tfrac{1}{4}y^2/a)Ci(\tfrac{1}{4}y^2/a)\Big]$
10.38	$\{\tfrac{1}{2}-S(ax^2)\}\cos(ax^2)$ $-\{\tfrac{1}{2}-C(ax^2)\}\sin(ax^2)$	$-\tfrac{1}{2}(2\pi a)^{-\tfrac{1}{2}}\Big(\sin(\tfrac{1}{4}y^2 a)si(\tfrac{1}{4}y^2/a)$ $+\cos(\tfrac{1}{4}y^2/a)Ci(\tfrac{1}{4}y^2/a)\Big]$
10.39	$\{\tfrac{1}{2}-C(ax^2)\}\cos(ax^2)$ $+\{\tfrac{1}{2}-S(ax^2)\}\sin(ax^2)$	$\tfrac{1}{2}(2\pi a)^{-\tfrac{1}{2}}\Big(\sin(\tfrac{1}{4}y^2/a)Ci(\tfrac{1}{4}y^2/a)$ $-\cos(\tfrac{1}{4}y^2/a)si(\tfrac{1}{4}y^2/a)\Big]$
10.40	$x^{-1}\{C(ax^2)\cos(ax^2)$ $+S(ax^2)\sin(ax^2)\}$	$\tfrac{1}{2}\pi\Big(\tfrac{1}{2}-S(\tfrac{1}{4}y^2/a)\Big]$
10.41	$x^{-1}\{C(ax^2)\sin(ax^2)$ $-S(ax^2)\cos(ax^2)\}$	$\tfrac{1}{2}\pi\Big(\tfrac{1}{2}-C(\tfrac{1}{4}y^2/a)\Big]$
10.42	$S(a/x)$	$\tfrac{1}{4}y^{-1}\{\sin z-\cos z+e^{-z}\};\qquad z=2(ay)^{\tfrac{1}{2}}$
10.43	$C(a/x)$	$\tfrac{1}{4}y^{-1}\{\sin z+\cos z-e^{-z}\};\qquad z=2(ay)^{\tfrac{1}{2}}$
10.44	$\tfrac{1}{2}-S(ax^{\tfrac{1}{2}})$	$-\tfrac{1}{4}\pi^{\tfrac{1}{2}}y^{-3/2}a\cos(z-\tfrac{7}{8}\pi)J_{\tfrac{1}{4}}(z)\ ;\qquad z=\tfrac{1}{8}a^2/y$
10.45	$\tfrac{1}{2}-C(ax^{\tfrac{1}{2}})$	$-\tfrac{1}{4}\pi^{\tfrac{1}{2}}y^{-3/2}a\cos(z-\tfrac{5}{8}\pi)J_{-\tfrac{1}{4}}(z)\qquad z=\tfrac{1}{8}a^2/y$
10.46	$x^{-1}\{\sin uC(u)$ $-\cos uS(u)\};u=a^2/x$	$\tfrac{1}{4}\pi\{J_o(z)+I_o(z)+\mathbf{H}_o(z)-\mathbf{L}_o(z)\}\ ;\quad z=2ay^{\tfrac{1}{2}}$
10.47	$x^{-1}\{\sin uS(u)$ $+\cos uC(u)\};u=a^2/x$	$\tfrac{1}{4}\pi\Big(\mathbf{H}_o(z)+\mathbf{L}_o(z)-J_o(z)-I_o(z)\Big]\ ;\quad z=2(ay)^{\tfrac{1}{2}}$
10.48	$u^{-3/2}\{\sin(bu)C(bu)$ $-\cos(bu)S(bu)\}$ $u=(a^2+x^2)^{\tfrac{1}{2}}$	$0\qquad\qquad y<b$

	$f(x)$	$g_c(y)$
10.49	$(z/a)^{-\frac{1}{2}}\{\sin xS(z)$ $+\cos zC(z)\}$;$z=a \sinh x$	$-\frac{1}{8}i\pi(\pi a)^{\frac{1}{2}}\text{cosech}(\frac{1}{2}\pi y)\left[I_{-\frac{1}{4}-\frac{1}{2}iy}(\frac{1}{2}a)I_{\frac{1}{4}-\frac{1}{2}iy}(\frac{1}{2}a)\right.$ $\left.-I_{\frac{1}{4}+\frac{1}{2}iy}(\frac{1}{2}a)I_{-\frac{1}{4}+\frac{1}{2}iy}(\frac{1}{2}a)\right]$
10.50	$(z/a)^{-\frac{1}{2}}\{\sin zS(z)$ $+\cos zC(z)\}$;$z=a \cosh x$	$\frac{1}{8}i\pi(\pi a)^{\frac{1}{2}}\text{cosech}(\frac{1}{2}\pi y)\left[J_{\frac{1}{4}+\frac{1}{2}iy}(\frac{1}{2}a)J_{-\frac{1}{4}+\frac{1}{2}iy}(\frac{1}{2}a)\right.$ $\left.-J_{\frac{1}{4}-\frac{1}{2}iy}(\frac{1}{2}a)J_{-\frac{1}{4}-\frac{1}{2}iy}(\frac{1}{2}a)\right]$
10.51	$(z/a)^{-\frac{1}{2}}\sin zC(z)$ $-\cos zS(z)$;$z=a \cosh x$	$\frac{1}{8}\pi(\pi a)^{\frac{1}{2}}\text{sech}(\frac{1}{2}\pi y)\left[J_{\frac{1}{4}+\frac{1}{2}iy}(\frac{1}{2}a)J_{-\frac{1}{4}+\frac{1}{2}iy}(\frac{1}{2}a)\right.$ $\left.+J_{\frac{1}{4}-\frac{1}{2}iy}(\frac{1}{2}a)J_{-\frac{1}{4}-\frac{1}{2}iy}(\frac{1}{2}a)\right]$
10.52	$\cos zC(z)+\sin zS(z)$ $z=a \cosh x$	$\frac{1}{2}\pi\left\{\frac{1}{4}\text{sech}(\frac{1}{2}\pi y)\{H^{(1)}_{iy}(a)+H^{(2)}_{-iy}(a)\}-\pi a\,\text{sech}(\pi y)\right.$ $\left.\cdot\left[\Gamma(\frac{1}{4}-\frac{1}{2}iy)\Gamma(\frac{1}{4}+\frac{1}{2}iy)\right]^{-1}S_{-\frac{1}{2},iy}(a)\right\}$
10.53	$\sin zC(z)-\cos zS(z)$ $z=a \cosh x$	$\frac{1}{2}\pi\left\{\frac{1}{4}\text{sech}(\frac{1}{2}\pi y)\{H^{(1)}_{iy}(a)+H^{(2)}_{-iy}(a)\}+\frac{1}{2}\pi a\,\text{sech}(\pi y)\right.$ $\left.\cdot\left[\Gamma(\frac{3}{4}-\frac{1}{2}iy)\Gamma(\frac{3}{4}+\frac{1}{2}iy)\right]^{-1}S_{\frac{1}{2},iy}(a)\right\}$
10.54	$\cos z\{\frac{1}{2}-S(z)\}$ $-\sin z\{\frac{1}{2}-C(z)\}$ $z=a \cosh x$	$\frac{1}{4}\pi^{-1}\Gamma(\frac{1}{4}-\frac{1}{2}iy)\Gamma(\frac{1}{4}+\frac{1}{2}iy)S_{\frac{1}{2},iy}(a)$
10.55	$\cos z\{\frac{1}{2}-C(z)\}$ $+\sin z\{\frac{1}{2}-S(z)\}$ $z=a \cosh x$	$\frac{1}{2}\pi^{-1}\Gamma(\frac{3}{4}-\frac{1}{2}iy)\Gamma(\frac{3}{4}+\frac{1}{2}iy)S_{-\frac{1}{2},iy}(a)$
10.56	$\sin zC(z)-\cos zS(z)$ $z=a \cosh^2 x$	$\frac{1}{4}\pi 2^{-\frac{1}{2}}\text{sech}(\frac{1}{2}\pi y)\left[\sin(\frac{1}{4}\pi+\frac{1}{2}a-\frac{1}{4}i\pi y)J_{\frac{1}{2}iy}(\frac{1}{2}a)\right.$ $\left.+\sin(\frac{1}{4}\pi+\frac{1}{2}a+\frac{1}{4}i\pi y)J_{-\frac{1}{2}iy}(\frac{1}{2}a)\right]$
10.57	$\cos zC(z)+\sin zS(z)$ $z=a \cos$	$\frac{1}{4}\pi 2^{-\frac{1}{2}}\text{sech}(\frac{1}{2}\pi y)\left[\cos(\frac{1}{4}\pi+\frac{1}{2}a-\frac{1}{4}i\pi y)J_{\frac{1}{2}iy}(\frac{1}{2}a)\right.$ $\left.+\cos(\frac{1}{4}\pi+\frac{1}{2}a+\frac{1}{4}i\pi y)J_{-\frac{1}{2}iy}(\frac{1}{2}a)\right]$
10.58	$\cos uC(v)+\sin uS(v)$ $u=a \cosh x$ $v=2a \cosh^2(\frac{1}{2}x)$	$\frac{1}{4}\pi\text{sech}(\pi y)\left[\cosh(\frac{1}{2}\pi y)\{J_{iy}(a)+J_{-iy}(a)\}\right.$ $\left.+i\sin h(\frac{1}{2}\pi y)\{J_{iy}(a)-J_{-iy}(a)\}\right]$
10.59	$\sin uC(v)-\cos uS(v)$ $u=a \cosh x$ $v=2a \cosh^2(\frac{1}{2}x)$	$\frac{1}{4}\pi\text{sech}(\pi y)\left[\cosh(\frac{1}{2}\pi y)\{J_{iy}(a)+J_{-iy}(a)\}\right.$ $\left.-i\sinh(\frac{1}{2}\pi y)\{J_{iy}(a)-J_{-iy}(a)\}\right]$

	$f(x)$	$g_c(y)$

1.11   The Exponential and Related Integrals

	$f(x)$	$g_c(y)$
11.1	$\mathrm{Ei}(-ax)$	$-y^{-1}\arctan(y/a)$
11.2	$\mathrm{Ei}(-bx)\quad x<a$ $0\qquad\quad x>a$	$y^{-1}\Big(\sin(ay)\,\mathrm{Ei}(-ab)-\arctan(y/b)$ $\quad -\tfrac12 i\,\mathrm{Ei}(-ab-iay)+\tfrac12 i\,\mathrm{Ei}(-ab+iay)\Big)$
11.3	$e^{ax}\mathrm{Ei}(-ax)$	$(a^2+y^2)^{-1}\Big(a\,\log(y/a)-\tfrac12\pi y\Big)$
11.4	$e^{-ax}\mathrm{Ei}(-bx),a\geq-b$	$-(a^2+y^2)^{-1}\Big(\tfrac12 a\,\log\{(a+b)^2+y^2\}$ $\quad -a\,\log b+y\,\arctan\{y/(a+b)\}\Big)$
11.5	$e^{-ax}\overline{\mathrm{Ei}}(ax)$	$-(a^2+y^2)^{-1}\{a\,\log(y/a)+\tfrac12\pi y\}$
11.6	$e^{-ax}\overline{\mathrm{Ei}}(bx)\quad,a\geq b$	$-(a^2+y^2)^{-1}\Big(\tfrac12 a\,\log\{(a-b)^2+y^2\}$ $\quad -a\,\log b+y\,\arctan\{y/a-b)\}\Big)$
11.7	$x^{-1}\Big(e^{-ax}\overline{\mathrm{Ei}}(ax)$ $\quad -e^{ax}\mathrm{Ei}(-ax)\Big)$	$\pi\arctan(a/y$
11.8	$\mathrm{Ei}(-ax)\overline{\mathrm{Ei}}(-ax)$	$\tfrac12\pi y^{-1}\log(1+y^2/a^2)$
11.9	$\mathrm{Ei}(-ax^2)$	$-\pi y^{-1}\mathrm{Erf}(\tfrac12 ya^{-\frac12})$
11.10	$e^{ax^2}\mathrm{Ei}(-ax^2)$	$-\tfrac12\pi(\pi/a)^{\frac12}\exp(\tfrac14 y^2/a)\mathrm{Erfc}(\tfrac12 ya^{-\frac12})$
11.11	$\exp(-ax^2)\overline{\mathrm{Ei}}(ax^2)$	$\tfrac12 i\pi(\pi/a)^{\frac12}\exp(-\tfrac14 y^2/a)\mathrm{Erf}(\tfrac12 iya^{-\frac12}$
11.12	$\exp(a\cosh x)$ $\mathrm{Ei}(-a\cosh x)$	$\tfrac12\pi^2\mathrm{cosech}^2(\pi y)\Big[I_{iy}(a)+I_{-iy}(a)$ $\quad -e^{\frac12\pi y}J_{iy}(ia)-e^{-\frac12\pi y}J_{-iy}(ia)\Big]$
11.13	$\exp(-a\cosh x)$ $\cdot\overline{\mathrm{Ei}}(a\cosh x)$	$\tfrac12\pi^2\mathrm{cosech}^2(\pi y)\Big(\cosh(\pi y)\{I_{iy}(a)+I_{-iy}(a)\}$ $\quad -e^{\frac12\pi y}J_{iy}(-ia)-e^{-\frac12\pi y}J_{-iy}(-ia)\Big)$
11.14	$e^{ax^2}\mathrm{Ei}\{-(b+ax^2)\}$	$-\tfrac12\pi(\pi/a)^{\frac12}\exp(\tfrac14 y^2/a)\mathrm{Erfc}(b^{\frac12}+\tfrac12 ya^{-\frac12})$
11.15	$e^{ax}\mathrm{Ei}(-u)+e^{-ax}\mathrm{Ei}(-v)$ $\genfrac{}{}{0pt}{}{u}{v}=a\{(b^2+x^2)^{\frac12}\pm x\}$	$-2a(a^2+y^2)^{-1}K_0\{b(a^2+y^2)^{\frac12}\}$

	$f(x)$	$g_c(y)$
11.16	$e^{au}Ei\{-a(x+u)\}$ $+e^{-au}Ei\{-a(x-u)\}$ $u=(x^2-b^2)^{\frac{1}{2}}$	$4ab(a^2+y^2)^{-\frac{1}{2}}S_{0,1}\{b(a^2+y^2)^{\frac{1}{2}}\}$
11.17	$si(ax)$	$-\frac{1}{2}y^{-1}\log\{(y+a)/(y-a)\}$ $y\neq a$
11.18	$e^{-bx}si(ax)$	$-\frac{1}{2}(b^2+y^2)^{-1}\left\{\frac{1}{2}y\ \log\left(\dfrac{b^2+(y+a)^2}{b^2+(y-a)^2}\right)\right.$ $\left.+\pi b-b\ arctan\{2ab(b^2+y^2-a^2)^{-1}\}\right\}$
11.19	$Si(bx)$ $x<a$ 0 $x>a$	$\frac{1}{2}y^{-1}\left(2\ \sin(ay)Si(ab)+Ci(ay+ab)-Ci(\|ay-ab\|)\right.$ $\left.+\log\|(y-b)/(y+b)\|\right]$,$y\neq b$ $\frac{1}{2}b^{-1}\{2\sin(ab)Si(ab)-Ci(2ab)-\gamma-\log(2ab)\},y=b$
11.20	$x^{-1}Si(ax)$	$\frac{1}{2}\pi\log(a/y)$ $y<a$ 0 $y>a$
11.21	$x(b^2+x^2)^{-1}Si(ax)$	$\frac{1}{4}\pi\left(e^{-by}\{\overline{Ei}(by)-Ei(-ab)\}\right.$ $\left.-e^{by}\{Ei(-by)-Ei(-ab)\}\right]$ $y<a$ $\frac{1}{4}\pi\left(e^{-by}\{Ei(-ab)-\overline{Ei}(ab)\}\right]$ $y>a$
11.22	$Ci(ax)$	0 $y<a$ $-\frac{1}{2}\pi y^{-1}$ $y>a$
11.23	$Ci(bx)$ $x<a$ 0 $x>a$	$\frac{1}{2}y^{-1}\left(2\sin(ay)Ci(ab)-Si(ay+ab)-Si(ay-ab)\right]$
11.24	$(b^2+x^2)^{-1}Ci(ax)$	$\frac{1}{2}\pi b^{-1}\cosh(by)Ei(-ab)$ $y\leq a$ $\frac{1}{4}\pi b^{-1}\left(e^{-by}\{\overline{Ei}(ab)+Ei(-ab)-\overline{Ei}(by)\}\right.$ $\left.+e^{by}Ei(-by)\right]$ $y\geq a$
11.25	$e^{-ax}Ci(bx)$	$-\frac{1}{4}(a^2+y^2)^{-1}\left(a\ \log\{(b^2+a^2-y^2)^2+4a^2y^2\}\right.$ $\left.-4a\log b+2y\ arctan\{2ay/(a^2+b^2-y^2)\}\right]$
11.26	$\sin(ax)Ci(ax)$ $-\cos(ax)si(ax)$	$a(y^2-a^2)^{-1}\log(y/a)$

	$f(x)$	$g_c(y)$	
11.27	$\sin(ax)\,si(ax)$ $+\cos(ax)\,Ci(ax)$	$-\tfrac{1}{2}\pi(a+y)^{-1}$	
11.28	$x^{-1}\{\cos(ax)\,Si(ax)$ $-\sin(ax)\,Ci(ax)\}$	$-\tfrac{1}{4}\pi\log\|(y+a)/(y-a)\|$	
11.29	$\{si(ax)\}^2$	$\tfrac{1}{2}\pi y^{-1}\log(1+y/a)$ $\tfrac{1}{2}\pi y^{-1}\log\|(y+a)/(y-a)\|$	$y<2a$ $y>2a$
11.30	$\{Ci(ax)\}^2$	$\tfrac{1}{2}\pi y^{-1}\log(y^2/a^2-1)$ $\tfrac{1}{2}\pi^{-1}\log(1+y/a)$	$y>2a$ $y<2a$
11.31	$si(ax^2)$	$\pi y^{-1}\{S(\tfrac{1}{4}y^2/a)-C(\tfrac{1}{4}y^2/a)\}$	
11.32	$Ci(ax^2)$	$-\pi y^{-1}\{C(\tfrac{1}{4}y^2/a)+S(\tfrac{1}{4}y^2/a)\}$	
11.33	$\cos(ax^2)\,si(ax^2)$ $-\sin(ax^2)\,Ci(ax^2)$	$\pi(\tfrac{1}{2}\pi/a)^{\frac{1}{2}}\Big(\sin z\{S(z)-\tfrac{1}{2}\}+\cos z\{C(z)-\tfrac{1}{2}\}\Big)$ $z=\tfrac{1}{4}y^2/a)$	
11.34	$\cos(ax^2)\,Ci(ax^2)$ $+\sin(ax^2)\,si(ax^2)$	$\pi(\tfrac{1}{2}\pi/a)^{\frac{1}{2}}\Big(\cos z\{S(z)-\tfrac{1}{2}\}-\sin z\{C(z)-\tfrac{1}{2}\}\Big)$ $z=\tfrac{1}{4}y^2/a$	
11.35	$x^{-1}si(a/x)$	$-\pi\displaystyle\int_c^\infty t^{-1}J_0(t)\,dt$ $\qquad c=2(ay)^{\frac{1}{2}}$	
11.36	$x^{-1}\{\sin(a/x)\,Ci(a/x)$ $-\cos(a/x)\,si(a/x)\}$	$\pi K_0\{2(ay)^{\frac{1}{2}}\}$	
11.37	$\sin z\,Ci(z)$ $-\cos z\,si(z)$ $z=a\cosh x$	$\tfrac{1}{2}\pi\,\text{sech}(\tfrac{1}{2}\pi y)\,S_{0,iy}(a)$	
11.38	$\cos z\,Ci(z)$ $+\sin z\,si(z)$ $z=a\cosh x$	$-\tfrac{1}{2}\pi y\,\text{cosech}(\tfrac{1}{2}\pi y)\,S_{-1,iy}(a)$	

1.12   Legendre Functions

	$f(x)$	$g_c(y)$
12.1	$P_\nu(x)$ $\quad x<1$ $0 \quad x>1$	$-\frac{1}{2}\pi^{\frac{1}{2}}\nu(\nu+1)\{\Gamma(\frac{3}{2}+\frac{3}{2}\nu)\Gamma(1-\frac{1}{2}\nu)\}^{-1}$ $y^{-\frac{1}{2}}S_{-\frac{1}{2},\frac{1}{2}+\nu}(y)$
12.2	$x^{\mu-1}P_\nu(x) \quad x<1$ $0 \quad x>1$ $\mathrm{Re}\,\mu>0$	$\pi^{\frac{1}{2}}2^{-\mu}\Gamma(\mu)\{\Gamma(1+\frac{1}{2}\mu+\frac{1}{2}\nu)\Gamma(\frac{1}{2}+\frac{1}{2}\mu-\frac{1}{2}\nu)\}^{-1}$ $\cdot\,_2F_3(\frac{1}{2}\mu,\frac{1}{2}+\frac{1}{2}\mu;\frac{1}{2}+\frac{1}{2}\mu-\frac{1}{2}\nu,\frac{1}{2}\nu+\frac{1}{2}\mu,\frac{1}{2};-\frac{1}{4}y^2)$
12.3	$(1-x^2)^{-\frac{1}{2}\mu}P_\nu^\mu(x) \quad x<1$ $0 \quad x>1$ $\mathrm{Re}\,\mu<1$	$\pi^{\frac{1}{2}}2^{\mu-1}\{\Gamma(\frac{3}{2}-\frac{3}{2}\mu+\frac{3}{2}\nu)\Gamma(1-\frac{1}{2}\mu-\frac{1}{2}\nu)\}^{-1}$ $\cdot(\mu+\nu)(\mu-\nu-1)y^{-\frac{1}{2}+\mu}S_{-\mu-\frac{1}{2},\nu+\frac{1}{2}}(y)$
12.4	$x^{\lambda-1}(1-x^2)^{-\frac{1}{2}\mu}P_\nu^\mu(x)$ $\quad,x<1$ $0 \quad x>1$ $\mathrm{Re}\,\lambda>0,\ \mathrm{Re}\,\mu<1$	$\pi^{\frac{1}{2}}2^{\mu-\lambda}\left[\Gamma(1+\frac{1}{2}\lambda-\frac{1}{2}\mu+\frac{1}{2}\nu)\Gamma(\frac{1}{2}+\frac{1}{2}\lambda-\frac{1}{2}\mu-\frac{1}{2}\nu)\right]^{-1}$ $\cdot\Gamma(\lambda)\,_2F_3(\frac{1}{2}\lambda,\frac{1}{2}+\frac{1}{2}\lambda;\frac{1}{2}+\frac{1}{2}\lambda-\frac{1}{2}\mu-\frac{1}{2}\nu,\frac{1}{2},1+\frac{1}{2}\lambda-\frac{1}{2}\mu$ $+\frac{1}{2}\nu;-\frac{1}{4}y^2)$
12.5	$0 \quad x<1$ $P_\nu(x) \quad x>1$ $-1<\mathrm{Re}\,\nu<0$	$(\frac{1}{2}\pi/y)^{\frac{1}{2}}\{\sin(\frac{1}{2}\pi\nu)Y_{\nu+\frac{1}{2}}(y)$ $-\cos(\frac{1}{2}\pi\nu)J_{\nu+\frac{1}{2}}(y)\}$
12.6	$(x^2-1)^{\frac{1}{2}\mu}P_\nu^\mu(x)$ $\mathrm{Re}(\mu+\nu)>0;\mathrm{Re}(\mu-\nu)<1$	$2^{\mu+1}\pi^{\frac{1}{2}}\{\Gamma(-\frac{1}{2}\nu-\frac{1}{2}\mu)\Gamma(\frac{1}{2}+\frac{1}{2}\nu-\frac{1}{2}\mu)\}^{-1}y^{-\frac{1}{2}-\mu}$ $\cdot S_{\mu-\frac{1}{2},\nu+\frac{1}{2}}(y)$
12.7	$0 \quad x<1$ $(x^2-1)^{-\frac{1}{2}\mu}P_\nu^\mu(x) \quad x>1$ $-\frac{1}{2}<\mathrm{Re}\,\mu<1,\mathrm{Re}\,\mu>\mathrm{Re}\,\nu>-1$ $\quad -\mathrm{Re}\,\mu$	$-(\frac{1}{2}\pi)^{\frac{1}{2}}y^{\mu-\frac{1}{2}}\left(\cos(\frac{1}{2}\pi\mu-\frac{1}{2}\pi\nu)J_{\nu+\frac{1}{2}}(y)\right.$ $\left.+\sin(\frac{1}{2}\pi\mu-\frac{1}{2}\pi\nu)Y_{\nu+\frac{1}{2}}(y)\right]$
12.8	$\{x(x+1)\}^{-\frac{1}{2}\mu}P_\nu^\mu(1+2x)$ $\mathrm{Re}\,\mu<1,-1-\mathrm{Re}\,\mu<\mathrm{Re}\,\nu<\mathrm{Re}\,\mu$	$-\frac{1}{2}\pi^{\frac{1}{2}}y^{\mu-\frac{1}{2}}\left(\cos(\frac{1}{2}y+\frac{1}{2}\pi\mu-\frac{1}{2}\pi\nu)J_{\nu+\frac{1}{2}}(\frac{1}{2}y)\right.$ $\left.+\sin(\frac{1}{2}y+\frac{1}{2}\pi\mu-\frac{1}{2}\pi\nu)Y_{\nu+\frac{1}{2}}(\frac{1}{2}y)\right]$
12.9	$Q_{2n+1}(x/a) \quad x<a$ $q_{2n+1}(x/a) \quad x>a$ $n=0,1,2,\ldots$	$(-1)^{n+1}(\frac{1}{2}\pi)^{3/2}(a/y)^{\frac{1}{2}}J_{2n+3/2}(ay)$

	$f(x)$		$g_c(y)$
12.10	$p_\nu(1+x^2)$ $-1<\mathrm{Re}\,\nu<0$		$-\pi^{-1}2^{\frac{1}{2}}\sin(\pi\nu)\{K_{\nu+\frac{1}{2}}(y2^{-\frac{1}{2}})\}^2$
12.11	$q_\nu(1+x^2)$ $\mathrm{Re}\,\nu>-1$		$\pi 2^{-\frac{1}{2}}I_{\nu+\frac{1}{2}}(y2^{-\frac{1}{2}})K_{\nu+\frac{1}{2}}(y2^{-\frac{1}{2}})$
12.12	$x^{-1}q_\nu(1+2a^2/x^2)$ $\mathrm{Re}\,\nu>-1$		$\frac{1}{2}\pi\Gamma(1+\nu)(ay)^{-1}W_{-\nu-\frac{1}{2},0}(ay)\mathrm{cosec}(\pi\nu)$ $\cdot\left[\cos(\pi\nu)M_{\nu+\frac{1}{2},0}(ay)-W_{-\nu-\frac{1}{2},0}(ay)\right]$
12.13	$p_\nu\{(x^2+a^2+b^2)/(2ab)\}$ $-1<\mathrm{Re}\,\nu<0$		$-2\pi^{-1}(ab)^{\frac{1}{2}}\sin(\pi\nu)K_{\nu+\frac{1}{2}}(ay)K_{\nu+\frac{1}{2}}(by)$
12.14	$q_\nu\{(x^2+a^2+b^2)/(2ab)\}$ $\mathrm{Re}\,\nu>-1$		$\pi(ab)^{\frac{1}{2}}I_{\nu+\frac{1}{2}}(by)K_{\nu+\frac{1}{2}}(ay)$ $\qquad a>b$
12.15	$P_\nu(2x^2/a^2-1)$ 0	$x<a$ $x>a$	$\frac{1}{2}\pi aJ_{\frac{1}{2}+\nu}(\frac{1}{2}ay)J_{-\frac{1}{2}-\nu}(\frac{1}{2}ay)$
12.16	0 $x^{-1}p_\nu(2x^2-1)$ $-1<\mathrm{Re}\,\nu<0$	$x<1$ $x>1$	$-\frac{1}{2}\pi\cos\mathrm{ec}(\pi\nu)_1F_1(1+\nu;1;iy)_1F_1(1+\nu;1;-iy)$
12.17	$P_\nu(x^2)$ $p_\nu(x^2)$ $-1<\mathrm{Re}\,\nu<1$	$x<1$ $x>1$	$-\pi^{-1}2^{-\frac{1}{2}}\sin(\pi\nu)K_{\nu+\frac{1}{2}}\{y(\frac{1}{2}i)^{\frac{1}{2}}\}K_{\nu+\frac{1}{2}}\{y(-\frac{1}{2}i)^{\frac{1}{2}}\}$
12.18	$P_\nu(\frac{1}{2}x^2/a^2-1)$ $p_\nu(\frac{1}{2}x^2/a^2-1)$ $-1<\mathrm{Re}\,\nu<0$	$x<2a$ $x>2a$	$-\frac{1}{2}\pi a\sin(\pi\nu)\left[\{J_{\nu+\frac{1}{2}}(ay)\}^2+\{Y_{\nu+\frac{1}{2}}(ay)\}^2\right]$
12.19	$Q_\nu(\frac{1}{2}x^2/a^2-1)$ $q_\nu(\frac{1}{2}x^2/a^2-1)$ $\mathrm{Re}\,\nu>-1$	$x<2a$ $x>2a$	$-\frac{1}{2}\pi^2 aJ_{\nu+\frac{1}{2}}(ay)Y_{-\nu-\frac{1}{2}}(ay)$

	$f(x)$	$g_c(y)$
12.20	$p_\nu\{(a^2+b^2-x^2)/(2ab)\}$ $\qquad\qquad x<a-b$ $\qquad 0 \qquad\qquad x>a-b$	$\tfrac{1}{2}\pi(ab)^{\frac{1}{2}}\Big(J_{\nu+\frac{1}{2}}(by)Y_{\nu+\frac{1}{2}}(ay)$ $-J_{\nu+\frac{1}{2}}(ay)Y_{\nu+\frac{1}{2}}(by)\Big)\qquad a>b$
12.21	$\qquad 0 \qquad\qquad x<a+b$ $p_\nu\{(x^2-a^2-b^2)/(2ab)\}$ $\qquad\qquad x>a+b$ $\qquad -1<\mathrm{Re}\,\nu<0$	$\tfrac{1}{2}\pi(ab)^{\frac{1}{2}}\Big(Y_{\nu+\frac{1}{2}}(ay)Y_{-\nu-\frac{1}{2}}by)$ $-J_{\nu+\frac{1}{2}}(ay)J_{-\nu-\frac{1}{2}}(by)\Big)$
12.22	$2\pi^{-1}\sin(\pi\nu)$ $q_\nu\{(a^2+b^2-x^2)/(2ab)\}$ $\qquad\qquad x<a-b$ $P_\nu\{(x^2-a^2-b^2)/(2ab)\}$ $\qquad\qquad a-b<x<a+b$ $\qquad 0 \qquad\qquad x>a+b$	$\pi(ab)^{\frac{1}{2}}J_{\nu+\frac{1}{2}}(by)J_{-\nu-\frac{1}{2}}(ay)\qquad a>b$
12.23	$z^{2n+1}Q_{2n+1}(zx),n=0,1,\ldots$ $z=(a^2+x^2)^{-\frac{1}{2}},\mathrm{Re}\,(\nu\pm\mu)<0$	$\tfrac{1}{2}\pi(-1)^{n+1}\{(2n+1)!\}^{-1}y^{2n+1}K_0(ay)$
12.24	$z^{-\nu}\{P_\nu^\mu(zx)+P_\nu^\mu(-zx)\}$ $z=(a^2+x^2)^{-\frac{1}{2}},\mathrm{Re}\,(\nu\pm\mu)<0$	$2\sin(\tfrac{1}{2}\pi\mu-\tfrac{1}{2}\pi\nu)\{\Gamma(-\nu-\mu)\}^{-1}y^{-\nu-1}K_\mu(ay)$
12.25	$z^{-\nu}\{P_\nu^\mu(zx)-P_\nu^\mu(-zx)\}$ $z=(a^2+x^2)^{-\frac{1}{2}},\mathrm{Re}\,(\nu\pm\mu)<0$	$-\pi^{\frac{1}{2}}2^\mu\{\Gamma(-\tfrac{1}{2}\nu-\tfrac{1}{2}\mu)\}^{-1}y^{-\nu-1}\Big(2^\nu\Gamma(\tfrac{1}{2}\nu+\tfrac{1}{2}\mu+\tfrac{1}{2})$ $\cdot\{I_\mu(ay)+I_{-\mu}(ay)+2\pi^{-1}\sin(\pi\nu)K_\mu(ay)\}$ $-4ie^{-\frac{1}{2}i\pi\nu}\{\Gamma(\tfrac{1}{2}+\tfrac{1}{2}\nu-\tfrac{1}{2}\mu)\}^{-1}s_{\nu,\mu}(iay)\Big)$
12.26	$P_\nu(-\cos x) \qquad x<\pi$ $\qquad 0 \qquad\qquad x>\pi$	$\pi^2\{\Gamma(\tfrac{1}{2}-\tfrac{1}{2}\nu-\tfrac{1}{2}y)\Gamma(\tfrac{1}{2}-\tfrac{1}{2}\nu+\tfrac{1}{2}y)\Gamma(1+\tfrac{1}{2}\nu+\tfrac{1}{2}y)$ $\cdot\Gamma(1+\tfrac{1}{2}\nu-\tfrac{1}{2}y)\}^{-1}$
12.27	$P_\nu(\cosh x)$ $\qquad -1<\mathrm{Re}\,\nu<0$	$-\tfrac{1}{4}\pi^{-2}\sin(\pi\nu)\Gamma(-\tfrac{1}{2}\nu+\tfrac{1}{2}iy)\Gamma(-\tfrac{1}{2}\nu-\tfrac{1}{2}iy)$ $\cdot\Gamma(\tfrac{1}{2}+\tfrac{1}{2}\nu+\tfrac{1}{2}iy)\Gamma(\tfrac{1}{2}+\tfrac{1}{2}\nu-\tfrac{1}{2}iy)\}^{-1}$
12.28	$p_\nu(a\cosh x)$ $\qquad -1<\mathrm{Re}\,\nu<0,a\geq1$	$-2^{\nu-3/2}(\pi a)^{-\frac{1}{2}}\sin(\pi\nu)\{\cosh(\pi y)-\cos(\pi\nu)\}^{-1}$ $\cdot\Gamma(\tfrac{1}{2}+\tfrac{1}{2}\nu+\tfrac{1}{2}iy)\Gamma(\tfrac{1}{2}+\tfrac{1}{2}\nu-\tfrac{1}{2}iy)$ $\cdot\Big(P_{-\frac{1}{2}+iy}^{-\nu-\frac{1}{2}}(z)+P_{-\frac{1}{2}+iy}^{-\nu-\frac{1}{2}}(-z)\Big);z=(1-a^{-2})^{\frac{1}{2}}$

	$f(x)$	$g_c(y)$
12.29	$q_\nu(a\cosh x)$ $a\geq 1,\ \mathrm{Re}\,\nu>-1$	$2^{\nu-1}(\tfrac12\pi/a)^{\frac12}\Gamma(\tfrac12+\tfrac12\nu+\tfrac12 iy)\,\Gamma(\tfrac12+\tfrac12\nu-\tfrac12 iy)$ $\cdot P^{-\frac12-\nu}_{-\frac12+iy}(z)\qquad;\ z=(1-a^{-2})^{\frac12}$
12.30	$(\sinh x)^\mu p^\mu_\nu(\cosh x)$ $\mathrm{Re}(1+\nu-\mu)\ 0,\ \mathrm{Re}(\nu+\mu)<0$	$\tfrac12\pi^{-1}2^{-\mu}\Gamma(\tfrac12+\tfrac12\nu-\tfrac12\mu+\tfrac12 iy)\,\Gamma(\tfrac12+\tfrac12\nu-\tfrac12\mu-\tfrac12 iy)$ $\cdot\Gamma(-\tfrac12\nu-\tfrac12\mu+\tfrac12 iy)\,\Gamma(-\tfrac12\nu-\tfrac12\mu-\tfrac12 iy)$ $\cdot\Big(\Gamma(-\nu-\mu)\,\Gamma(1+\nu-\mu)\,\Gamma(\tfrac12-\mu)\Big)^{-1}$
12.31	$\sinh x)^{-\nu-1}q^\mu_\nu(\coth x)$ $-1<\mathrm{Re}(\nu+\mu)<0$	$\tfrac14 e^{i\pi\mu}2^\nu\{\Gamma(1+\nu)\,\Gamma(1+\nu-\mu)\}^{-1}$ $\cdot\Gamma(\tfrac12+\tfrac12\nu-\tfrac12\mu+\tfrac12 iy)\,\Gamma(\tfrac12+\tfrac12\nu-\tfrac12\mu-\tfrac12 iy)$ $\cdot\Gamma(\tfrac12+\tfrac12\nu+\tfrac12\mu+\tfrac12 iy)\,\Gamma(\tfrac12+\tfrac12\nu+\tfrac12\mu-\tfrac12 iy)$
12.32	$(a^2\cosh^2 x-1)^{\frac12\nu}$ $p^\mu_\nu(a\cosh x)$ $\mathrm{Re}(\nu+\mu)<0;$ $\mathrm{Re}(1+\nu-\mu)>0$ $a>1$	$\tfrac12\pi^{-\frac12}a^{\mu-\nu-1}2^{-\mu}\Gamma(\tfrac12\nu-\tfrac12\mu+\tfrac12+\tfrac12 iy)\,\Gamma(-\tfrac12\nu-\tfrac12\mu-\tfrac12 iy)$ $\cdot\Gamma(\tfrac12+\tfrac12\nu+\tfrac12\mu-\tfrac12 iy)\,\Gamma(-\tfrac12\nu-\tfrac12\mu+\tfrac12 iy)$ $\cdot\Big(\Gamma(\tfrac12-\mu)\,\Gamma(1+\nu-\mu)\,\Gamma(-\nu-\mu)\Big)^{-1}$ $\cdot\,_2F_1\big(\tfrac12+\tfrac12\nu-\tfrac12\mu+\tfrac12 iy,\ \tfrac12+\tfrac12\nu-\tfrac12\mu-\tfrac12 iy;\ \tfrac12-\mu;\ 1-a^{-2}\big)$
12.33	$P_\nu(1-2a^2\cos^2 x)\quad x<\tfrac12\pi$ $0\qquad\qquad x>\tfrac12\pi$ $a\geq 1$	$\tfrac12\pi P^{\frac12 y}_\nu(z)P^{-\frac12 y}_\nu(z)\quad;\qquad z=(1-a^2)^{\frac12}$
12.34	$p_\nu(1+2a^2\cos^2 x)\quad x<\tfrac12\pi$ $0\qquad\qquad x>\tfrac12\pi$	$\tfrac12\pi p^{\frac12 y}_\nu(z)p^{-\frac12 y}_\nu(z)\quad;\qquad z=(1+a^2)^{\frac12}$
12.35	$p_\nu(1+2a^2\sinh^2 x)$ $-1<\mathrm{Re}\,\nu<0,\ a<1$	$\tfrac14\,\mathrm{sech}(\tfrac12\pi y)\Big(P^{-\frac12 iy}_\nu(s)\{Q^{\frac12 iy}_\nu(s)+Q^{\frac12 iy}_{-\nu-1}(s)\}$ $+P^{\frac12 iy}_\nu(s)\{Q^{-\frac12 iy}_\nu(s)+Q^{-\frac12 iy}_{-\nu-1}(s)\}\Big)$ $s=(1-a^2)^{\frac12}$
12.36	$P_\nu(1+2a^2\sinh^2 x)$ $a>1;\ -1<\mathrm{Re}\,\nu<0$ $s=(1-a^{-2})^{\frac12}$	$\tfrac14 a^{-1}\sec(\pi\nu)$ $\cdot\Big(P^{\frac12+\nu}_{-\frac12+\frac12 iy}(s)\{Q^{-\frac12-\nu}_{-\frac12+\frac12 iy}(s)+Q^{-\frac12-\nu}_{-\frac12-\frac12 iy}(s)\}$ $-P^{-\frac12-\nu}_{-\frac12+\frac12 iy}(s)\{Q^{\frac12+\nu}_{-\frac12+\frac12 iy}(s)+Q^{\frac12+\nu}_{-\frac12-\frac12 iy}(s)\}\Big)$
12.37	$q_\nu(1+2a^2\sinh^2 x)$ $a<1\ ;\ \mathrm{Re}\,\nu>-1$	$-\tfrac14 i\pi\,\mathrm{cosech}(\tfrac12\pi y)$ $\cdot\Big(P^{\frac12 iy}_\nu(s)Q^{-\frac12 iy}_\nu(s)-P^{-\frac12 iy}_\nu(s)Q^{\frac12 iy}_\nu(s)\Big)$ $s=(1-a^2)^{\frac12}$

	$f(x)$	$g_c(y)$
12.38	$q_\nu(1+2a^2\sinh^2 x)$ $a>1$; $-1<\mathrm{Re}\nu<0$	$-\tfrac{1}{4}\pi a^{-1}\cos\mathrm{ec}(\pi\nu)$ $\cdot\left[P^{-\frac{1}{2}-\nu}_{-\frac{1}{2}+\frac{1}{2}iy}(s)\{Q^{\frac{1}{2}+\nu}_{-\frac{1}{2}+\frac{1}{2}iy}(s)+Q^{\frac{1}{2}+\nu}_{-\frac{1}{2}-\frac{1}{2}iy}(s)\}\right]$ $s=(1-a^{-2})^{\frac{1}{2}}$
12.39	$p_\nu(1+2a^2\cosh^2 x)$ $-1<\mathrm{Re}\nu<0$ $s=(1+a^{-2})^{\frac{1}{2}}$	$\tfrac{1}{4}a^{-1}\tan(\pi\nu)$ $\cdot\Big[\Gamma(1+\nu+\tfrac{1}{2}iy)\,\Gamma(1+\nu-\tfrac{1}{2}iy)\{p^{-\frac{1}{2}-\nu}_{-\frac{1}{2}+\frac{1}{2}iy}(s)\}^2$ $-\Gamma(-\nu+\tfrac{1}{2}iy)\,\Gamma(-\nu-\tfrac{1}{2}iy)\{p^{\frac{1}{2}+\nu}_{-\frac{1}{2}+\frac{1}{2}iy}(s)\}^2\Big]$
12.40	$q_\nu(1+2a^2\cosh^2 x)$ $\mathrm{Re}\nu>-1$	$\tfrac{1}{4}\pi a^{-1}\Gamma(1+\nu+\tfrac{1}{2}iy)\,\Gamma(1+\nu-\tfrac{1}{2}iy)\{p^{-\frac{1}{2}-\nu}_{-\frac{1}{2}+\frac{1}{2}iy}(s)\}^2$ $s=(1+a^{-2})^{\frac{1}{2}}$
12.41	$p_\nu(2a^2\cosh^2 x-1)$ $-1<\mathrm{Re}\nu<0$, $a<1$	$-\tfrac{1}{2}i\,\sin(\pi\nu)\cosech(\pi\nu)$ $\Big(P^{-\frac{1}{2}iy}_{\nu}(s)\{Q^{\frac{1}{2}iy}_{\nu}(s)+Q^{\frac{1}{2}iy}_{-\nu-1}(s)$ $-P^{\frac{1}{2}iy}_{\nu}(s)\{Q^{-\frac{1}{2}iy}_{\nu}(s)+Q^{-\frac{1}{2}iy}_{-\nu-1}(s)\}\Big]$, $s=(1-a^2)^{\frac{1}{2}}$
12.42	$p_\nu(2a^2\cosh^2 x-1)$ $-1<\mathrm{Re}\nu<0$, $a>1$ $s=(1-a^{-2})^{\frac{1}{2}}$	$\tfrac{1}{4}a^{-1}\mathrm{sech}(\tfrac{1}{2}\pi y)$ $\cdot\Big[P^{-\frac{1}{2}-\nu}_{-\frac{1}{2}+\frac{1}{2}iy}(s)\{Q^{\nu+\frac{1}{2}}_{-\frac{1}{2}+\frac{1}{2}iy}(s)+Q^{\nu+\frac{1}{2}}_{-\frac{1}{2}-\frac{1}{2}iy}(s)\}$ $+P^{\nu+\frac{1}{2}}_{-\frac{1}{2}+\frac{1}{2}iy}(s)\{Q^{-\nu-\frac{1}{2}}_{-\frac{1}{2}+\frac{1}{2}iy}(s)+Q^{-\nu-\frac{1}{2}}_{-\frac{1}{2}-\frac{1}{2}iy}(s)\}\Big]$
12.43	$q_\nu(2a^2\cosh^2 x-1)$ $\mathrm{Re}\nu>-1$, $a>1$	$\tfrac{1}{4}\pi a^{-1}\Gamma(1+\nu+\tfrac{1}{2}iy)\,\Gamma(1+\nu-\tfrac{1}{2}iy)$ $\{p^{-\nu-\frac{1}{2}}_{-\frac{1}{2}+\frac{1}{2}iy}(s)\}^2$; $s=(1-a^{-2})^{\frac{1}{2}}$
12.44	$P_\nu(2a^2\sinh^2 x-1)$, $\sinh x<\frac{1}{a}$ 0 , $\sinh x>\frac{1}{a}$	$\tfrac{1}{2}\pi a^{-1}p^{\nu+\frac{1}{2}}_{-\frac{1}{2}+\frac{1}{2}iy}(s)\,p^{-\nu-\frac{1}{2}}_{-\frac{1}{2}-\frac{1}{2}iy}(s)$; $s=(1+a^{-2})^{\frac{1}{2}}$
12.45	0 , $\sinh x<\frac{1}{a}$ $p_\nu(2a^2\sinh^2 x-1)$, $\sinh x>\frac{1}{a}$ $-1<\mathrm{Re}\nu<0$	$(2\pi a)^{-1}\Big(q^{-\nu-\frac{1}{2}}_{-\frac{1}{2}+\frac{1}{2}iy}(s)+q^{\nu+\frac{1}{2}}_{-\frac{1}{2}+\frac{1}{2}iy}(s)$ $+q^{-\nu-\frac{1}{2}}_{-\frac{1}{2}-\frac{1}{2}iy}(s)\,q^{\nu+\frac{1}{2}}_{-\frac{1}{2}-\frac{1}{2}iy}(s)\Big]$ $s=(1+a^{-2})^{\frac{1}{2}}$
12.46	$(\mathrm{sech}\,x)^{2n+2}Q_{2n+1}(\tanh x)$	$(-1)^{n+1}2^{2n-1}\{(2n+1)!\}^{-2}\cosh(\tfrac{1}{2}\pi y)$ $\left(\Gamma(n+1+\tfrac{1}{2}iy)\,\Gamma(n+1-\tfrac{1}{2}iy)\right)^2$

	$f(x)$	$g_c(y)$
12.47	$(\sinh x)^{-\mu} q^{\mu}_{\nu}(\cosh x)$ $\mathrm{Re}(\nu+\mu+1)>0, \ \mathrm{Re}\,\mu<\tfrac{1}{2}$	$e^{i\pi\mu}\pi^{\frac{1}{2}}2^{\mu-2}\Gamma(\tfrac{1}{2}-\mu)$ $\cdot \Gamma(\tfrac{1}{2}+\tfrac{1}{2}\nu+\tfrac{1}{2}\mu+\tfrac{1}{2}iy)\,\Gamma(\tfrac{1}{2}+\tfrac{1}{2}\nu+\tfrac{1}{2}\mu-\tfrac{1}{2}iy)$ $\cdot\left[\Gamma(1+\tfrac{1}{2}\nu-\tfrac{1}{2}\mu+\tfrac{1}{2}iy)\,\Gamma(1+\tfrac{1}{2}\nu-\tfrac{1}{2}\mu-\tfrac{1}{2}iy)\right]^{-1}$
12.48	$(a^2\cosh^2 x-1)^{\frac{1}{2}\mu}$ $\cdot q^{-\mu}_{\nu}(a\cosh x) \qquad a\geq 1$ $\mathrm{Re}(\mu-\nu-1)<0$	$\tfrac{1}{4}\pi^{\frac{1}{2}}e^{-i\pi\mu}2^{-\mu}a^{\mu-\nu-1}\{\Gamma(\tfrac{3}{2}+\nu)\}^{-1}$ $\cdot \Gamma(\tfrac{1}{2}+\tfrac{1}{2}\nu-\tfrac{1}{2}\mu+\tfrac{1}{2}iy)\,\Gamma(\tfrac{1}{2}+\tfrac{1}{2}\nu-\tfrac{1}{2}\mu-\tfrac{1}{2}iy)$ $\cdot {}_2F_1(\tfrac{1}{2}+\tfrac{1}{2}\nu-\tfrac{1}{2}\mu+\tfrac{1}{2}iy,\tfrac{1}{2}+\tfrac{1}{2}\nu-\tfrac{1}{2}\mu-\tfrac{1}{2}iy;3/2+\nu;a^{-2})$
12.49	$z^{-\mu-1}Q_{2n+1}(az^{-\frac{1}{2}}\sinh x)$ $n=0,1,2,\ldots$ $z=b^2+a^2\sin\ x$	$\tfrac{1}{2}(-1)^{n+1}2^{2n}a^{-2n-2}\{(2n+1)!\}^{-2}\cos h(\tfrac{1}{2}\pi y)$ $\cdot\left[\Gamma(n+1+\tfrac{1}{2}iy)\,\Gamma(n+1-\tfrac{1}{2}iy)\right]^2$ $\cdot {}_2F_1(1+\mu+\tfrac{1}{2}iy,1+\mu-\tfrac{1}{2}iy;2n+2;1-b^2/a^2)$
12.50	$\log(\sinh x)q_{\nu}(\cosh x)$ $\mathrm{Re}\,\nu>-1$	$\tfrac{1}{4}\pi\Gamma(\tfrac{1}{2}+\tfrac{1}{2}\nu+\tfrac{1}{2}iy)\,\Gamma(\tfrac{1}{2}+\tfrac{1}{2}\nu-\tfrac{1}{2}iy)$ $\cdot\left[\Gamma(1+\tfrac{1}{2}\nu+\tfrac{1}{2}iy)\,\Gamma(1+\tfrac{1}{2}\nu-\tfrac{1}{2}iy)\right]^{-1}\Big(-\gamma-\log 4$ $+\Psi(\nu+1)-\tfrac{1}{2}\Psi(\tfrac{1}{2}+\tfrac{1}{2}\nu+\tfrac{1}{2}iy)-\tfrac{1}{2}\Psi(\tfrac{1}{2}+\tfrac{1}{2}\nu-\tfrac{1}{2}iy)$ $-\tfrac{1}{2}\Psi(1+\tfrac{1}{2}\nu+\tfrac{1}{2}iy)-\tfrac{1}{2}\Psi(1+\tfrac{1}{2}\nu-\tfrac{1}{2}iy)\Big]$
12.51	$(b^2+a^2\sinh^2 x)^{\frac{1}{2}\nu}$ $\cdot\left[P^{\mu}_{\nu}(z)+P^{\mu}_{\nu}(-z)\right]$ $z=a\sinh x(b^2+a^2\sinh^2 x)^{-\frac{1}{2}}$ $\mathrm{Re}(\mu+\nu<0$	$\tfrac{1}{2}\pi^{-1}2^{-\nu}\sin(\tfrac{1}{2}\pi\mu-\tfrac{1}{2}\pi\nu)a^{\nu-\mu}\{\Gamma(-\nu)\,\Gamma(-\nu-\mu)\}^{-1}$ $\cdot\cosh(\tfrac{1}{2}\pi y)\,\Gamma(\tfrac{1}{2}\mu-\tfrac{1}{2}\nu+\tfrac{1}{2}iy)\,\Gamma(\tfrac{1}{2}\mu-\tfrac{1}{2}\nu-\tfrac{1}{2}iy)$ $\cdot\Gamma(-\tfrac{1}{2}\mu-\tfrac{1}{2}\nu+\tfrac{1}{2}iy)\,\Gamma(-\tfrac{1}{2}\mu-\tfrac{1}{2}\nu-\tfrac{1}{2}iy)$ $\cdot {}_2F_1(\tfrac{1}{2}\mu-\tfrac{1}{2}\nu+\tfrac{1}{2}iy,\tfrac{1}{2}\mu-\tfrac{1}{2}\nu-\tfrac{1}{2}iy;-\nu;1-b^2/a^2)$
12.52	$z^{-\frac{1}{2}\nu-\frac{1}{2}}P^{\mu}_{\nu}(bz^{-\frac{1}{2}}\cosh x)$ $\mathrm{Re}(\nu-\mu+1)<0$ $,z=a^2+b^2\cosh^2 x$	$2^{\nu-1}a^{-\mu}\Gamma(\tfrac{1}{2}+\tfrac{1}{2}\nu-\tfrac{1}{2}\mu+\tfrac{1}{2}iy)\,\Gamma(\tfrac{1}{2}+\tfrac{1}{2}\nu-\tfrac{1}{2}\mu-\tfrac{1}{2}iy)$ $\cdot\{\Gamma(1-\mu)\,\Gamma(1+\nu-\mu)\}^{-1}y^{\mu-x-1}{}_2F_1(\tfrac{1}{2}+\tfrac{1}{2}\nu-\tfrac{1}{2}\mu+\tfrac{1}{2}iy,$ $\tfrac{1}{2}+\tfrac{1}{2}\nu-\tfrac{1}{2}\mu-\tfrac{1}{2}iy;1-\mu;-a^2/b^2)$
12.53	$P^{\mu}_{-\frac{1}{2}+x}(\cos a) \qquad a<\pi$ $\mathrm{Re}\,\mu<\tfrac{1}{2}$	$(\tfrac{1}{2}\pi)^{\frac{1}{2}}\{\Gamma(1-\mu)\}^{-1}(\sin a)^{\mu}(\cos y-\cos a)^{-\mu-\frac{1}{2}}$ $\qquad\qquad\qquad\qquad y<a$ $\qquad\qquad 0 \qquad\qquad\qquad y>a$
12.54	$\cos(\pi x)q_{-\frac{1}{2}+x}(z)q_{-\frac{1}{2}-x}(z)$	$\tfrac{1}{2}\pi(z^2-\sin^2\tfrac{1}{2}y)^{-\frac{1}{2}}\mathbf{K}\left(\cos(\tfrac{1}{2}y)(z^2-\sin^2\tfrac{1}{2}y)^{-\frac{1}{2}}\right)$ $\qquad\qquad\qquad\qquad y<\pi$ $\qquad\qquad 0 \qquad\qquad\qquad y>\pi$

	$f(x)$	$g_c(y)$
12.55	$\left[\Gamma(\frac{1}{2}+\mu+x)\Gamma(\frac{1}{2}+\mu-x)\right]^{-1}$ $\cdot q^{\mu}_{-\frac{1}{2}+x}(z)\,q^{\mu}_{-\frac{1}{2}-x}(z)$ $z>1$	$\frac{1}{4}\pi e^{i2\pi\mu}(z^2-1)^{-\frac{1}{2}}$ $\cdot P_{\mu-\frac{1}{2}}\left(1+2(z^2-1)^{-1}\cos^2(\frac{1}{2}y)\right)\quad y<\pi$ $\qquad\qquad\qquad\qquad 0\qquad\qquad\quad y>\pi$
12.56	$P^{x}_{\nu}(a)\,P^{-x}_{\nu}(a)\ ,\ a<1$	$\frac{1}{2}P_{\nu}\left(1-2(1-a^2)\cos^2(\frac{1}{2}y)\right)\qquad y<\pi$ $\qquad\qquad\qquad\qquad 0\qquad\qquad\qquad y>\pi$
12.57	$P^{x}_{\nu}(z)\,P^{-x}_{\nu}(z)\qquad z>1$ $\qquad\qquad\qquad\qquad z>1$	$\frac{1}{2}P_{\nu}\left(1-2(1-z^2)\cos^2(\frac{1}{2}y)\right)\qquad y<\pi$ $\qquad\qquad\qquad\qquad 0\qquad\qquad\qquad y>\pi$
12.58	$\operatorname{sech}(\pi x)\,P_{-\frac{1}{2}+ix}(a)$	$2^{-\frac{1}{2}}(a+\cosh y)^{-\frac{1}{2}}\qquad\qquad\qquad -1<a<1$
12.59	$\tanh(x\pi)\operatorname{cosech}(bx)$ $\cdot P_{-\frac{1}{2}+ix}(a)$ $b\geq\pi\ ,\ -1<a<1$	$b^{-1}\sum_{n=0}^{\infty}(-1)^{n}\varepsilon_{n}\cos(n\pi y/b)\,Q_{\frac{1}{2}n\pi-\frac{1}{2}}(a)$ $\qquad\qquad\quad -b\leq y\leq b$
12.60	$\operatorname{sech}(\pi x)\,P^{\mu}_{-\frac{1}{2}+ix}(a)$ $-1<a<1$	$2^{-\mu-\frac{1}{2}}(1+a)^{\frac{1}{2}\mu}(a+\cosh y)^{-\frac{1}{2}\mu-\frac{1}{2}}$ $\cdot P^{\mu}_{\mu}\{(1+\cosh y)^{\frac{1}{2}}(a+\cosh y)^{-\frac{1}{2}}\}$
12.61	$\operatorname{sech}^{2}(\pi x)\,P_{-\frac{1}{2}+ix}(a)$ $-1<a<1$	$2^{-\frac{1}{2}}\pi^{-1}(\cosh y-a)^{-\frac{1}{2}}$ $\left[\log\{(1+\cosh y)^{\frac{1}{2}}+(\cosh y-a)^{\frac{1}{2}}\}\right.$ $\left.-\log\{(1+\cosh y)^{\frac{1}{2}}-(\cosh y+a)^{\frac{1}{2}}\}\right]$
12.62	$x^{-1}\operatorname{sech}(\pi x)$ $\cdot\{P_{\frac{1}{2}+ix}(a)-P_{\frac{1}{2}-ix}(a)\}$ $-1<a<1$	$i2^{3/2}\{(a+\cosh y)^{\frac{1}{2}}-(1+\cosh y)^{\frac{1}{2}}\}$
12.63	$\Gamma(\frac{1}{4}+ix)\Gamma(\frac{1}{4}-ix)P_{-\frac{1}{2}+ix}(a)$ $a<1$	$2(2\pi)^{\frac{1}{2}}\{(1-a^2)^{\frac{1}{2}}+\cosh y\}^{-\frac{1}{2}}$ $\cdot K\left(2^{\frac{1}{2}}\{1+(1-a^2)^{-\frac{1}{2}}\cosh y\}^{-\frac{1}{2}}\right)$
12.64	$\{\Psi(\frac{1}{2}+ix)+\Psi(\frac{1}{2}-ix)\}$ $\cdot\operatorname{sech}(\pi x)\,P_{-\frac{1}{2}+ix}(a)$ $-1<a<1$	$(2a)^{-\frac{1}{2}}(1+\cosh y)^{-\frac{1}{2}}\left(-2\gamma+\log(1+a)-4\log 2\right.$ $2\log\{(1+\cosh y)^{\frac{1}{2}}+(a+\cosh y)^{\frac{1}{2}}\}$ $\left.-2\log(a+\cosh y)\right)$

	$f(x)$	$g_c(y)$
12.65	$\Gamma(\tfrac{1}{4}-\tfrac{1}{2}\mu+\tfrac{1}{2}ix)\,\Gamma(\tfrac{1}{4}-\tfrac{1}{2}\mu-\tfrac{1}{2}ix)$ $\cdot P^{\mu}_{-\frac{1}{2}+ix}(a) \qquad 0<a<1$	$\pi^{\frac{1}{2}}2^{1+\mu}(1-a^2)^{-\frac{1}{4}}q_{-\mu-\frac{1}{2}}\!\left((1-a^2)^{-\frac{1}{2}}\cosh\,y\right)$
12.66	$\Gamma(\mu+ix)\,\Gamma(\mu-ix)\,P^{\frac{1}{2}-\mu}_{-\frac{1}{2}+ix}(a)$ $-1<a<1 \; ,\mathrm{Re}\mu>0$	$(\tfrac{1}{2}\pi)^{\frac{1}{2}}\Gamma(\mu)(1-a^2)^{\frac{1}{2}\mu-\frac{1}{4}}(a+\cosh\,y)^{-\mu}$
12.67	$\Gamma(\mu+ix)\,\Gamma(\mu-ix)$ $\cdot\,\mathrm{sech}(\pi x)\,P^{\frac{1}{2}-\mu}_{-\frac{1}{2}+ix}(a)$ $0<a<1,\; \mathrm{Re}\mu>0$	$2^{1-\mu}(1-a^2)^{\frac{1}{2}\mu-\frac{1}{4}}\Gamma(2\mu)e^{i\pi\mu-\frac{1}{2}i\pi}$ $\cdot\,(\cosh\,y-a)^{-\frac{1}{2}\mu-\frac{1}{4}}q^{\frac{1}{2}-\mu}_{\mu-\frac{1}{2}}(z)$ $z=2^{\frac{1}{2}}\cosh(\tfrac{1}{2}y)(\cosh\,y-a)^{-\frac{1}{2}}$
12.68	$\Gamma(\tfrac{1}{2}-\mu+ix)\,\Gamma(\tfrac{1}{2}-\mu-ix)$ $\cdot\,\cosh(\tfrac{1}{2}\pi x)\,P^{\mu}_{-\frac{1}{2}+ix}(a)$ $0<a<1 \; ,\mathrm{Re}\mu<\tfrac{1}{2}$	$(\tfrac{1}{2}\pi)^{\frac{1}{2}}(1-a^2)^{-\frac{1}{2}\mu}\Gamma(\tfrac{1}{2}-\mu)(a^2+\sinh^2\,y)^{\frac{1}{2}\mu-\frac{1}{4}}$ $\cdot\cos\!\left((\tfrac{1}{2}-\mu)\arctan(a^{-1}\sinh\,y)\right)$
12.69	$\Gamma(\tfrac{1}{4}-\tfrac{1}{2}\mu+\tfrac{1}{2}ix)\,\Gamma(\tfrac{1}{4}-\tfrac{1}{2}\mu-\tfrac{1}{2}ix)$ $\cdot\left(\cosh(\pi x)+\sin(\pi\mu)\right)^{-1}$ $\cdot\,\{P^{\mu}_{-\frac{1}{2}+ix}(a)+P^{\mu}_{-\frac{1}{2}+ix}(-a)\}$ $-1<a<1 \; ,-\tfrac{1}{2}<\mathrm{Re}\mu<\tfrac{1}{2}$	$2^{1+\mu}\pi^{3/2}\sec(\pi\mu)(1-a^2)^{-\frac{1}{4}}$ $\cdot P_{\mu-\frac{1}{2}}\{(1-a^2)^{-\frac{1}{2}}\cosh\,y\}$
12.70	$\mathrm{sech}(\pi x)\left(P_{-\frac{1}{2}+ix}(a)\right)^2$ $0<a\leq1$	$\pi^{-1}K\{(1-a^2)^{\frac{1}{2}}\mathrm{sech}(\tfrac{1}{2}y)\}\mathrm{sech}(\tfrac{1}{2}y)$
12.71	$\mathrm{sech}(\pi x)$ $\cdot P_{-\frac{1}{2}+ix}(a)P_{-\frac{1}{2}+ix}(-a)$ $-1<a<1$	$\pi^{-1}\{\cosh^2(\tfrac{1}{2}y)-a^2\}^{-\frac{1}{2}}K(z)$ $z=(1-a^2)^{\frac{1}{2}}\{\cosh^2(\tfrac{1}{2}y)-a^2\}^{-\frac{1}{2}}$
12.72	$\{\mathrm{sech}(\pi x)\}^2$ $\cdot P_{-\frac{1}{2}+ix}(a)P_{-\frac{1}{2}+ix}(-a)$ $-1<a<1$	$\pi^{-1}\mathrm{sech}(\tfrac{1}{2}y)K\!\left(\{1-(1-a^2)\mathrm{sech}^2(\tfrac{1}{2}y)\}^{\frac{1}{2}}\right)$

	$f(x)$	$g_c(y)$
12.73	$\Gamma(\frac{1}{2}-\mu+ix)\Gamma(\frac{1}{2}-\mu-ix)$ $\cdot\left(P^{\mu}_{-\frac{1}{2}+ix}(a)\right)^2$ $0\leq a<1$, $\mathrm{Re}\mu<\frac{1}{2}$	$(1-a^2)^{-\frac{1}{2}}q_{-\mu-\frac{1}{2}}\left(2(1-a^2)^{-1}\cosh^2(\frac{1}{2}y)-1\right)$
12.74	$Q_{-\frac{1}{2}+ix}(a)+Q_{-\frac{1}{2}-ix}(a)$ $-1<a<1$	$\pi 2^{-\frac{1}{2}})\cosh y-a)^{-\frac{1}{2}}$
12.75	$P_{-\frac{1}{2}+ix}(z)$ $z>1$	$2^{-\frac{1}{2}}(z-\cosh y)^{-\frac{1}{2}}$ $\cosh y<z$ 0 $\cosh y>z$
12.76	$\mathrm{sech}(\pi x)P_{-\frac{1}{2}+ix}(z)$, $z<1$	$2^{-\frac{1}{2}}(z+\cosh y)^{-\frac{1}{2}}$
12.77	$\{\mathrm{sech}(\pi x)\}^2 P_{-\frac{1}{2}+ix}(a)$ $a>1$	$\pi^{-1}2^{\frac{1}{2}}(a-\cosh y)^{-\frac{1}{2}}\arctan\left(\frac{a-\cosh y}{1+\cosh y}\right)^{\frac{1}{2}}$ $\cosh y<a$ $\pi^{-1}2^{-\frac{1}{2}}(\cosh y-a)^{-\frac{1}{2}}\left(-\log(1+a)\right.$ $+2\log\{(1+\cosh y)^{\frac{1}{2}}(\cosh y-a)^{\frac{1}{2}}\}\left.\right]$, $\cosh y>a$
12.78	$\Gamma(\frac{1}{2}-\mu+ix)\Gamma(\frac{1}{2}-\mu-ix)$ $\cdot\left(P^{\mu}_{-\frac{1}{2}+ix}(a)\right)^2$ $a<1$ $\mathrm{Re}\mu<\frac{1}{2}$	$(a^2-1)^{-\frac{1}{2}}q_{-\frac{1}{2}-\mu}\left(1+2(a^2-1)^{-1}\cosh^2(\frac{1}{2}y)\right)$
12.79	$\left(\Psi(\frac{1}{2}+ix)+\Psi(\frac{1}{2}-ix)\right)$ $\cdot P_{-\frac{1}{2}+ix}(\cosh a)$	$(\frac{1}{2}\cosh a-\frac{1}{2}\cosh y)^{-\frac{1}{2}}\{-\gamma-\log 4+\log(\sinh a)$ $-\log(\cosh a-\cosh y)\}$ $y<a$ $2^{-\frac{1}{2}}(\cosh y-\cosh a)^{-\frac{1}{2}}$ $y>a$
12.80	$\mathrm{sech}(\pi x)$ $\{\Psi(\frac{1}{2}+ix)q_{-\frac{1}{2}+ix}(\cosh a)$ $-\Psi(\frac{1}{2}-ix)q_{-\frac{1}{2}-ix}(\cosh a)\}$	$i\pi(2\cosh a+2\cosh y)^{-\frac{1}{2}}\{-\gamma-\log 4$ $+\log(\sinh a)-\log(\cosh a+\cosh y)\}$
12.81	$\{P_{-\frac{1}{2}+ix}(a)\}^2$, $a>1$	$\pi^{-1}(a^2-1)^{-\frac{1}{2}}K\{(a^2-1)^{-\frac{1}{2}}(a^2-\cosh^2\frac{1}{2}y)^{\frac{1}{2}}\}$ $\cosh\frac{1}{2}y<a$ 0 $\cosh\frac{1}{2}y>a$

	$f(x)$	$g_c(y)$
12.82	$\operatorname{sech}(\pi x)\{p_{-\frac12+ix}(a)\}^2$, $a>1$	$\pi^{-1}z^{-\frac12}K\{(a^2-1)^{\frac12}z^{-\frac12}\}$, $z=a^2+\sinh^2(\tfrac12 y)$
12.83	$\operatorname{sech}(\pi x)p_{-\frac12+ix}(a)$ $\{q_{-\frac12+ix}(a)+q_{-\frac12-ix}(a)\}$ $a>1$	$z^{-\frac12}K(z^{-\frac12}\cosh\tfrac12 y)$, $z=a^2+\sinh^2(\tfrac12 y)$
12.84	$\{q_{-\frac12+ix}(a)\}^2$ $+\{q_{-\frac12-ix}(a)\}^2$, $a>1$	$\pi\operatorname{cosec}(\tfrac12 y)K\{1-(a^2-1)\operatorname{cosech}^2(\tfrac12 y)\}$ $\qquad\qquad \sinh(\tfrac12 y)>(a^2-1)^{\frac12}$ $\qquad 0 \qquad \sinh(\tfrac12 y)<(a^2-1)^{\frac12}$
12.85	$p^{\mu}_{-\frac12+ix}(a)$, $a>1$, $\operatorname{Re}\mu<\tfrac12$	$(\tfrac12\pi)^{\frac12}\{\Gamma(\tfrac12-\mu)\}^{-1}(a^2-1)^{\frac12\mu}(a-\cosh y)^{-\mu-\frac12}$ $\qquad\qquad \cosh y<a$ $\qquad 0 \qquad \cosh y>a$
12.86	$\operatorname{sech}(\pi x)p^{\mu}_{-\frac12+ix}(a)$, $a>1$	$2^{-\mu-\frac12}(a+1)^{\frac12\mu}(a+\cosh y)^{-\frac12-\frac12\mu}$ $p^{\mu}_{\mu}\left((1+\cosh y)^{\frac12}(a+\cosh y)^{-\frac12}\right)$
12.87	$\Gamma(\mu+ix)\Gamma(\mu-ix)$ $\cdot p^{\frac12-\mu}_{-\frac12+ix}(a)$ $a>1$, $\operatorname{Re}\mu>0$	$(\tfrac12\pi)^{\frac12}\Gamma(\mu)(a^2-1)^{\frac12\mu-\frac14}(a+\cosh y)^{-\mu}$
12.88	$\operatorname{sech}(\pi x)\Gamma(\mu+ix)\Gamma(\mu-ix)$ $\cdot p^{\frac12-\mu}_{-\frac12+ix}(a)$, $a>1$, $\operatorname{Re}\mu>0$	$2^{\frac12-\mu}\Gamma(2\mu)\pi^{\frac12}(a^2-1)^{-\frac14}(a-1)^{\frac12\mu}(a-\cosh y)^{-\frac12\mu}$ $p^{-\mu}_{-\mu}\left((\tfrac12+\tfrac12 a)^{-\frac12}\cosh(\tfrac12 y)\right)$ $\cosh y<a$ $2^{1-\mu}\Gamma(2\mu)(a-1)^{\frac12\mu-\frac14}(\cosh y-a)^{-\frac12\mu-\frac14}$ $\cdot e^{i\pi\mu-\frac12 i\pi}q^{\frac12-\mu}_{\mu-\frac12}\left((\tfrac12\cosh y-\tfrac12 a)^{-\frac12}\cosh(\tfrac12 y)\right)$ $\cosh y>a$
12.89	$q^{\mu}_{-\frac12+ix}(z)+q^{\mu}_{-\frac12-ix}(z)$ $\operatorname{Re}\mu<\tfrac12$, $z=\cosh a$	$2^{-\frac12}\pi^{3/2}\{\Gamma(\tfrac12-\mu)\}^{-1}e^{i\pi\mu}\sinh^{\mu}a$ $\cdot(\cosh y-\cosh a)^{-\mu-\frac12}$, $y>a$ $\qquad\qquad 0 \qquad,\quad y<a$

	$f(x)$	$g_c(y)$
12.90	$\text{sech}(\pi x)$ $\cdot \{q_{-\frac{1}{2}+ix}^{\mu-\frac{1}{2}}(z) - q_{-\frac{1}{2}-ix}^{\mu-\frac{1}{2}}(z)\}$ $z = \cosh a, \qquad \text{Re}\,\mu > 0$	$(\tfrac{1}{2}\pi)^{\frac{1}{2}} e^{i\pi\mu} \Gamma(\mu)(\sinh a)^{\mu-\frac{1}{2}}$ $\cdot (\cosh y + \cosh a)^{-\mu}$
12.91	$p_{-\frac{1}{2}+ix}^{\mu}(a)\, p_{-\frac{1}{2}-ix}^{-\mu}(a), a>1$	$\tfrac{1}{2}(a^2-1)^{-\frac{1}{2}} P_{\mu-\frac{1}{2}}\{2(a^2-1)^{-1}\sinh^2(\tfrac{1}{2}y)-1\}$ $\qquad\qquad\qquad\qquad \sinh(\tfrac{1}{2}y) < (a^2-1)^{\frac{1}{2}}$ $\qquad 0 \qquad\qquad\qquad \sinh(\tfrac{1}{2}y) > (a^2-1)^{\frac{1}{2}}$

1.13 Bessel Functions of Arguments x, x^2 and $1/x$

	$f(x)$	$g_c(y)$
13.1	$J_0(ax)$	$(a^2-y^2)^{-\frac{1}{2}} \qquad\qquad\qquad\qquad\qquad\qquad , y<a$ $0 \qquad\qquad\qquad\qquad\qquad , y>a$
13.2	$x^{-\frac{1}{2}} J_0(ax)$	$(\pi a)^{-\frac{1}{2}}\left[K\{(\tfrac{1}{2}-\tfrac{1}{2}y/a)^{\frac{1}{2}}\}+K\{(\tfrac{1}{2}+\tfrac{1}{2}y/a)^{\frac{1}{2}}\}\right] \quad, y<a$ $(\tfrac{1}{2}\pi)^{-\frac{1}{2}}(a+y)^{-\frac{1}{2}}K\{(\tfrac{1}{2}+\tfrac{1}{2}y/a)^{-\frac{1}{2}}\} \qquad\qquad , y>a$
13.3	$x^{-\frac{1}{2}}\log x\, J_0(ax)$	$-(\pi a)^{-\frac{1}{2}}\left[K\{(\tfrac{1}{2}-\tfrac{1}{2}y/a)^{\frac{1}{2}}\}+K\{(\tfrac{1}{2}+\tfrac{1}{2}y/a)^{\frac{1}{2}}\}\right]$ $\cdot\{\gamma+\log 4+\tfrac{1}{2}\log(a^2-y^2)\} \qquad\qquad , y<a$ $-(2\pi)^{-\frac{1}{2}}\{\pi+2\gamma+4\log 2+\log(y^2-a^2)\}$ $\cdot(a+y)^{-\frac{1}{2}}K\{(\tfrac{1}{2}+\tfrac{1}{2}y/a)^{-\frac{1}{2}}\} \qquad , y>a$
13.4	$\log(bx)\,J_0(ax)$	$-(a^2-y^2)^{-\frac{1}{2}}\{\gamma-\log(\tfrac{1}{2}ab)+\log(a^2-y^2)\} \;, y<a$ $-\tfrac{1}{2}\pi(y^2-a^2)^{-\frac{1}{2}} \qquad\qquad\qquad\qquad , y>a$
13.5	$J_{2n}(ax)$ $n=0,1,2,\ldots$	$(-1)^n(a^2-y^2)^{-\frac{1}{2}}T_{2n}(y/a) \qquad\qquad , y<a$ $0 \qquad\qquad\qquad\qquad , y>a$
13.6	$x^{-\frac{1}{2}}J_{2n+1}(ax)$ $n=0,1,2,\ldots$	$(-1)^n(\tfrac{1}{2}\pi/a)^{\frac{1}{2}}P_{2n}(y/a) \qquad\qquad , y<a$ $0 \qquad\qquad\qquad\qquad , y>a$
13.7	$J_\nu(ax)$ $\text{Re}\,\nu > -1$	$(a^2-y^2)^{-\frac{1}{2}}\cos\{\nu\arcsin(y/a)\} \qquad , y<a$ $-a^\nu \sin(\tfrac{1}{2}\pi\nu)(y^2-a^2)^{-\frac{1}{2}}\{y+(y^2-a^2)^{\frac{1}{2}}\}^{-\nu}, y>a$

	$f(x)$	$g_c(y)$	
13.8	$x^{-1}J_\nu(ax)$ $\mathrm{Re}\,\nu>0$	$\nu^{-1}\cos\{\nu\,\arcsin(y/a)\}$ $\nu^{-1}a^\nu\cos(\tfrac{1}{2}\pi\nu)\{y+(y^2-a^2)^{\frac{1}{2}}\}^{-\nu}$	$y<a$ $y>a$
13.9	$x^{-2}J_\nu(ax)$ $\mathrm{Re}\,\nu>1$	$\tfrac{1}{2}\{\nu(\nu-1)\}^{-1}a\,\cos\{(\nu-1)\arcsin(y/a)\}$ $+\tfrac{1}{2}\{\nu(\nu+1)\}^{-1}a\,\cos\{(\nu+1)\arcsin(y/a)\}$ $\tfrac{1}{2}\{\nu(\nu-1)\}^{-1}a^\nu\sin(\tfrac{1}{2}\pi\nu)\{y+(y^2-a^2)^{\frac{1}{2}}\}^{1-\nu}$ $-\tfrac{1}{2}\{\nu(\nu+1)\}^{-1}a^{\nu+2}\sin(\tfrac{1}{2}\pi\nu)\{y+(y^2-a^2)^{\frac{1}{2}}\}^{-\nu-1}$	 $,y<a$ $,y>a$
13.10	$x^\nu J_\nu(ax)$ $-\tfrac{1}{2}<\mathrm{Re}\,\nu<\tfrac{1}{2}$	$\pi^{-\frac{1}{2}}\Gamma(\tfrac{1}{2}+\nu)(2a)^\nu(a^2-y^2)^{-\nu-\frac{1}{2}}$ $-\pi^{-\frac{1}{2}}\Gamma(\tfrac{1}{2}+\nu)(2a)^\nu\sin(\pi\nu)(y^2-a^2)^{-\nu-\frac{1}{2}}$	$,y<a$ $,y>a$
13.11	$x^{-\nu}J_\nu(ax)$ $\mathrm{Re}\,\nu>-\tfrac{1}{2}$	$\pi^{\frac{1}{2}}(2a)^{-\nu}\{\Gamma(\tfrac{1}{2}+\nu)\}^{-1}(a^2-y^2)^{\nu-\frac{1}{2}}$ 0	$,y<a$ $,y>a$
13.12	$x^{-\nu}J_{\nu+2n}(ax)$ $\mathrm{Re}\,\nu>\tfrac{1}{2}$; $n=0,1,2,.$	$(-1)^n\tfrac{1}{2}(2n)!\,(\tfrac{1}{2}a)^{-\nu}\Gamma(\nu)\{\Gamma(2\nu+2n)\}^{-1}$ $\cdot(a^2-y^2)^{\nu-\frac{1}{2}}C_{2n}^\nu(y/a)$ 0	 $,y<a$ $,y>a$
13.13	$x^{1+\nu}J_\nu(ax)$ $-1<\mathrm{Re}\,\nu<-\tfrac{1}{2}$	0 $(2a)^\nu 2\pi^{\frac{1}{2}}\{\Gamma(-\tfrac{1}{2}-\nu)\}^{-1}y(y^2-a^2)^{-\nu-3/2}$	$,y<a$ $,y>a$
13.14	$x^{1-\nu}J_\nu(ax)$ $\mathrm{Re}\,\nu>\tfrac{1}{2}$	$2^{1-\nu}a^{\nu-2}\{\Gamma(\nu)\}^{-1}{}_2F_1(1,1-\nu;\tfrac{1}{2};y^2/a^2)$ $-(\tfrac{1}{2}a)^\nu\{\Gamma(1+\nu)\}^{-1}y^{-2}{}_2F_1(1,\tfrac{3}{2};1+\nu;a^2/y^2)$	$y<a$ $y>a$
13.15	$x^{-\frac{1}{2}}J_\nu(ax)$ $\mathrm{Re}\,\nu>-\tfrac{1}{2}$	$(\tfrac{1}{2}\pi/a)^{\frac{1}{2}}\sec(\pi\nu)\cos(\tfrac{1}{2}\pi\nu+\tfrac{1}{4}\pi)$ $\cdot\{P_{\nu-\frac{1}{2}}(y/a)+P_{\nu-\frac{1}{2}}(-1/a)$ $(\tfrac{1}{2}\pi a)^{-\frac{1}{2}}\sin(\tfrac{1}{2}\pi\nu-\tfrac{1}{4}\pi)q_{\nu-\frac{1}{2}}(y/a)$	 $y<a$ $y>a$
13.16	$(b^2+x^2)^{-1}J_0(ax)$	$\tfrac{1}{2}\pi b^{-1}e^{-by}I_0(ab)$	$y>a$
13.17	$x(b^2+x^2)^{-1}J_0(ax)$	$\cosh(by)K_0(ab)$	$y<a$
13.18	$x^{-\nu}(b^2+x^2)^{-1}J_\nu(ax)$ $\mathrm{Re}\,\nu>-3/2$	$\tfrac{1}{2}\pi b^{-\nu-1}e^{-by}I_\nu(ab)$	$y>a$

	$f(x)$	$g_c(y)$
13.19	$x^{\nu+1}(b^2+x^2)^{-1}J_\nu(ax)$ $-1<\mathrm{Re}\,\nu<3/2$	$b^\nu\cosh(by)K_\nu(ab)$ $\qquad\qquad y<a$
13.20	$x^{2n-\nu}(b^2+x^2)^{-1}$ $\cdot J_\nu(ax)$ $\mathrm{Re}\,\nu>2n-\frac{5}{2},n=0,1,2,\ldots$	$(-1)^n\tfrac{1}{2}\pi b^{2n-\nu-1}e^{-by}I_\nu(ab)\qquad y>a$
13.21	$x^{\nu+2n+1}(b^2+x^2)^{-1}$ $\cdot J_\nu(ax)$ $-1<\mathrm{Re}\,(\nu+n)<3/2-n$	$(-1)^n b^{\nu+2n}\cosh(by)K_\nu(ab)\qquad y<a$
13.22	$(b^2+x^2)^{-1}$ $\{J_\nu(ax)+J_{-\nu}(ax)\}$ $-1<\mathrm{Re}\,\nu<1$	$-2^{-1}\sin(\tfrac{1}{2}\pi\nu)\cosh(by)K_\nu(ab)\qquad y<a$
13.23	$x^\mu J_\nu(ax)$ $-1-\mathrm{Re}\,\nu<\mathrm{Re}\,\mu<\tfrac{1}{2}$	$(2\pi a)^{-\frac{1}{2}}\cos(\tfrac{1}{2}\pi\nu-\tfrac{1}{2}\pi\mu)\,\Gamma(1+\nu+\mu)\,\Gamma(1+\mu-\nu)$ $\cdot(a^2-y^2)^{-\frac{1}{2}\mu-\frac{1}{4}}\{P_{\nu-\frac{1}{2}}^{-\mu-\frac{1}{2}}(y/a)+P_{\nu-\frac{1}{2}}^{-\mu-\frac{1}{2}}(-y/a)\}$ $\qquad\qquad\qquad\qquad y<a$ $(\tfrac{1}{2}\pi a)^{-\frac{1}{2}}\sin(\tfrac{1}{2}\pi\mu+\tfrac{1}{2}\pi\nu)(y^2-a^2)^{-\frac{1}{2}\mu-\frac{1}{4}}$ $\cdot e^{-i\pi(\frac{1}{2}+\mu)}q_{\nu-\frac{1}{2}}^{\mu+\frac{1}{2}}(y/a)\qquad\qquad y>a$
13.24	$x^{-\frac{1}{2}}\log xJ_\nu(ax)$ $\mathrm{Re}\,\nu>-\tfrac{1}{2}$	$(\tfrac{1}{2}\pi/a)^{\frac{1}{2}}\sec(\pi\nu)\Big[\{P_{\nu-\frac{1}{2}}(z)+P_{\nu-\frac{1}{2}}(-z)\}$ $\cdot\{\tfrac{1}{2}\pi\sin(\tfrac{1}{2}\pi\nu+\tfrac{1}{4}\pi)-\tfrac{1}{2}\cos(\tfrac{1}{2}\pi\nu+\tfrac{1}{4}\pi)\log(a^2-y^2)$ $+\Psi(\tfrac{1}{2}+\nu)\cos(\tfrac{1}{2}\pi\nu+\tfrac{1}{4}\pi)\}-\cos(\tfrac{1}{2}\pi\nu+\tfrac{1}{4}\pi)$ $\cdot\{Q_{-\nu-\frac{1}{2}}(z)+Q_{-\nu-\frac{1}{2}}(-z)\}\Big]\ ;\ z=y/a\ ,\qquad y<a$ $(\tfrac{1}{2}\pi a)^{-\frac{1}{2}}q_{\nu-\frac{1}{2}}(z)\{\tfrac{1}{2}\sin(\tfrac{1}{2}\pi\nu-\tfrac{1}{4}\pi)\log(y^2-a^2)$ $-\tfrac{1}{2}\pi\cos(\tfrac{1}{2}\pi\nu+\tfrac{1}{4}\pi)-\Psi(\tfrac{1}{2}+\nu)\sin(\tfrac{1}{2}\pi\nu-\tfrac{1}{4}\pi)\};z=\tfrac{y}{a},$ $\qquad\qquad\qquad\qquad y>a$
13.25	$x^\nu\sin xJ_\nu(x)$ $-1<\mathrm{Re}\,\nu<\tfrac{1}{2}$	$\pi^{\frac{1}{2}}2^{\nu-1}\{\Gamma(\tfrac{1}{2}-\nu)\}^{-1}(y^2+2y)^{-\nu-\frac{1}{2}}\qquad y<2$ $\pi^{\frac{1}{2}}2^{\nu-1}\{\Gamma(\tfrac{1}{2}-\nu)\}^{-1}\{(y^2+2y)^{-\nu-\frac{1}{2}}-(y^2-2y)^{-\nu-\frac{1}{2}}\}$ $\qquad\qquad\qquad\qquad y>2$

	$f(x)$	$g_c(y)$
13.26	$x^{-\nu}\cos x\, J_\nu(x)$ $\text{Re}\,\nu>-\tfrac{1}{2}$	$(\tfrac{1}{2}\pi)^{\frac{1}{2}}2^{-\nu}\{\Gamma(\tfrac{1}{2}+\nu)\}^{-1}(2y-y^2)^{\nu-\frac{1}{2}}$ $\quad$, $y<2$ 0 $\qquad\qquad$, $y>2$
13.27	$x^{-\nu}\sin x\, J_{\nu+1}(x)$ $\text{Re}\,\nu>-\tfrac{1}{2}$	$\tfrac{1}{2}\pi^{\frac{1}{2}}2^{-\nu}\{\Gamma(\tfrac{1}{2}+\nu)\}^{-1}(1-y)(2y-y^2)^{\nu-\frac{1}{2}}$ $\quad$, $y<2$ 0 $\qquad\qquad$, $y>2$
13.28	$x^{-1}\{\text{si}(ax)+\tfrac{1}{2}\pi J_0(ax)\}$	$-\pi\log\left[\tfrac{1}{2}y^{-\frac{1}{2}}\{(y+a)^{\frac{1}{2}}+(y-a)^{\frac{1}{2}}\}\right]$ $\quad$, $y>a$ 0 $\qquad\qquad$, $y<a$
13.29	$\{J_0(ax)\}^2$	$(\pi a)^{-1}\mathbf{K}\{(1-\tfrac{1}{4}y^2/a^2)^{\frac{1}{2}}\}$ $\quad$, $y<2a$ 0 $\qquad\qquad$, $y>2a$
13.30	$x^{-\frac{1}{2}}\{J_0(ax)\}^2$	$(\tfrac{1}{2}\pi/y)^{\frac{1}{2}}\left[P_{-3/2}\{(1-4a^2/y^2)^{\frac{1}{2}}\}\right]^2$ $\quad$, $y>2a$
13.31	$J_0(ax)\,J_0(bx)$	$2\pi^{-1}z^{-\frac{1}{2}}\mathbf{K}\{2(ab)^{\frac{1}{2}}z^{-\frac{1}{2}}\}$ $\qquad$ $y<\lvert a-b\rvert$ $\pi^{-1}(ab)^{-\frac{1}{2}}\mathbf{K}\{\tfrac{1}{2}(ab)^{-\frac{1}{2}}z^{\frac{1}{2}}\}$ $\qquad$ $\lvert a-b\rvert<y<a+b$ 0 $\qquad\qquad$ $y>a+b$ $z=(a+b)^2-y^2$
13.32	$x^{-\frac{1}{2}}\{J_\nu(ax)\}^2$ $\text{Re}\,\nu>-\tfrac{1}{4}$	$(\tfrac{1}{2}\pi/y)^{\frac{1}{2}}\Gamma(\tfrac{1}{4}+\nu)\{\Gamma(\tfrac{3}{4}-\nu)\}^{-1}$ $\cdot\left[P_{-\frac{1}{4}}^{-\nu}\{(1-4a^2/y^2)^{\frac{1}{2}}\}\right]^2$ $\qquad$ $y>2a$
13.33	$x^{-2\nu}\{J_\nu(ax)\}^2$ $\text{Re}\,\nu>0$	$2^{1-\nu}\{\Gamma(\tfrac{1}{2}-\nu)\}^{-2}\displaystyle\int_z^{\frac{1}{2}\pi}(4a^2-y^2)^{\nu-\frac{1}{2}}\cos^{2\nu}t\,dt$ $\,$, $y<2a$ 0 $\qquad\qquad$, $y>2a$ $z=\arcsin(\tfrac{1}{2}y/a)$
13.34	$x^{-\frac{1}{2}}J_\nu(ax)\,J_{-\nu}(ax)$	$(\tfrac{1}{2}\pi)^{\frac{1}{2}}y^{-\frac{1}{2}}P_{-\frac{1}{4}}^{\nu}(z)\,P_{-\frac{1}{4}}^{-\nu}(z)$; $z=(1-4a^2/y)^{\frac{1}{2}}$ $\,$, $y>2a$
13.35	$J_\nu(ax)\,J_{-\nu}(ax)$	$\tfrac{1}{2}a^{-1}P_{\nu-\frac{1}{2}}(\tfrac{1}{2}y^2/a^2-1)$ $\qquad$, $y<2a$ 0 $\qquad\qquad$, $y>2a$
13.36	$x^{\frac{1}{2}}J_{\nu-\frac{1}{4}}(ax)\,J_{-\nu-\frac{1}{4}}(ax)$	$(\tfrac{1}{2}\pi y)^{-\frac{1}{2}}(4a^2-y^2)^{-\frac{1}{2}}\cos\{2\nu\arccos(\tfrac{1}{2}y/a)\}$, $y<2a$ 0

	$f(x)$	$g_c(y)$
13.37	$J_\nu(ax)J_\nu(bx)$ $\text{Re}\,\nu > -\tfrac{1}{2}$	$\pi^{-1}(ab)^{-\frac{1}{2}}q_{\nu-\frac{1}{2}}(z)$ $y < \lvert a-b \rvert$ $\pi^{-1}(ab)^{-\frac{1}{2}}Q_{\nu-\frac{1}{2}}(z)$ $\lvert a-b \rvert < y < a+b$ $-\sin(\pi\nu)\pi^{-1}(ab)^{-\frac{1}{2}}q_{\nu-\frac{1}{2}}(-z)$ $y > a+b$ $z = (2ab)^{-1}(a^2+b^2-y^2)$
13.38	$J_\nu(bx)J_{-\nu}(ax)$ $a > b$	$\pi^{-1}\cos(\pi\nu)(ab)^{-\frac{1}{2}}q_{\nu-\frac{1}{2}}(z)$ $y < a-b$ $\tfrac{1}{2}(ab)^{-\frac{1}{2}}P_{\nu-\frac{1}{2}}(-z)$ $a-b < y < a+b$ 0 $y > a+b$ $z = (2ab)^{-1}(a^2+b^2-y^2)$
13.39	$x^{\nu-\mu+1}J_\mu(ax)J_\nu(bx)$ $-1 < \text{Re}\,\nu < \text{Re}\,\mu\ ,\ b > a$	0 $y < b-a$
13.40	$x^{\nu-\mu-1}J_\mu(ax)J_\nu(bx)$ $0 < \text{Re}\,\nu < 2+\text{Re}\,\mu\ ,\ b > a$	$2^{\nu-\mu-1}b^{-\nu}a^\mu\Gamma(\nu)\{\Gamma(1+\nu)\}^{-1}$ $y < b-a$
13.41	$J_\mu(ax)J_\nu(bx)$	Watson,G.N.J.Lond.Math.Soc.,Vol.9,1936,p.21
13.42	$x^\lambda J_\mu(ax)J_\nu(bx)$	Bailey,W.N.Proc.Lond.Math.Soc.,Vol.40,1936
13.43	$\{x^\nu J_\nu(ax)\}^2$ $-\tfrac{1}{2} < \text{Re}\,\nu < \tfrac{1}{2}$	$2^{3\nu}\{\Gamma(\tfrac{1}{2}-\nu)\}^{-1}(2\pi ay)^{-\frac{1}{2}}(y/a)^{-\nu}(4a^2-y^2)^{-\nu}$ $\cdot\left(\pi p^\nu_{\nu-\frac{1}{2}}(z)-2e^{-i\pi\nu}\sin(\pi\nu)q^\nu_{\nu-\frac{1}{2}}(z)\right)$ $,y < 2a$ $-2^{3\nu}\{\Gamma(\tfrac{1}{2}-\nu)\}^{-1}\sin(\pi\nu)(\tfrac{1}{2}\pi ay)^{-\frac{1}{2}}(y/a)^{-\nu}$ $\cdot(y^2-4a^2)^{-\nu}e^{-i\pi\nu}q^\nu_{\nu-\frac{1}{2}}(z)$ $,y > 2a$ $z = (4ay)^{-1}(y^2+4a^2)$
13.44	$x^{-\nu-\mu}J_\nu(ax)J_\mu(ax)$ $\text{Re}\,(\nu+\mu) > -1$	$\pi^{-\frac{1}{2}}a^{\nu+\mu-1}\{\Gamma(\tfrac{1}{2}+\nu+\mu)\}^{-1}$ $\int_0^z (\cos t)^{-\nu-\mu}\cos\{(\mu-\nu)t\}(\cos^2 t-\tfrac{1}{4}y^2/a^2)^{\nu+\mu-\frac{1}{2}}dt$ $y < 2a$ 0 $y > 2a$ $z = \arccos(\tfrac{1}{2}y/a$
13.45	$Y_o(ax)$	0 $y < a$ $-(y^2-a^2)^{-\frac{1}{2}}$ $y > a$

	$f(x)$	$g_c(y)$	
13.46	$Y_0(ax)\log(bx)$	$\frac{1}{2}\pi(a^2-y^2)^{-\frac{1}{2}}$	$y<a$
		$(y^2-a^2)^{-\frac{1}{2}}\{\gamma+\log(y^2-a^2)-\log(\frac{1}{2}ab)\}$	$y>a$
13.47	$(b^2+x^2)^{-1}\{\frac{1}{2}\pi Y_0(ax)$ $-J_0(ax)\log x\}$	$-\frac{1}{2}\pi b^{-1}e^{-by}\{I_0(ab)\log b+K_0(ab)\}$	$y>a$
13.48	$(b^2+x^2)^{-1}\{-\frac{1}{2}\pi J_0(ax)$ $+Y_0(ax)\log x\}$	$-b^{-1}\log b\,\cosh(by)K_0(ab$	$y<a$
13.49	$J_0(ax)Y_0(ax)$	$-(\pi a)^{-1}K(\frac{1}{2}y/a)$	$y<2a$
		$-2(\pi y)^{-1}K(2a/y)$	$y>2a$
13.50	$J_0(bx)Y_0(ax)$ $a\geq b$	0	$y<a-b$
		$-\pi^{-1}(ab)^{-\frac{1}{2}}K\{\frac{1}{2}(ab/z)^{-\frac{1}{2}}\}$	$a-b<y<a+b$
		$-2\pi^{-1}z^{-\frac{1}{2}}K\{2(ab/z)^{\frac{1}{2}}\}$	$y>a+b$
		$z=y^2-(a-b)^2$	
13.51	$J_0(ax)Y_0(bx)$ $a\geq b$	$-4\pi^{-1}v^{-\frac{1}{2}}K\{(-u/v)^{\frac{1}{2}}\}$	$y<a-b$
		$-\pi^{-1}(ab)^{-\frac{1}{2}}K\{\frac{1}{2}(ab/u)^{-\frac{1}{2}}\}$	$a-b<y<a+b$
		$-2\pi^{-1}u^{-\frac{1}{2}}K\{2(ab/u)^{\frac{1}{2}}\}$	$y>a+b$
		$u=y^2-(a-b)^2$, $v=(a+b)^2-y^2$	
13.52	$\{Y_0(ax)\}^2$	$(\pi a)^{-1}K\{(1-\frac{1}{4}y^2/a^2)^{\frac{1}{2}}\}$	$y<2a$
		$4(\pi y)^{-1}K\{(1-4a^2/y^2)^{\frac{1}{2}}\}$	$y>2a$
13.53	$\{J_0(ax)\}^2+\{Y_0(ax)\}^2$	$2(\pi a)^{-1}K\{(1-\frac{1}{4}y^2/a^2)^{\frac{1}{2}}\}$	$y<2a$
		$4(\pi y)^{-1}K\{(1-4a^2/y^2)^{\frac{1}{2}}\}$	$y>2a$
13.54	$Y_0(ax)Y_0(bx)$	$2\pi^{-1}u^{-\frac{1}{2}}K\{2(ab/u)^{\frac{1}{2}}\}$	$y<|a-b|$
		$\pi^{-1}(ab)^{-\frac{1}{2}}K\{\frac{1}{2}(ab/u)^{-\frac{1}{2}}\}$	$|a-b|<y>a+b$
		$4\pi^{-1}v^{-\frac{1}{2}}K\{(-u/v)^{\frac{1}{2}}\}$	
		$u=(a+b)^2-y^2$; $v=y^2-(a-b)^2$	$y>a+b$
13.55	$Y_\nu(ax)$ $-1<\mathrm{Re}\,\nu<1$	$-\tan(\frac{1}{2}\pi\nu)(a^2-y^2)^{-\frac{1}{2}}\cos\{\nu\arcsin(y/a)\}$,	$y<a$
		$-\sin(\frac{1}{2}\pi\nu)(y^2-a^2)^{-\frac{1}{2}}\Big(a^{-\nu}\{y-(y^2-a^2)^{\frac{1}{2}}\}^\nu\cot(\pi\nu)$	
		$+a^\nu\{y-(y^2-a^2)^{\frac{1}{2}}\}^{-\nu}\mathrm{cosec}(\pi\nu)\}\Big)$,	$y>a$

	$f(x)$	$g_c(y)$	
13.56	$x^\nu Y_\nu(ax)$ $-\tfrac{1}{2}<\mathrm{Re}\nu<\tfrac{1}{2}$	0 $-(2a)^\nu \pi^{\frac{1}{2}}\{\Gamma(\tfrac{1}{2}-\nu)\}^{-1}(y^2-a^2)^{-\nu-\frac{1}{2}}$	$y<a$ $y>a$
13.57	$x^{-\nu}Y_\nu(ax)$ $-\tfrac{1}{2}<\mathrm{Re}\nu<\tfrac{1}{2}$	$-(2a)^{-\nu}\pi^{\frac{1}{2}}\Gamma(\tfrac{1}{2}-\nu)\sin(\pi\nu)(a^2-y^2)^{\nu-\frac{1}{2}}$ $-(2a)^{-\nu}\pi^{-\frac{1}{2}}\Gamma(\tfrac{1}{2}-\nu)(y^2-a^2)^{\nu-\frac{1}{2}}$	$y<a$ $y>a$
13.58	$Y_\nu(ax)\cos(\tfrac{1}{2}\pi\nu)$ $+J_\nu(ax)\sin(\tfrac{1}{2}\pi\nu)$ $-1<\mathrm{Re}\nu<1$	0 $-\tfrac{1}{2}a^{-\nu}(y^2-a^2)^{-\frac{1}{2}}\Big(\{y+(y^2-a^2)^{\frac{1}{2}}\}^\nu+\{y-(y^2-a^2)^{\frac{1}{2}}\}^\nu\Big)$	$y<a$ $y>a$
13.59	$x^\nu\{Y_\nu(ax)\cos(ax)$ $-J_\nu(ax)\sin(ax)\}$ $-\tfrac{1}{2}<\mathrm{Re}\nu<\tfrac{1}{2}$	$-\pi^{\frac{1}{2}}(2a)^\nu\{\Gamma(\tfrac{1}{2}-\nu)\}^{-1}(y^2+2ay)^{-\nu-\frac{1}{2}}$	
13.60	$x^\nu\{J_\nu(ax)\sin(ax)$ $+Y_\nu(ax)\cos(ax)\}$ $-\tfrac{1}{2}<\mathrm{Re}\nu<\tfrac{1}{2}$	0 $-\pi^{\frac{1}{2}}(2a)^\nu\{\Gamma(\tfrac{1}{2}-\nu)\}^{-1}(y^2-2ay)^{-\nu-\frac{1}{2}}$	$y<2a$ $,\ y>2a$
13.61	$\sin(ax-\tfrac{1}{2}\pi\nu)J_\nu(ax)$ $-\cos(ax-\tfrac{1}{2}\pi\nu)Y_\nu(ax)$ $-1<\mathrm{Re}\nu<1$	$\tfrac{1}{2}a^{-\nu}(y^2+2ay)^{-\frac{1}{2}}\Big(\{y+a+(y^2+2ay)^{\frac{1}{2}}\}^\nu$ $+\{y+a-(y^2+2ay)^{\frac{1}{2}}\}^\nu\Big)$	
13.62	$x^\mu Y_\nu(ax)$ $\mathrm{Re}\,\mu<\tfrac{1}{2},\ \mathrm{Re}(\mu\pm\nu)>-1$	$(2a\pi)^{-\frac{1}{2}}\sin(\tfrac{1}{2}\pi\mu-\tfrac{1}{2}\pi\nu)\Gamma(1+\mu+\nu)\Gamma(1+\mu-\nu)$ $\cdot(a^2-y^2)^{-\frac{1}{2}\mu-\frac{1}{4}}\Big(P_{\nu-\frac{1}{2}}^{-\mu-\frac{1}{2}}(z)+P_{\nu-\frac{1}{2}}^{-\mu-\frac{1}{2}}(-z)\Big);z=y/a,y<a$ $-(\tfrac{1}{2}\pi a)^{-\frac{1}{2}}(y^2-a^2)^{-\frac{1}{2}\mu-\frac{1}{4}}\Big(\cos(\tfrac{1}{2}\pi\mu+\tfrac{1}{2}\pi\nu)e^{-\frac{1}{2}i\pi-i\pi\mu}$ $\cdot q_{\nu-\frac{1}{2}}^{\mu+\frac{1}{2}}(z)-\sin(\tfrac{1}{2}\pi\mu-\tfrac{1}{2}\pi\nu)\Gamma(1+\mu+\nu)\Gamma(1+\mu-\nu)$ $\cdot P_{\nu-\frac{1}{2}}^{-\mu-\frac{1}{2}}(z)\Big)\ ,\ z=y/a\ ,$	 $y>a$
13.63	$x^\nu(b^2+x^2)^{-1}Y_\nu(ax)$ $-\tfrac{1}{2}<\mathrm{Re}\nu<5/2$	$-b^{\nu-1}\cosh(by)K_\nu(ab)$	$y\le a$
13.64	$(b^2+x^2)^{-1}$ $\{Y_\nu(ax)+Y_{-\nu}(ax)\}$ $-1<\mathrm{Re}\nu<1$	$-2b^{-1}\cos(\tfrac{1}{2}\pi\nu)\cosh(by)K_\nu(ab)$	$y\le a$

	$f(x)$	$g_c(y)$	
13.65	$J_\nu(ax)Y_{-\nu}(ax)$ $\mathrm{Re}\,\nu>-\tfrac{1}{2}$	$-(\pi a)^{-1}Q_{\nu-\frac{1}{2}}(\tfrac{1}{2}y^2/a^2-1)$	$y<2a$
		$-(\pi a)^{-1}q_{\nu-\frac{1}{2}}(\tfrac{1}{2}y^2/a^2-1)$	$y>2a$
13.66	$J_\nu(ax)Y_\nu(bx)$ $a\geq b$, $\mathrm{Re}\,\nu>-\tfrac{1}{2}$	$-(ab)^{-\frac{1}{2}}p_{\nu-\frac{1}{2}}(z)$	$y<a-b$
		$-(\tfrac{1}{2}ab)^{-\frac{1}{2}}P_{\nu-\frac{1}{2}}(z)$	$a-b<y<a+b$
		$-\pi^{-1}(ab)^{-\frac{1}{2}}\cos(\pi\nu)q_{\nu-\frac{1}{2}}(-z)$	$y>a+b$
		$z=(a^2+b^2-y^2)/(2ab)$	
13.67	$J_\nu(bx)Y_\nu(ax)$ $a\geq b$, $\mathrm{Re}\,\nu>-\tfrac{1}{2}$	0	$y<a-b$
		$-\tfrac{1}{2}(ab)^{-\frac{1}{2}}P_{\nu-\frac{1}{2}}(z)$	$a-b<y<a+b$
		$-\pi^{-1}(ab)^{-\frac{1}{2}}\cos(\pi\nu)q_{\nu-\frac{1}{2}}(-z)$	$y>a+b$
		$z=(a^2+b^2-y^2)/(2ab)$	
13.68	$J_\nu(bx)Y_{-\nu}(ax)$ $a>b$, $\mathrm{Re}\,\nu>-\tfrac{1}{2}$	$\pi^{-1}\sin(\pi\nu)(ab)^{-\frac{1}{2}}q_{\nu-\frac{1}{2}}(z)$	$y<a-b$
		$-\pi^{-1}(ab)^{-\frac{1}{2}}Q_{\nu-\frac{1}{2}}(-z)$	$a-b<y<a+b$
		$-\pi^{-1}(ab)^{-\frac{1}{2}}q_{\nu-\frac{1}{2}}(-z)$	$y>a+b$
		$z=(a^2+b^2-y^2)/(2ab)$	
13.69	$J_\nu(ax)Y_{-\nu}(bx)$ $a>b$, $\mathrm{Re}\,\nu>-\tfrac{1}{2}$	$(ab)^{-\frac{1}{2}}\left[\pi^{-1}\sin(\pi\nu)q_{\nu-\frac{1}{2}}(z)-\cos(\pi\nu)p_{\nu-\frac{1}{2}}(z)\right]$ $y<a-b$	
		$-\pi^{-1}(ab)^{-\frac{1}{2}}Q_{\nu-\frac{1}{2}}(-z)$	$a-b<y<a+b$
		$-\pi^{-1}(ab)^{-\frac{1}{2}}q_{\nu-\frac{1}{2}}(-z)$	$y>a+b$
		$z=(a^2+b^2-y^2)/(2ab)$	
13.70	$J_\nu^2(ax)+Y_\nu^2(ax)$ $-\tfrac{1}{2}<\mathrm{Re}\,\nu<\tfrac{1}{2}$	$a^{-1}\sec(\pi\nu)P_{\nu-\frac{1}{2}}(\tfrac{1}{2}y^2/a^2-1)$	$y<2a$
		$a^{-1}\sec(\pi\nu)p_{\nu-\frac{1}{2}}(\tfrac{1}{2}y^2/a^2-1)$	$y>2a$
13.71	$Y_\nu(ax)Y_\nu(bx)$ $-\tfrac{1}{2}<\mathrm{Re}\,\nu<\tfrac{1}{2}$	$(ab)^{-\frac{1}{2}}\left[\pi^{-1}q_{\nu-\frac{1}{2}}(z)+\tan(\pi\nu)p_{\nu-\frac{1}{2}}(z)\right], y<|a-b|$	
		$(ab)^{-\frac{1}{2}}\left[\pi^{-1}Q_{\nu-\frac{1}{2}}(z)+\tan(\pi\nu)P_{\nu-\frac{1}{2}}(z)\right], |a-b|<y<a+b$	
		$(ab)^{-\frac{1}{2}}\left[\pi^{-1}\sin(\pi\nu)q_{\nu-\frac{1}{2}}(-z)\right.$	
		$\left. +\sec(\pi\nu)p_{\nu-\frac{1}{2}}(-z)\right.]$, $y>a+b$	
		$z=(a^2+b^2-y^2)/(2ab)$	

	$f(x)$	$g_c(y)$
13.72	$J_0(ax^2)$	$\frac{1}{8}\pi a^{-1}y\{J_{-\frac{1}{4}}^2(z)-J_{\frac{1}{4}}^2(z)\}$; $z=\frac{1}{8}y^2/a$
13.73	$x^{\frac{1}{2}}J_{\frac{1}{4}}(ax^2)$	$-\frac{1}{4}a^{-1}(\pi y)^{\frac{1}{2}}\{J_{-\frac{1}{4}}^2(z)+H_{-\frac{1}{4}}^2(z)\}$; $z=\frac{1}{4}y^2/a$
13.74	$x^{\frac{1}{2}}J_{-\frac{1}{4}}^2(ax^2)$	$\frac{1}{2}a^{-1}(\frac{1}{2}\pi y)^{\frac{1}{2}}J_{-\frac{1}{4}}(\frac{1}{2}y^2/a)$
13.75	$x^{3/2}J_{\frac{1}{4}}(ax^2)$	$\frac{1}{4}a^{-2}y(\frac{1}{2}\pi y)^{\frac{1}{2}}J_{-\frac{3}{4}}(\frac{1}{4}y^2/a)$
13.76	$x^{3/2}J_{-\frac{3}{4}}(ax^2)$	$\frac{1}{4}a^{-2}y(\frac{1}{2}\pi y)^{\frac{1}{2}}J_{\frac{1}{4}}(\frac{1}{4}y^2/a)$
13.77	$x^{\frac{1}{2}}e^{-ax^2}J_{-\frac{1}{4}}(bx^2)$	$\frac{1}{2}(\frac{1}{2}\pi y)^{\frac{1}{2}}(a^2+b^2)^{-\frac{1}{2}}e^{-az}J_{-\frac{1}{4}}(bz)$; $z=\frac{1}{4}y^2/(a^2+b^2)$
13.78	$x^{\frac{1}{2}}\cos(ax^2)J_{-\frac{1}{4}}(ax^2)$	$\frac{1}{2}(ay)^{-\frac{1}{2}}\cos(\frac{1}{8}\pi-\frac{1}{8}y^2/a)$
13.79	$x^{\frac{1}{2}}\sin(ax^2)J_{-\frac{1}{4}}(ax^2)$	$\frac{1}{2}(ay)^{-\frac{1}{2}}\sin(\frac{1}{8}\pi-\frac{1}{8}y^2/a)$
13.80	$x^{1/3}\sin x^2 J_{-\frac{1}{3}}(x^2)$	$\frac{1}{8}\pi^{\frac{1}{2}}(\frac{1}{2}y^2)^{1/6}\sin(\frac{\pi}{12}-\frac{y^2}{16})$
13.81	$x^{1/3}\cos x^2 J_{-\frac{1}{3}}(x^2)$	$\frac{1}{8}\pi^{\frac{1}{2}}(\frac{1}{2}y^2)^{1/6}\cos(\frac{\pi}{12}-\frac{y^2}{16})$
13.82	$x\,J_{\frac{1}{4}}^2(ax^2)$	$-\frac{1}{4}a^{-1}\{J_0(\frac{1}{8}y^2/a)+Y_0(\frac{1}{8}y^2/a)\}$
13.83	$x\,J_{-\frac{1}{4}}^2(ax^2)$	$\frac{1}{4}a^{-1}\{J_0(\frac{1}{8}y^2/a)-Y_0(\frac{1}{8}y^2/a)\}$
13.84	$x^{\frac{1}{2}}J_{-\frac{1}{8}}^2(ax^2)$	$-\frac{1}{4}a^{-1}(\frac{1}{2}\pi y)^{\frac{1}{2}}J_{-\frac{1}{8}}(z)Y_{\frac{1}{8}}(z)$; $z=y^2/(16a)$
13.85	$x^{\frac{1}{2}}J_{-\frac{1}{8}}(ax^2)J_{\frac{1}{8}}(ax^2)$	$-\frac{1}{4}a^{-1}(\frac{1}{2}\pi)^{\frac{1}{2}}J_{-\frac{1}{8}}(z)$ $\cdot\left(\sin(\pi/8)J_{-\frac{1}{8}}(z)+\cos(\pi/8)Y_{-\frac{1}{8}}(z)\right)$; $z=y^2/(16a)$

	$f(x)$	$g_c(y)$
13.86	$x^{\frac{1}{2}}J_{-\nu-\frac{1}{8}}(ax^2)$	$y^{-1}(\tfrac{1}{2}\pi y)^{-\frac{1}{2}}\Big(e^{-i\pi/8}W_{\nu,-\frac{1}{8}}(-z)W_{-\nu,-\frac{1}{8}}(-z)$
	$\cdot J_{\nu-\frac{1}{8}}(ax^2)$	$+e^{i\pi/8}W_{\nu,-\frac{1}{8}}(z)W_{-\nu,-\frac{1}{8}}(z)\Big)$; $z=\tfrac{1}{8}iy^2/a$
13.87	$x^{2\nu}J_{\nu}(ax^2)$ $-\tfrac{1}{4}<\mathrm{Re}\,\nu<1$	$\tfrac{1}{4}(\tfrac{1}{2}a)^{-\nu-\frac{1}{2}}\Big(\{\Gamma\tfrac{3}{4})\}^{-1}\Gamma(\tfrac{1}{4}+\nu)\,_1F_2(\tfrac{1}{4}+\nu;\tfrac{1}{2},\tfrac{3}{4};-\dfrac{y^2}{64a^2})$
		$-\tfrac{1}{4}\{\Gamma(\tfrac{5}{4})\}^{-1}\Gamma(\tfrac{3}{4}+\nu)y^2a^{-1}\,_1F_2(\tfrac{3}{4}+\nu;\tfrac{3}{2},\tfrac{5}{4};-\dfrac{y^2}{64a^2})\Big)$
13.88	$x^{-2\nu}J_{\nu}(ax^2)$ $\mathrm{Re}\,\nu>-1$	$2^{-\nu-1}a^{\nu-\frac{1}{2}}\Big(\cos(\tfrac{1}{4}\pi+\pi\nu)\{\Gamma(\tfrac{3}{4})\}^{-1}\Gamma(\tfrac{1}{4}-\nu)$
		$\cdot\,_1F_2(\tfrac{1}{4}-\nu;\tfrac{1}{2},\tfrac{3}{4};-\dfrac{y^2}{64a^2})-\tfrac{1}{4}\sin(\tfrac{1}{4}\pi+\pi\nu)\{\Gamma(\tfrac{5}{4})\}^{-1}$
		$\cdot\Gamma(\tfrac{3}{4}-\nu)y^2a^{-1}\,_1F_2(\tfrac{3}{4}-\nu;\tfrac{3}{2},\tfrac{5}{4};-\dfrac{y^2}{64a^2})\Big)$
13.89	$Y_o(ax^2)$	$-\tfrac{1}{8}\pi y a^{-1}\{J^2_{\frac{1}{4}}(z)+J^2_{-\frac{1}{4}}(z)\}$; $z=\tfrac{1}{8}y^2/a$
13.90	$x^{\frac{1}{2}}Y_{-\frac{1}{4}}(ax^2)$	$-\tfrac{1}{2}a^{-1}(\tfrac{1}{2}\pi y)^{\frac{1}{2}}\mathbf{H}_{-\frac{1}{4}}(\tfrac{1}{8}y^2/a)$
13.91	$x^{3/2}Y_{\frac{1}{4}}(ax^2)$	$-\tfrac{1}{2}a^{-2}y(\tfrac{1}{2}\pi y)^{\frac{1}{2}}\mathbf{H}_{-\frac{3}{4}}(\tfrac{1}{8}y^2/a)$
13.92	$xJ_{\frac{1}{4}}(ax^2)Y_{\frac{1}{4}}ax^2)$	$-\tfrac{1}{4}a^{-1}\{J_o(z)+\mathbf{H}_o(z)\}$; $z=\tfrac{1}{8}y^2/a$
13.93	$x^{\frac{1}{2}}J_{-\frac{1}{8}}(ax^2)Y_{-\frac{1}{8}}(ax^2)$	$-\tfrac{1}{2}a^{-1}(\tfrac{1}{2}\pi y)^{\frac{1}{2}}J^2_{-\frac{1}{8}}(\tfrac{1}{8}y^2/a)$
13.94	$xY^2_{\frac{1}{4}}(ax^2)$	$\tfrac{1}{4}a^{-1}\{J_o(z)+Y_o(z)-2\mathbf{H}_o(z)\}$; $z=\tfrac{1}{8}y^2/a$
13.95	$xY^2_{-\frac{1}{4}}(ax^2)$	$-\tfrac{1}{4}a^{-1}\{J_o(z)+Y_o(z)+2\mathbf{H}_o(z)\}$; $z=\tfrac{1}{8}y^2/a$
13.96	$x^{-1}J_{\nu}(a/x)$ $\mathrm{Re}\,\nu>0$	$2\{J_{\nu}(u)K_{\nu}(u)+J_{\nu}(v)K_{\nu}(v)\}$; $\dfrac{u}{v}=(\pm 2iay)^{\frac{1}{2}}$
13.97	$x^{-1}e^{-a/x}J_{\nu}(b/x)$ $\mathrm{Re}\,\nu>-1$	$J_{\nu}(ui^{\frac{1}{2}})K_{\nu}(vi^{\frac{1}{2}})+J_{\nu}\{u(-i)^{\frac{1}{2}}\}K_{\nu}\{v(-i)^{\frac{1}{2}}\}$ $\dfrac{v}{u}=(2y)^{\frac{1}{2}}\{(a^2+b^2)^{\frac{1}{2}}\pm a\}$

	$f(x)$	$g_c(y)$
13.98	$x^{-1}\sin(a/x)J_\nu(b/x)$ $\mathrm{Re}\,\nu>-2$	$\tfrac{1}{2}\pi J_\nu(cy^{\frac{1}{2}})\{J_\nu(dy^{\frac{1}{2}})\cos(\tfrac{1}{2}\pi\nu)-Y_\nu(dy^{\frac{1}{2}})\sin(\tfrac{1}{2}\pi\nu)\}$ $+\sin(\tfrac{1}{2}\pi\nu)I_\nu(cy^{\frac{1}{2}})K_\nu(dy^{\frac{1}{2}})$ $\begin{matrix}c\\d\end{matrix}=(a+b)^{\frac{1}{2}}\pm(a-b)^{\frac{1}{2}}\,,\qquad a\geq b$
13.99	$x^{-1}\cos(a/x)J_\nu(b/x)$ $\mathrm{Re}\,\nu>-1$	$-\tfrac{1}{2}\pi J_\nu(cy^{\frac{1}{2}})\{J_\nu(dy^{\frac{1}{2}})\sin(\tfrac{1}{2}\pi\nu)+Y_\nu(dy^{\frac{1}{2}})\cos(\tfrac{1}{2}\pi\nu)\}$ $+\cos(\tfrac{1}{2}\pi\nu)I_\nu(cy^{\frac{1}{2}})K_\nu(dy^{\frac{1}{2}})\,,\qquad a\geq b$ $\begin{matrix}c\\d\end{matrix}=(a+b)^{\frac{1}{2}}\pm(a-b)^{\frac{1}{2}}\,,\qquad a\geq b$
13.100	$x^{-\frac{1}{2}}\cos\!\left(\dfrac{a}{x}\right)J_{2n-\frac{1}{2}}\!\left(\dfrac{a}{x}\right)$ $n=1,2,3,\ldots$	$(-1)^n\tfrac{1}{2}(\pi/y)^{\frac{1}{2}}J_{4n-1}\{2(2ay)^{\frac{1}{2}}\}$
13.101	$x^{-\frac{1}{2}}\sin\!\left(\dfrac{a}{x}\right)J_{2n+\frac{1}{2}}\!\left(\dfrac{a}{x}\right)$ $n=0,1,2,\ldots$	$(-1)^n\tfrac{1}{2}(\pi/y)^{\frac{1}{2}}J_{4n+1}\{2(2ay)^{\frac{1}{2}}\}$
13.102	$x^{-1}e^{-a/x}Y_\nu(b/x)$ $-1<\mathrm{Re}\,\nu<1$	$Y_\nu(ui^{\frac{1}{2}})K_\nu(vi^{\frac{1}{2}})+Y_\nu\{u(-i)^{\frac{1}{2}}\}K_\nu\{v(-i)\}^{\frac{1}{2}}$ $\begin{matrix}v\\u\end{matrix}=(2y)^{\frac{1}{2}}\{(a^2+b^2)^{\frac{1}{2}}\pm a\}\,;\qquad a\geq b$
13.103	$x^{-1}\sin(a/x)Y_\nu(b/x)$ $-2<\mathrm{Re}\,\nu<2$	$\tfrac{1}{2}\pi Y_\nu(cy^{\frac{1}{2}})\{Y_\nu(dy^{\frac{1}{2}})\cos(\tfrac{1}{2}\pi\nu)-Y_\nu(dy^{\frac{1}{2}})\sin(\tfrac{1}{2}\pi\nu)\}$ $+K_\nu(dy^{\frac{1}{2}})\{I_\nu(cy^{\frac{1}{2}})\cos(\tfrac{1}{2}\pi\nu)+\dfrac{2}{\pi}K_\nu(cy^{\frac{1}{2}})\sin(\tfrac{1}{2}\pi\nu)\}$ $\begin{matrix}c\\d\end{matrix}=(a+b)^{\frac{1}{2}}\pm(a-b)^{\frac{1}{2}}\,,\qquad a\geq b$
13.104	$x^{-1}\cos(a/x)Y_\nu(b/x)$ $-1<\mathrm{Re}\,\nu<1$	$-\tfrac{1}{2}\pi Y_\nu(cy^{\frac{1}{2}})\{J_\nu(dy^{\frac{1}{2}})\sin(\tfrac{1}{2}\pi\nu)+Y_\nu(dy^{\frac{1}{2}})\cos(\tfrac{1}{2}\pi\nu)\}$ $-K_\nu(dy^{\frac{1}{2}})\{I_\nu(cy^{\frac{1}{2}})\sin(\tfrac{1}{2}\pi\nu)+\dfrac{2}{\pi}K_\nu(cy^{\frac{1}{2}})\cos(\tfrac{1}{2}\pi\nu)\}$ $\begin{matrix}c\\d\end{matrix}=(a+b)^{\frac{1}{2}}\pm(a-b)^{\frac{1}{2}}\,,\qquad a\geq b$
13.105	$x^{2\mu}J_{2\nu}(a/x)$ $-\tfrac{1}{4}<\mathrm{Re}\,\mu<\mathrm{Re}\,\nu-\tfrac{1}{2}$	$\pi^{\frac{1}{2}}4^{\mu-2\nu}\Gamma(\tfrac{1}{2}+\mu-\nu)\{\Gamma(1+2\nu)\Gamma(\nu-\mu)\}^{-1}y^{2\nu-2\mu-1}$ $\cdot\,_0F_3\!\left(1+2\nu,\tfrac{1}{2}+\nu-\mu,\nu-\mu;(\tfrac{1}{4}ay)^2\right)$ $+\tfrac{1}{4}4^{-\mu}a^{1+2\mu}\Gamma(\nu-\mu-\tfrac{1}{2})$ $\cdot\{\Gamma(\nu+\mu+3/2)\}^{-1}{}_0F_3\!\left(\tfrac{1}{2},\mu-\nu+3/2,\mu+\nu+3/2;(\tfrac{1}{4}ay)^2\right)$

1.14   Bessel Functions of Argument $(ax^2+bx+c)^{\frac{1}{2}}$

	$f(x)$	$g_c(y)$
14.1	$J_0(ax^{\frac{1}{2}})$	$y^{-1}\sin(\tfrac{1}{4}a^2/y)$
14.2	$x^{-\frac{1}{2}}J_1(ax^{\frac{1}{2}})$	$4a^{-1}\sin^2(\tfrac{1}{8}a^2/y)$
14.3	$J_0(ax^{\frac{1}{2}})\log(bx)$	$2y^{-1}\Big(\sin z\{\log(\tfrac{1}{2}ab^{\frac{1}{2}}/y)-\tfrac{1}{2}\mathrm{Ci}(z)\}+\tfrac{1}{2}\cos z\,\mathrm{si}(z)\Big)$ $z=\tfrac{1}{4}a^2/y$
14.4	$J_\nu(ax^{\frac{1}{2}})$ $\mathrm{Re}\,\nu>-2$	$-\tfrac{1}{4}\pi^{\frac{1}{2}}ay^{-3/2}\Big(\sin(\tfrac{1}{8}a^2/y-\tfrac{1}{4}\pi\nu)J_{\frac{1}{2}\nu-\frac{1}{2}}(\tfrac{1}{8}a^2/y)$ $+\cos(\tfrac{1}{4}\pi\nu-\tfrac{1}{8}a^2/y)J_{\frac{1}{2}\nu+\frac{1}{2}}(\tfrac{1}{8}a^2/y)\Big)$
14.5	$x^{-\frac{1}{2}}J_\nu(ax^{\frac{1}{2}}),\ \mathrm{Re}\,\nu>-1$	$(\pi/y)^{\frac{1}{2}}\cos(\tfrac{1}{4}\pi+\tfrac{1}{4}\pi\nu-\tfrac{1}{8}a^2/y)J_{\frac{1}{2}\nu}(\tfrac{1}{8}a^2/y)$
14.6	$x^{\frac{1}{2}\nu}J_\nu(ax^{\frac{1}{2}})$ $-1<\mathrm{Re}\,\nu<\tfrac{1}{2}$	$y^{-1}(\tfrac{1}{2}a/y)^\nu\sin(\tfrac{1}{4}a^2/y-\tfrac{1}{2}\nu\pi)$
14.7	$x^{\frac{1}{2}\nu}e^{-ax}J_\nu\{2(bx)^{\frac{1}{2}}\}$ $\mathrm{Re}\,\nu>-1$	$b^{\frac{1}{2}\nu}s^{\frac{1}{2}\nu+\frac{1}{2}}e^{-abs}\cos\{bys-(\nu+1)\arctan(y/a)\}$ $s=(a^2+y^2)^{-1}$
14.8	$J_\nu(ax)J_{2\nu}(bx^{\frac{1}{2}})$ $\mathrm{Re}\,\nu>-\tfrac{1}{2}$	$s^{\frac{1}{2}}\cos(\tfrac{1}{4}b^2ys)J_\nu(\tfrac{1}{4}ab^2s);\ s=(a^2-y^2)^{-1};\quad y<a$ $s^{\frac{1}{2}}\sin(\tfrac{1}{4}b^2yS-\pi\nu)J_\nu(\tfrac{1}{4}ab^2S);\ S=(y^2-a^2)^{-1},\ y>a$
14.9	$J_\nu(ax^{\frac{1}{2}})J_\nu(bx^{\frac{1}{2}})$ $\mathrm{Re}\,\nu>-1$	$y^{-1}J_\nu(\tfrac{1}{2}ab/y)\sin\{\tfrac{1}{4}(a^2+b^2)/y-\tfrac{1}{2}\pi\nu\}$
14.10	$Y_0(ax^{\frac{1}{2}})$	$(\pi y)^{-1}\Big(\mathrm{Ci}(z)\sin z-\{\pi+\mathrm{si}(z)\}\cos z\Big),\ z=\tfrac{1}{4}a^2/y$
14.11	$x^{-\frac{1}{2}}Y_0(ax^{\frac{1}{2}})$	$\tfrac{1}{2}(\pi/y)^{\frac{1}{2}}\Big(J_0(z)\sin(z-\tfrac{\pi}{4})+Y_0(z)\cos(z-\tfrac{\pi}{4})\Big);\ z=\tfrac{1}{8}a^2/y$
14.12	$J_0(ax^{\frac{1}{2}})Y_0(ax^{\frac{1}{2}})$	$\tfrac{1}{2}y^{-1}\Big(\sin z\,Y_0(z)-\cos z\,J_0(z)\Big)\qquad ;z=\tfrac{1}{4}a^2/y$
14.13	$J_0(bx^{\frac{1}{2}})Y_0(ax)$ $+2J_0(ax)Y_0(bx^{\frac{1}{2}})$	$s^{\frac{1}{2}}\cos(\tfrac{1}{4}b^2ys)Y_0(\tfrac{1}{4}ab^2s)\ ;\ s=(a^2-y^2)^{-1}\ ;\ y<a$ $s^{\frac{1}{2}}\sin(\tfrac{1}{4}b^2yS)Y_0(\tfrac{1}{4}ab^2S)\ ;\ S=(y^2-a^2)^{-1}\ ;\ y>a$

	$f(x)$	$g_c(y)$	
14.14	$J_0(ax^{\frac{1}{2}})J_0(bx^{\frac{1}{2}})$ $+J_0(bx^{\frac{1}{2}})Y_0(ax^{\frac{1}{2}})$	$y^{-1}\Big(\sin\{\tfrac{1}{4}(a^2+b^2)/y\}Y_0(\tfrac{1}{2}ab/y)$ $\qquad -\cos\{\tfrac{1}{4}(a^2+b^2)/y\}J_0(\tfrac{1}{2}ab/y)\Big]$	
14.15	$J_\nu(ax^{\frac{1}{2}})Y_{-\nu}(bx^{\frac{1}{2}})$ $+J_{-\nu}(bx^{\frac{1}{2}})Y_\nu(ax^{\frac{1}{2}})$ $-1<\mathrm{Re}\nu<1$	$y^{-1}\Big(\sin\{\tfrac{1}{2}\pi\nu+\tfrac{1}{4}(a^2+b^2)/y\}Y_\nu(\tfrac{1}{2}ab/y)$ $\qquad -\cos\{\tfrac{1}{2}\pi\nu+\tfrac{1}{4}(a^2+b^2)/y\}J_\nu(\tfrac{1}{2}ab/y)\Big]$	
14.16	$J_0\{b(a^2+x^2)^{\frac{1}{2}}\}$	$(b^2-y^2)^{-\frac{1}{2}}\cos\{a(b^2-y^2)^{\frac{1}{2}}\}$ 0	$y<b$ $y>b$
14.17	$\log(a^2+x^2)$ $\cdot J_0\{b(a^2+x^2)^{\frac{1}{2}}\}$	$-2s^{-1}\Big(\sin(as)\,\mathrm{si}(as)-\cos(as)\{\log(ab/s)$ $\qquad -\mathrm{Ci}(2as)\}\Big]$ $-\pi S^{-1}e^{-aS}$ $\qquad s=(b^2-y^2)^{\frac{1}{2}} \; ; \; S=(y^2-b^2)^{\frac{1}{2}}$	$y<b$ $y>b$
14.18	$Y_0\{b(a^2+x^2)^{\frac{1}{2}}\}$	$(b^2-y^2)^{-\frac{1}{2}}\sin\{a(b^2-y^2)^{\frac{1}{2}}\}$ $-(y^2-b^2)^{-\frac{1}{2}}\exp\{-a(y^2-b^2)^{\frac{1}{2}}\}$	$y<b$ $y>b$
14.19	$\log(a^2+x^2)$ $\cdot Y_0\{b(a^2+x^2)^{\frac{1}{2}}\}$	$2s^{-1}\Big(\cos(as)\,\mathrm{si}(2as)+\sin(as)\{\log(ab/s)$ $\qquad\qquad +\mathrm{Ci}(2as)\}\Big]$ $-2S^{-1}\Big(e^{-aS}\log(ab/S)-e^{aS}\mathrm{Ei}(-2aS)\Big]$	$,\ y<b$ $,\ y>b$
14.20	$J_{2n}(br)$ $n=1,2,3,\ldots$	$(-1)^n n(\tfrac{1}{2}\pi a)^{\frac{1}{2}}\sum\limits_{0}^{n}(-1)^k(n+k-1)!\{k!(n-k)!\}^{-1}$ $\cdot(\tfrac{1}{2}ab^2)^{-k}s^{k-\frac{1}{2}}J_{k-\frac{1}{2}}(as)$ 0	$,\ y<b$ $,\ y>b$
14.21	$r^{-1}J_\nu(br)$	$\cos(\tfrac{1}{2}\pi\nu)I_{\frac{1}{2}\nu}(\tfrac{1}{2}ay-\tfrac{1}{2}aS)K_{\frac{1}{2}\nu}(\tfrac{1}{2}ay+\tfrac{1}{2}aS)$	$y>b$
14.22	$r^\nu J_\nu(br)$ $\mathrm{Re}\nu<\tfrac{1}{2}$	$-(\tfrac{1}{2}\pi a)^{\frac{1}{2}}(ab)^\nu s^{-\nu-\frac{1}{2}}Y_{\nu+\frac{1}{2}}(as)$ $-(2a/\pi)^{\frac{1}{2}}\sin(\pi\nu)(ab)^\nu S^{-\nu-\frac{1}{2}}K_{\nu+\frac{1}{2}}(aS)$	$,\ y<b$ $,\ y>b$
14.23	$r^{-\nu}J_\nu(br)$ $\mathrm{Re}\nu>-\tfrac{1}{2}$	$(\tfrac{1}{2}\pi a)^{\frac{1}{2}}(ab)^{-\nu}s^{\nu-\frac{1}{2}}J_{\nu-\frac{1}{2}}(as)$ 0	$,\ y<b$ $,\ y>b$

$$r=(a^2+x^2)^{\frac{1}{2}} \qquad s=(b^2-y^2)^{\frac{1}{2}} , \qquad S=(y^2-b^2)^{\frac{1}{2}}$$

	$f(x)$	$g_c(y)$
14.24	$Y_{2n}(br)$ $n=1,2,3,\ldots$	$(-1)^n n (\tfrac{1}{2}\pi a)^{\frac{1}{2}} \sum\limits_{k=0}^{n} (-1)^k (n+k-1)! \{k!(n-k)!\}^{-1}$ $\quad \cdot (\tfrac{1}{2}ab^2)^{-k} s^{k-\frac{1}{2}} Y_{k-\frac{1}{2}}(as) \quad , \; y<b$ $(-1)^{n+1} n (2a/\pi)^{\frac{1}{2}} \sum\limits_{k=0}^{n} (-1)^k (n+k-1)! \{k!(n-k)!\}^{-1}$ $\quad \cdot (\tfrac{1}{2}ab^2)^{-k} S^{k-\frac{1}{2}} K_{k-\frac{1}{2}}(aS) \quad\quad , \; y>b$
14.25	$r^{-1} Y_\nu(br)$	$-K_{\frac{1}{2}\nu}(\tfrac{1}{2}ay+\tfrac{1}{2}aS)$ $\cdot \left(\sin(\tfrac{1}{2}\pi\nu) I_{\frac{1}{2}\nu}(\tfrac{1}{2}ay-\tfrac{1}{2}aS) + \tfrac{1}{\pi} K_{\frac{1}{2}\nu}(\tfrac{1}{2}ay-\tfrac{1}{2}aS) \right) , \; y>b$
14.26	$r^\nu Y_\nu(br)$ $\mathrm{Re}\,\nu<\tfrac{1}{2}$	$(\tfrac{1}{2}\pi a)^{\frac{1}{2}} (ab)^\nu s^{-\nu-\frac{1}{2}} J_{\nu+\frac{1}{2}}(as) \quad\quad , \; y<b$ $-(2a/\pi)^{\frac{1}{2}} (ab)^\nu \cos(\pi\nu) S^{-\nu-\frac{1}{2}} K_{\nu+\frac{1}{2}}(aS) , \; y>b$
14.27	$r^{-\nu} Y_\nu(br)$ $\mathrm{Re}\,\nu>-\tfrac{1}{2}$	$(\tfrac{1}{2}\pi a)^{\frac{1}{2}} (ab)^{-\nu} s^{\nu-\frac{1}{2}} Y_{\nu-\frac{1}{2}}(as) \quad\quad , \; y<b$ $-(2a/\pi)^{\frac{1}{2}} (ab)^{-\nu} S^{\nu-\frac{1}{2}} K_{\nu-\frac{1}{2}}(aS) \quad , \; y>b$
14.28	$(x^2+2ax)^{-\frac{1}{2}}$ $\cdot J_{2\nu}\{b(x^2+2ax)^{\frac{1}{2}}\}$ $\mathrm{Re}\,\nu>-\tfrac{1}{2}$	$\tfrac{1}{2}\pi J_\nu(\tfrac{1}{2}ay-\tfrac{1}{2}aS) \left[J_\nu(\tfrac{1}{2}ay+\tfrac{1}{2}aS)\sin(ay-\pi\nu) \right.$ $\left. -Y_\nu(\tfrac{1}{2}ay+\tfrac{1}{2}aS)\cos(ay-\pi\nu) \right] \quad , \; y>b$
14.29	$(x^2+2ax)^{\frac{1}{2}\nu}$ $\cdot J_\nu\{b(x^2+2ax)^{\frac{1}{2}}\}$ $-1<\mathrm{Re}\,\nu<\tfrac{1}{2}$	$(2a/\pi)^{\frac{1}{2}} (ab)^\nu \cos(ay) s^{-\nu-\frac{1}{2}} K_{\nu+\frac{1}{2}}(as) \quad , \; y<b$ $(\tfrac{1}{2}\pi a)^{\frac{1}{2}} (ab)^\nu S^{-\nu-\frac{1}{2}}$ $\cdot \left[Y_{\nu+\frac{1}{2}}(aS)\sin(\pi\nu-ay) - J_{\nu+\frac{1}{2}}(aS)\cos(\pi\nu-ay) \right]$ $\quad\quad , \; y>b$
14.30	$r^{-\frac{1}{2}} P_{2n}(\tfrac{x}{r}) J_{2n+\frac{1}{2}}(br)$ $n=0,1,2,\ldots$	$(-1)^n (\tfrac{1}{2}\pi a)^{\frac{1}{2}} P_{2n}(y/b) J_0(as) \quad , \; y<b$ $0 \quad\quad , \; y>b$
14.31	$T_{2n}(a/r) J_{2n}(br)$ $n=0,1,2,\ldots$	$(-1)^n s^{-1} \cos(as) T_{2n}(s/b) \quad\quad , \; y<b$ $0 \quad\quad , \; y>b$

$$r=(a^2+x^2)^{\frac{1}{2}} \qquad\qquad s=(b^2-y^2)^{\frac{1}{2}} \quad , \quad S=(y^2-b^2)^{\frac{1}{2}}$$

	$f(x)$	$g_c(y)$	
14.32	$T_{2n+1}\left(\frac{a}{r}\right)J_{2n+1}(br)$ $n=0,1,2,\ldots$	$(-1)^n s^{-1}\sin(as)T_{2n+1}(s/b)$ 0	$y<b$ $y>b$
14.33	$T_{2n}\left(\frac{x}{r}\right)J_{2n}(br)$ $n=0,1,\ldots$	$(-1)^n s^{-1}\cos(as)T_{2n}(y/b)$ 0	$y<b$ $y>b$
14.34	$x^{-1}U_{2n}\left(\frac{x}{r}\right)J_{2n+1}(br)$ $n=0,1,\ldots$	$(-1)^n (ab)^{-1}U_{2n}(y/b)\sin(as)$ 0	$y<b$ $y>b$
14.35	$r^{-\nu}C_{2n}^{\nu}(x/r)$ $\cdot J_{\nu+2n}(x/r);n=0,1,\ldots$	$(-1)^n (\tfrac{1}{2}\pi a)^{\frac{1}{2}}(ab)^{-\nu}s^{\nu-\frac{1}{2}}C_{2n}^{\nu}\left(\frac{y}{b}\right)J_{\nu-\frac{1}{2}}(as)$ 0	$y<b$ $y>b$
14.36	$(c^2+x^2)^{-1}x^{2n}r^{-\nu}$ $\cdot J_{\nu}(br)$ $n=0,1,\ldots;\mathrm{Re}\,\nu>2n-\frac{5}{2}$	$\tfrac{1}{2}\pi(-1)^n c^{2n-1}e^{-cy}(a^2-c^2)^{-\frac{1}{2}\nu}J_{\nu}\{b(a^2-c^2)^{\frac{1}{2}}\}$ 0	$,y>b$ $,y<b$
14.37	$x^{\frac{1}{2}}J_{-\frac{1}{4}}(\tfrac{1}{2}ar+\tfrac{1}{2}ab)$ $\cdot J_{-\frac{1}{4}}(\tfrac{1}{2}ar-\tfrac{1}{2}ab)$	$(\tfrac{1}{2}\pi y)^{-\frac{1}{2}}s^{-1}\cos(as)$ 0	$y<b$ $y>b$
14.38	$x^{\frac{1}{2}}J_{-\frac{1}{4}}(\tfrac{1}{2}ar-\tfrac{1}{2}ab)$ $\cdot Y_{-\frac{1}{4}}(\tfrac{1}{2}ar+\tfrac{1}{2}ab)$	$(\tfrac{1}{2}\pi y)^{-\frac{1}{2}}s^{-1}\sin(as)$ $-(\tfrac{1}{2}\pi y)^{-\frac{1}{2}}S^{-1}e^{-aS}$	$y<b$ $y>b$
14.39	$J_{\nu}(\tfrac{1}{2}bx+\tfrac{1}{2}br)$ $\cdot J_{\nu}(\tfrac{1}{2}bx-\tfrac{1}{2}br)$ $\mathrm{Re}\,\nu>-\tfrac{1}{2}$	$s^{-1}J_{2\nu}(as)$ 0	$y<b$ $y>b$
14.40	$J_{\nu}(w)J_{\nu}(z)$ $\mathrm{Re}\,\nu>-\tfrac{1}{2}$ $\genfrac{}{}{0pt}{}{w}{z}=\tfrac{1}{2}b\{x\pm(x^2-b^2)^{\frac{1}{2}}\}$	$s^{-1}I_{2\nu}(as)$ 0	$y<b$ $y>b$

$$r=(a^2+x^2)^{\frac{1}{2}} \qquad\qquad s=(b^2-y^2)^{\frac{1}{2}}\ ;\ S=(y^2-b^2)^{\frac{1}{2}}$$

	$f(x)$		$g_c(y)$	
14.41	$Y_\nu(\frac{1}{2}br+\frac{1}{2}bx)$		$-s^{-1}J_{2\nu}(as)$	$y<b$
	$\cdot Y_\nu(\frac{1}{2}br-\frac{1}{2}bx)$		$4\pi^{-1}\cos(\pi\nu)S^{-1}K_{2\nu}(S)$	$y>b$
	$-\frac{1}{2}<\mathrm{Re}\nu<\frac{1}{2}$			
14.42	$J_\nu(w)Y_\nu(z)$		$2s^{-1}Y_{2\nu}(as)$	$y<b$
	$+Y_\nu(w)J_\nu(z)$		$4\pi^{-1}\sin(\pi\nu)S^{-1}K_{2\nu}(aS)$	$y>b$
	$-\frac{1}{2}<\mathrm{Re}\nu<\frac{1}{2}$			
	$\genfrac{}{}{0pt}{}{w}{z}=\frac{1}{2}br\pm\frac{1}{2}bx$			
14.43	$J_0(bu)$	$x<a$	$q^{-1}\sin(aq)$	
	0	$x>a$		
14.44	$Y_0(bu)$	$x<a$	$(\pi q)^{-1}\Big[\sin(aq)\{Ci(aq+ay)+Ci(aq-ay)\}$	
	0	$x>a$	$\quad -\cos(aq)\{Si(aq+ay)+Si(aq-ay)\}\Big]$	
14.45	$\log u\, J_0(bu)$	$x<a$	$q^{-1}\Big[\sin(aq)\{-\frac{1}{2}Ci(aq+ay)+Ci(2aq)-\frac{1}{2}Ci(aq-ay)$	
	0	$x>a$	$\quad +\log(ab)-\log q\}-\cos(aq)\{Si(2aq)$	
			$\quad -\frac{1}{2}Si(aq+ay)-\frac{1}{2}Si(aq-ay)\}\Big]$	
14.46	$u^{-1}J_\nu(bu)$	$x<a$	$\frac{1}{2}\pi J_{\frac{1}{2}\nu}(\frac{1}{2}aq+\frac{1}{2}ay)J_{\frac{1}{2}\nu}(\frac{1}{2}aq-\frac{1}{2}ay)$	
	0	$x>a$		
	$\mathrm{Re}\nu>-1$			
14.47	$u^\nu J_\nu(bu)$	$x<a$	$(\frac{1}{2}\pi a)^{\frac{1}{2}}(ab)^\nu q^{-\nu-\frac{1}{2}}J_{\nu+\frac{1}{2}}(aq)$	
	0	$x>a$		
	$\mathrm{Re}\nu>-1$			
14.48	$u^{-1}Y_\nu(bu)$	$x<a$	$\frac{1}{4}\pi\Big[J_{\frac{1}{2}\nu}(\frac{1}{2}aq+\frac{1}{2}ay)Y_{\frac{1}{2}\nu}(\frac{1}{2}aq-\frac{1}{2}ay)+J_{\frac{1}{2}\nu}(\frac{1}{2}aq-\frac{1}{2}ay)$	
	0	$x>a$	$\cdot Y_{\frac{1}{2}\nu}(\frac{1}{2}aq+\frac{1}{2}ay)-\tan(\frac{1}{2}\pi\nu)\{J_{\frac{1}{2}\nu}(\frac{1}{2}aq+\frac{1}{2}ay)$	
	$-1<\mathrm{Re}\nu<1$		$\cdot J_{\frac{1}{2}\nu}(\frac{1}{2}aq-\frac{1}{2}ay)+Y_{\frac{1}{2}\nu}(\frac{1}{2}aq+\frac{1}{2}ay)Y_{\frac{1}{2}\nu}(\frac{1}{2}aq-\frac{1}{2}ay)\}\Big]$	
14.49	$P_{2n}(\frac{x}{a})J_0(bu)$	$x<a$	$(-1)^n(\frac{1}{2}\pi a)^{\frac{1}{2}}P_{2n}(y/q)q^{-\frac{1}{2}}J_{2n+\frac{1}{2}}(aq)$	
	0	$x>a$		

$r=(a^2+x^2)^{\frac{1}{2}};u=(a^2-x^2)^{\frac{1}{2}}\qquad q=(b^2+y^2)^{\frac{1}{2}}\ ;\ s=(b^2-y^2)^{\frac{1}{2}},S=(y^2-b^2)^{\frac{1}{2}}$

	$f(x)$	$g_c(y)$
14.50	$u^{\nu-\frac{1}{2}}C_{2n}^{\nu}(x/a)$ $\cdot J_{\nu-\frac{1}{2}}(bu)$ $x<a$ 0 $x>a$	$(-1)^n(\frac{1}{2}\pi/b)^{\frac{1}{2}}(ab)^{\nu}u^{-\nu}C_{2n}^{\nu}(y/q)J_{\nu+2n}(aq)$
14.51	$J_0(bv)$ $x<a$ 0 $x>a$	$2\cos(\frac{1}{2}ay)q^{-1}\sin(\frac{1}{2}aq)$
14.52	$Y_0(bv)$ $x<a$ 0 $x>a$	$2\pi^{-1}w^{-1}\cos(\frac{1}{2}ay)\left[\sin(\frac{1}{2}aq)\{Ci(\frac{1}{2}aq+\frac{1}{2}ay)\right.$ $+Ci)\frac{1}{2}aq-\frac{1}{2}ay)\}-\cos(\frac{1}{2}aq)\{Si(\frac{1}{2}aq+\frac{1}{2}ay)$ $\left.+Si(\frac{1}{2}aq-\frac{1}{2}ay)\}\right]$
14.53	$v^{\nu}J_{\nu}(bv)$, $x<a$ 0 , $x>a$ $Re\nu>-1$	$(\pi a)^{\frac{1}{2}}(\frac{1}{2}ab)^{\nu}\cos(\frac{1}{2}ay)r^{-\nu-\frac{1}{2}}J_{\nu+\frac{1}{2}}(aq)$
14.54	$v^{-1}Y_{2\nu}(bv)$, $x<a$ 0 , $x<a$ $-\frac{1}{2}<Re\nu<\frac{1}{2}$	$\frac{1}{2}\pi\cos(\frac{1}{2}ay)\left[J_{\nu}(\frac{1}{4}aq+\frac{1}{4}ay)Y_{\nu}(\frac{1}{4}aq-\frac{1}{4}ay)\right.$ $+J_{\nu}(\frac{1}{4}aq-\frac{1}{4}ay)Y_{\nu}(\frac{1}{4}aq+\frac{1}{4}ay)$ $-\tan(\pi\nu)\{J_{\nu}(\frac{1}{4}aq+\frac{1}{4}ay)$ $\left.\cdot J_{\nu}(\frac{1}{4}aq-\frac{1}{4}ay)+Y_{\nu}(\frac{1}{4}aq+\frac{1}{4}ay)Y_{\nu}(\frac{1}{4}aq-\frac{1}{4}ay)\}\right]$
14.55	0 $x<a$ $J_0(bU)$ $x>a$	$s^{-1}e^{-as}$ $y<b$ $-S^{-1}\sin(aS)$ $y>b$
14.56	0 $x<a$ $u^{-1}J_{\nu}(bU)$ $x>a$ $Re\nu>-1$	$-\frac{1}{2}\pi J_{\frac{1}{2}\nu}(\frac{1}{2}ay-\frac{1}{2}aS)Y_{-\frac{1}{2}\nu}(\frac{1}{2}ay+\frac{1}{2}aS)$ $y>b$
14.57	0 $x<a$ $u^{\nu}J_{\nu}(bU)$ $x>a$ $-1<Re\nu<\frac{1}{2}$	$(2a/\pi)^{\frac{1}{2}}(ab)^{\nu}s^{-\nu-\frac{1}{2}}K_{\nu+\frac{1}{2}}(as)$ $y<b$ $-(\frac{1}{2}\pi a)^{\frac{1}{2}}(ab)^{\nu}S^{-\nu-\frac{1}{2}}Y_{-\nu-\frac{1}{2}}(aS)$ $y>b$
14.58	0 $x<a$ $u^{-1}Y_{2\nu}(U)$ $x>a$ $-\frac{1}{2}<Re\nu<\frac{1}{2}$	$-\frac{1}{4}\pi\sec(\pi\nu)\left[Y_{\nu}(\frac{1}{4}ay+\frac{1}{4}aS)Y_{\nu}(\frac{1}{4}ay-\frac{1}{4}aS)\right.$ $\left.+2\cos(2\pi\nu)J_{\nu}(\frac{1}{4}ay+\frac{1}{4}aS)J_{\nu}(\frac{1}{4}ay-\frac{1}{4}aS)\right]$ $y>b$

$v=(ax-x^2)^{\frac{1}{2}}, u=(a^2-x^2)^{\frac{1}{2}}$ $q=(b^2+y^2)^{\frac{1}{2}}, s=(b^2-y^2)^{\frac{1}{2}}, S=(y^2-b^2)^{\frac{1}{2}}$

	$f(x)$	$g_c(y)$
14.59	$\begin{aligned}0 \quad &x<a\\ Y_0(bU) \quad &x>a\end{aligned}$	$\begin{aligned}-(\pi S)^{-1}&\Big[\sin(aS)\{Ci(ay-aS)+Ci(ay+aS)\}\\ &+\cos(aS)\{\pi+si(ay-aS)-si(ay+aS)\}\Big] \quad y>b\end{aligned}$
14.60	$\begin{aligned}\left(\frac{x-a}{x+a}\right)^{\frac{1}{2}\nu}J_\nu(bU) \quad &x>a\\ 0 \quad &x<a\\ \operatorname{Re}\nu>-1\end{aligned}$	$\begin{aligned}s^{-1}e^{-as}\cos\{\nu\arcsin(y/b)\} \quad &y<b\\ -s^{-1}b^\nu(y+S)^{-\nu}\sin(\tfrac{1}{2}\pi\nu+aS) \quad &y>b\end{aligned}$
14.61	$\begin{aligned}0 \quad &x<a\\ x(c^2+x^2)^{-1}U^\nu\\ J_\nu(bU) \quad &x>a\\ -1\operatorname{Re}\nu<3/2\end{aligned}$	$(a^2+c^2)^{\frac{1}{2}\nu}\cosh(cy)\,K_\nu\{b(a^2+c^2)^{\frac{1}{2}}\} \qquad y<b$
14.62	$\begin{aligned}0 \quad &x<a\\ x(c^2+x^2)^{-1}U^{\nu+2n-2}\\ J_\nu(bU) \quad &x>a\\ -n<\operatorname{Re}\nu<7/2-2n\\ n=0,1,2,\ldots\end{aligned}$	$\begin{aligned}(-1)^{n+1}&\cosh(cy)(a^2+c^2)^{\frac{1}{2}\nu+n-1}\\ &\cdot K_\nu\{b(a^2+c^2)^{\frac{1}{2}}\} \qquad y<b\end{aligned}$
14.63	$\begin{aligned}0 \quad &x<a\\ \log(x+a)J_0(bU) \quad &x>a\end{aligned}$	$\begin{aligned}s^{-1}&\Big(e^{-as}\log(ab/s)-e^{as}Ei(-2as)\Big) \quad y<b\\ -s^{-1}&\Big[\sin(aS)[\log(1+y/S)+Ci(2aS)]\\ &-\cos(aS)si(2aS)\Big] \quad y>b\end{aligned}$
14.64	$\begin{aligned}\log\left(\frac{x-a}{x+a}\right)J_0(bU) \quad &x>a\\ 0 \quad &x<a\\[4pt] U-(x^2-a^2)^{\frac{1}{2}}\end{aligned}$	$\begin{aligned}s^{-1}&\Big(\sin(aS)\{Ci(ay-aS)+Ci(ay+aS\\ &+2\log(\tfrac{y}{b}+\tfrac{S}{b})\}+\cos(aS)\{si(ay-aS)\\ &\qquad\qquad -si(ay+aS)\}\Big) \ , \ y>b\\ s=(b^2-y^2)^{\frac{1}{2}} \ , \ S=(y^2-b^2)^{\frac{1}{2}}\end{aligned}$

1.15 Bessel Functions of Trigonometric and Hyperbolic Arguments

	$f(x)$	$g_c(y)$
15.1	$\begin{aligned}J_{2\nu}\{2a\cos(\tfrac{1}{2}x)\} \quad &x<\pi\\ 0 \quad &x>\pi\\ \operatorname{Re}\nu>-\tfrac{1}{2}\end{aligned}$	$\pi J_{\nu-y}(a)J_{\nu+y}(a)$
15.2	$\begin{aligned}Y_{2\nu}\{2a\cos(\tfrac{1}{2}x)\} \quad &x<\pi\\ 0 \quad &x>\pi\\ -\tfrac{1}{2}<\operatorname{Re}\nu<\tfrac{1}{2}\end{aligned}$	$\begin{aligned}\pi\operatorname{cosec}(2\pi\nu)&\{\cos(2\pi\nu)J_{\nu+y}(a)J_{\nu-y}(a)\\ &-J_{y-\nu}(a)J_{-y-\nu}(a)\}\end{aligned}$

	$f(x)$	$g_c(y)$
15.3	$J_\nu(2a \sin x$ $\quad x<\pi$ $\qquad\qquad 0 \quad x>\pi$ $\mathrm{Re}\,\nu>-1$	$\pi\cos(\tfrac{1}{2}\pi y) J_{\frac{1}{2}\nu-\frac{1}{2}y}(a) J_{\frac{1}{2}\nu+\frac{1}{2}y}(a)$
15.4	$Y_\nu(2a \sin x)$ $\quad x<\pi$ $\qquad\qquad 0 \quad x>\pi$ $-1<\mathrm{Re}\,\nu<1$	$-\pi\,\mathrm{cosec}(\pi\nu)\cos(\tfrac{1}{2}\pi y)\{J_{-\frac{1}{2}\nu-\frac{1}{2}y}(a) J_{-\frac{1}{2}\nu+\frac{1}{2}y}(a)$ $-\cos(\pi\nu)J_{\frac{1}{2}\nu-\frac{1}{2}y}(a) J_{\frac{1}{2}\nu+\frac{1}{2}y}(a)\}$
15.5	$(\sec x)^m J_m(a \cos x)$ $\qquad\qquad\qquad x<\frac{1}{2}\pi$ $\qquad\qquad 0 \quad x>\frac{1}{2}\pi$ $m=0,1,\ldots$	$\tfrac{1}{2}\pi(\tfrac{1}{2}a)^m m!\sum_{n=0}^{m}\varepsilon_n\{(m+n)!(m-n)!\}^{-1}$ $\cdot J_{n-\frac{1}{2}y}(\tfrac{1}{2}a) J_{n+\frac{1}{2}y}(\tfrac{1}{2}a)$
15.6	$(\mathrm{cosec}\,x)^m J_m(a \sin x)$ $\qquad\qquad\qquad x<\pi$ $\qquad\qquad 0 \quad x>\pi$ $m=0,1,\ldots$	$\pi(\tfrac{1}{2}a)^m m!\cos(\tfrac{1}{2}\pi y)\sum_{n=o}^{m}[(m+n)!(m-n)!]^{-1}$ $\cdot J_{n-\frac{1}{2}y}(\tfrac{1}{2}a) J_{n+\frac{1}{2}y}(\tfrac{1}{2}a)$
15.7	$\mathrm{cosec}\,x J_{2\nu}(2a \sin x)$ $\qquad\qquad\qquad x<\pi$ $\qquad\qquad 0 \quad x>\pi$ $\mathrm{Re}\,\nu>0$	$\tfrac{1}{2}\pi a\nu^{-1}\cos(\tfrac{1}{2}\pi y)\{J_{-\frac{1}{2}-\frac{1}{2}y+\nu}(a) J_{-\frac{1}{2}+\frac{1}{2}y+\nu}(a)$ $+J_{\frac{1}{2}-\frac{1}{2}y+\nu}(a) J_{\frac{1}{2}+\frac{1}{2}y+\nu}(a)\}$
15.8	$\sec(\tfrac{1}{2}x)$ $\cdot J_{2\nu}[2a \cos(\tfrac{1}{2}x)]\,x<\pi$ $\qquad\qquad 0 \quad x>\pi$ $\mathrm{Re}\,\nu>1$	$\tfrac{1}{2}\pi a\nu^{-1}\{J_{\nu-\frac{1}{2}-y}(a) J_{\nu-\frac{1}{2}+y}(a)$ $+J_{\nu+\frac{1}{2}-y}(a) J_{\nu+\frac{1}{2}+y}(a)\}$
15.9	$J_0\{(a^2+b^2-2ab \cos x)^{\frac{1}{2}}\}$ $\qquad\qquad\qquad x<\pi$ $\qquad\qquad 0 \quad x>\pi$	$y\sin(\pi y)\sum_{n=o}^{\infty}(-1)^n\varepsilon_n(y^2-n^2)^{-1}J_n(a)J_n(b)$
15.10	$Y_0\{(a^2+b^2-2ab \cos x)^{\frac{1}{2}}\}$ $\qquad\qquad\qquad x<\pi$ $\qquad\qquad 0 \quad x>\pi$	$y\sin(\pi y)\sum_{n=o}^{\infty}(-1)^n\varepsilon_n(y^2-n^2)^{-1}J_n(a)Y_n(b)$ $\qquad\qquad\qquad a\leq b$
15.11	$J_0\{2a \sinh(\tfrac{1}{2}x)\}$	$\{I_{iy}(a)+I_{-iy}(a)\}K_{iy}(a)$

	$f(x)$	$g_c(y)$
15.12	$J_{2\nu}\{2a\ \sinh(\tfrac{1}{2}x)\}$ $\mathrm{Re}\,\nu > -\tfrac{1}{2}$	$I_{\nu-iy}(a)K_{\nu+iy}(a)+I_{\nu+iy}(a)K_{\nu-iy}(a)$
15.13	$Y_o\{2a\ \sinh(\tfrac{1}{2}x)\}$	$-2\pi^{-1}\cosh(\pi y)\{K_{iy}(a)\}^2$
15.14	$Y_{2\nu}[2a\ \sinh(\tfrac{1}{2}x)]$ $-\tfrac{1}{2}<\mathrm{Re}\,\nu<\tfrac{1}{2}$	$\operatorname{cosec}(2\pi\nu)\Big(\cos(2\pi\nu)\,[I_{\nu-iy}(a)K_{\nu+iy}(a)$ $+I_{\nu+iy}(a)K_{\nu-iy}(a)]-K_{\nu-iy}(a)I_{-\nu-iy}(a)$ $-I_{-\nu+iy}(a)K_{\nu+iy}(a)\Big)$
15.15	$J_{2\nu}\{2a\ \cosh(\tfrac{1}{2}x)\}$	$-\tfrac{1}{2}\pi\{J_{\nu+iy}(a)Y_{\nu-iy}(a)+J_{\nu-iy}(a)Y_{\nu+iy}(a)\}$
15.16	$Y_{2\nu}\{2a\ \cosh(\tfrac{1}{2}x)\}$	$\tfrac{1}{2}\pi\{J_{\nu+iy}(a)J_{\nu-iy}(a)-Y_{\nu+iy}(a)Y_{\nu-iy}(a)\}$
15.17	$J_o\{a(2\ \sinh x)^{\frac{1}{2}}\}$	$\{J_{iy}(a)+J_{-iy}(a)\}K_{iy}(a)$
15.18	$J_o\{a(2\ \cosh x)^{\frac{1}{2}}\}$	$i\pi^{-1}\Big(K_{iy}(ae^{\frac{1}{4}i\pi})K_{iy}(ae^{\frac{3}{4}i\pi})$ $-K_{iy}(ae^{-\frac{1}{4}i\pi})K_{iy}(ae^{-\frac{3}{4}i\pi})\Big)$
15.19	$\{\operatorname{cosech}(\tfrac{1}{2}x)\}^m$ $J_m\{2a\ \sinh(\tfrac{1}{2}x)\}$ $m=0,1,2,\ldots$	$a^m m!\ \sum\limits_{n=o}^{m}\varepsilon_n\{(m+n)!(m-n)!\}^{-1}$ $\cdot\Big(I_{n-iy}(a)K_{n+iy}(a)+I_{n+iy})a)K_{n-iy}(a)\Big)$
15.20	$\{\operatorname{sech}(\tfrac{1}{2}x)\}^m$ $J_m\{2a\ \cosh(\tfrac{1}{2}x)\}$ $m=0,1,2,\ldots$	$-\tfrac{1}{2}\pi m!\,a^m\ \sum\limits_{n=o}^{m}\varepsilon_n\{(m+n)!(m-n)!\}^{-1}$ $\cdot\Big(J_{n+iy}(a)Y_{n-iy}(a)+J_{n-iy}(a)Y_{n+iy}(a)\Big)$
15.21	$\{\operatorname{sech}(\tfrac{1}{2}x)\}^m$ $Y_m\{2a\ \cosh(\tfrac{1}{2}x)\}$ $m=0,1,2,\ldots$	$\tfrac{1}{2}\pi m!\,a^m\ \sum\limits_{n=o}^{m}\varepsilon_n[(m+n)!(m-n)!]^{-1}$ $\cdot\Big(J_{n+iy}(a)J_{n-iy}(a)-Y_{n+iy}(a)Y_{n-iy}(a)\Big)$
15.22	$J_o\{(a^2+b^2+2ab\ \cosh x)^{\frac{1}{2}}\}$	$-\tfrac{1}{2}\pi\{J_{iy}(a)Y_{-iy}(b)+Y_{iy}(a)J_{-iy}(b)\}$
15.23	$Y_o\{(a^2+b^2+2ab\ \cosh x)^{\frac{1}{2}}\}$	$\tfrac{1}{2}\pi\{J_{iy}(a)J_{-iy}(b)-Y_{iy}(a)Y_{-iy}(b)\}$

	$f(x)$	$g_c(y)$
15.24	$J_0\{(2ab \cosh x - a^2 - b^2)^{\frac{1}{2}}\}$ $\cosh x > \frac{1}{2}(a^2+b^2)/(ab)$ $0, \cosh x < \frac{1}{2}(a^2+b^2)/(ab)$	$K_{ix}(a)\{I_{ix}(b)+I_{-ix}(b)\}$
15.25	$0, \cosh x < b/a$ $J_0\{(a \cosh x - b)^{\frac{1}{2}}\}$ $\cosh x > b/a$	$\{I_{iy}(u)+I_{-iy}(u)\}K_{iy}(v)$ ${v \atop u} = \frac{1}{2}\{(b+a)^{\frac{1}{2}} \pm (b-a)^{\frac{1}{2}}\}$ $b>a$
15.26	$Y_0\{a(2 \cosh x)^{\frac{1}{2}}\}$	$-\pi^{-1}\left(K_{iy}(ae^{\frac{1}{4}i\pi})K_{iy}(ae^{\frac{3}{4}i\pi})\right.$ $\left. + K_{iy}(ae^{-\frac{1}{4}i\pi})K_{iy}(ae^{-\frac{3}{4}i\pi})\right)$
15.27	$J_0\{(b - a \cosh x)^{\frac{1}{2}}\}$ $\cosh x < b/a$ $0, \cosh x > b/a$	$\frac{1}{2}\pi\{J_{iy}(u)Y_{iy}(v) - J_{iy}(v)Y_{iy}(u)\}$ ${u \atop v} = \frac{1}{2}\{(b+a)^{\frac{1}{2}} \pm (b-a)^{\frac{1}{2}}\},$ $b>a$
15.28	$(\mathrm{sech}\ x)^{\mu}J_{\nu}(a\ \mathrm{sech}\ x)$ $\mathrm{Re}(\nu+\mu)>0$	$2^{\mu-2}a^{\nu}\{\Gamma(1+\nu)\Gamma(\mu+\nu)\}^{-1}\Gamma(\frac{1}{2}\mu+\frac{1}{2}\nu+\frac{1}{2}iy)$ $\cdot \Gamma(\frac{1}{2}\mu+\frac{1}{2}\nu-\frac{1}{2}iy)\,{}_2F_3(\frac{1}{2}\mu+\frac{1}{2}\nu+\frac{1}{2}iy, \frac{1}{2}\mu+\frac{1}{2}\nu-\frac{1}{2}iy;$ $\frac{1}{2}\mu+\frac{1}{2}\nu, \frac{1}{2}+\frac{1}{2}\mu+\frac{1}{2}\nu, 1+\nu; -\frac{1}{4}a^2)$

1.16 Bessel Functions of Variable Order

	$f(x)$	$g_c(y)$
16.1	$J_{\nu-x}(a)J_{\nu+x}(a)$ $\mathrm{Re}\,\nu > -\frac{1}{2}$	$\frac{1}{2}J_{2\nu}(2a \cos \frac{1}{2}y)$ $y < \pi$ 0 $y > \pi$
16.2	$\mathrm{sech}(\frac{1}{2}\pi x)$ $\cdot \{J_{ix}(a)+J_{-ix}(a)\}$ $= i\ \mathrm{cosech}(\frac{1}{2}\pi x)$ $\cdot \{Y_{ix}(a)-Y_{-ix}(a)\}$	$2 \sin u$
16.3	$\mathrm{cosech}(\frac{1}{2}\pi x)$ $\cdot \{J_{ix}(a)-J_{-ix}(a)\}$ $= -i\ \mathrm{sech}(\frac{1}{2}\pi x)$ $\cdot \{Y_{ix}(a)+Y_{-ix}(a)\}$	$-2i \cos u$

	$f(x)$	$g_c(y)$
16.4	$\cosh(\tfrac{1}{2}\pi x)\,\mathrm{sech}(\pi x)$ $\cdot\{J_{ix}(a)+J_{-ix}(a)\}$	$\cos u\{C(v)-S(v)\}+\sin u\{C(v)+S(v)\}$
16.5	$\sinh(\tfrac{1}{2}\pi x)\,\mathrm{sech}(\pi x)$ $\cdot\{J_{ix}(a)-J_{-ix}(a)\}$	$i\sin u\{C(v)\}-S(v)\}-i\cos u\{C(v)+S(v)\}$
16.6	$\mathrm{sech}(\tfrac{1}{2}\pi x)\,\mathrm{sech}(\pi x)$ $\cdot\{J_{ix}(a)+J_{-ix}(a)\}$	$2\cos u\{C(v)-S(v)\}+2\sin u\{C(v)+S(v)-1\}$
16.7	$\mathrm{cosech}(\tfrac{1}{2}\pi x)\,\mathrm{sech}(\pi x)$ $\cdot\{J_{ix}(a)-J_{-ix}(a)\}$	$2i\cos u\{C(v)+S(v)-1\}-2i\sin u\{C(v)-S(v)\}$
16.8	$\mathrm{sech}(\tfrac{1}{2}\pi x)\,\mathrm{sech}(\pi x)$ $\cdot\{Y_{ix}(a)+Y_{-ix}(a)\}$	$2\cos u\{C(v)+S(v)-1\}-2\sin u\{C(v)-S(v)\}$
16.9	$\mathrm{cosech}(\tfrac{1}{2}\pi x)\,\mathrm{sech}(\pi x)$ $\cdot\{Y_{ix}(a)-Y_{-ix}(a)\}$	$2i\cos u\{S(v)-C(v)\}+2i\sin u\{1-C(v)-S(v)\}$
16.10	$\exp(\tfrac{1}{2}\pi x)H_{ix}^{(2)}(a)$	$i\exp(-iu)$
16.11	$\exp(-\tfrac{1}{2}\pi x)H_{ix}^{(1)}$	$-i\exp(iu)$
16.12	$\mathrm{sech}(\pi x)J_{ix}(a)J_{-ix}(a)$	$\pi^{-1}\int_{o}^{2a}(t^2+4a^2\sinh^2\tfrac{1}{2}y)^{-\tfrac{1}{2}}\cos t\,dt$
16.13	$\{J_{ix}(a)\}^2+\{Y_{ix}(a)\}^2$	$2\pi^{-1}K_o\{2a\,\sin h(\tfrac{1}{2}y)\}$
16.14	$\mathrm{sech}(\pi x)\Big(\{J_{ix}(a)\}^2$ $+\{Y_{ix}(a)\}^2\Big)$	$\mathbf{H}_o(2a\cos\tfrac{1}{2}y)-Y_o(2a\cosh\tfrac{1}{2}y)$
16.15	$J_{ix}(a)Y_{-ix}(b)$ $+J_{-ix}(b)Y_{ix}(a)$	$-J_o\{(a^2+b^2+2bu)^{\tfrac{1}{2}}\}$

$$u=a\cosh y,\quad v=a+u=2a\cosh^2(\tfrac{1}{2}y)$$

	$f(x)$	$g_c(y)$
16.16	$J_{ix}(a)J_{-ix}(b)$ $+Y_{ix}(a)Y_{-ix}(b)$	$Y_0\{(a^2+b^2+bu)^{\frac{1}{2}}\}$
16.17	$\exp(\pi x)H_{ix}^{(2)}(a)H_{ix}^{(2)}(b)$	$iH_0^{(2)}\{(a^2+b^2+2bu)^{\frac{1}{2}}\}$
16.18	$\mathrm{sech}(\pi x)\left[\{J_{ix}(a)\}^2\right.$ $+\{J_{-ix}(a)\}^2$	$\mathbf{H}_0\{(2a\cosh(\frac{1}{2}y)\}$
16.19	$J_{ix}(a)J_{ix}(b)$ $+Y_{ix}(a)Y_{ix}(b)$	$-Y_0\{(a^2+b^2-2ab\cosh y)^{\frac{1}{2}}\},\cosh y<\frac{a^2+b^2}{2ab}$ $2\pi^{-1}K_0[(2ab\cosh y-a^2-b^2)^{\frac{1}{2}}],\cosh y>\frac{a^2+b^2}{2ab}$
16.20	$J_{ix}(a)Y_{ix}(b)$ $-J_{ix}(b)Y_{ix}(a)$ $=i\,\mathrm{cosech}(\pi x)$ $\{J_{ix}(a)J_{-ix}(b)$ $-J_{ix}(b)J_{-ix}(a)\}$	$-J_0\{(a^2+b^2-2ab\cosh y)^{\frac{1}{2}}\},\cosh y<\frac{a^2+b^2}{2ab}$ $0\qquad\qquad,\cosh y>\frac{a^2+b^2}{2ab}$
16.21	$\mathrm{cosech}(2\pi x)$ $\cdot\{J_{\nu+ix}(a)Y_{\nu-ix}(a)$ $-J_{\nu-ix}(a)Y_{\nu+ix}(a)$	$i\,\mathrm{cosec}(2\pi\nu)\{\mathbf{J}_{2\nu}(2a\cos\frac{1}{2}y)$ $\qquad\qquad -J_{2\nu}(2a\cosh\frac{1}{2}y)\}$
16.22	$\mathrm{sech}(\pi x)$ $\mathrm{cosech}(\pi x)$ $\cdot\{J_{\nu+ix}(a)J_{-\nu+ix}(a)$ $\{\pm\}J_{\nu-ix}(a)J_{-\nu-ix}(a)\}$	$\frac{1}{2}\left\{\begin{array}{c}\mathrm{cosec}(\pi\nu)\\-i\,\sec(\pi\nu)\end{array}\right\}\cdot\{\{\mp\}\}$ $\cdot[\mathbf{J}_{2\nu}(2a\cosh\frac{1}{2}y)\{\mathbf{J}_{-2\nu}(2a\cosh\frac{1}{2}y\}]$

1.17   Modified Bessel Functions of Arguments x, x^2 and $1/x$

	$f(x)$	$g_c(y)$
17.1	$e^{-ax}I_o(ax)$	$(2y)^{-\frac{1}{2}}\left[(y^2+4a^2)^{\frac{1}{2}}-y\right]^{\frac{1}{2}}(y^2+4a^2)^{-\frac{1}{2}}$
17.2	$e^{-bx}I_o(ax)$ $b\geq a$	$(2z)^{-\frac{1}{2}}\{z^{\frac{1}{2}}+b^2-a^2-y^2\}^{\frac{1}{2}}$ $z=(b^2-a^2-y^2)^2+4b^2y^2$
17.3	$x^\nu e^{-bx}I_\nu(ax)$ $b>a$ $\mathrm{Re}\,\nu>-\frac{1}{2}$	$(z^2+4b^2y^2)^{-\frac{1}{4}-\frac{1}{2}\nu}\cos\{(\frac{1}{2}+\nu)\arctan(2by/z)\}$ $\cdot\pi^{-\frac{1}{2}}(2a)^\nu\Gamma(\frac{1}{2}+\nu),\ z=b^2-a^2-y^2$
17.4	$K_o(ax)$	$\frac{1}{2}\pi(a^2+y^2)^{-\frac{1}{2}}$
17.5	$\log(bx)K_o(ax)$	$\frac{1}{2}\pi(a^2+y^2)^{-\frac{1}{2}}\{\log(\frac{1}{2}ab)-\gamma-\log(a^2+y^2)\}$
17.6	$x^{-\frac{1}{2}}K_o(ax)$	$(\frac{1}{2}\pi)^{\frac{1}{2}}(z/y)^{\frac{1}{2}}\left[K\{(\frac{1}{2}-\frac{1}{2}z)^{\frac{1}{2}}\}+K\{(\frac{1}{2}+\frac{1}{2}z)^{\frac{1}{2}}\}\right],$ $z=(1+\dfrac{a^2}{y^2})^{-\frac{1}{2}}$
17.7	$x^{\frac{1}{2}}K_o(ax)$	$\frac{1}{4}(2\pi)^{\frac{1}{2}}(z/y)^{3/2}\left[2E\{(\frac{1}{2}-\frac{1}{2}z)^{\frac{1}{2}}\}+2E\{(\frac{1}{2}+\frac{1}{2}z)^{\frac{1}{2}}\}\right.$ $\left.-K\{(\frac{1}{2}-\frac{1}{2}z)^{\frac{1}{2}}\}-K\{(\frac{1}{2}+\frac{1}{2}z)^{\frac{1}{2}}\}\right],\ z=y(a^2+y^2)^{-\frac{1}{2}}$
17.8	$x^{2n}K_o(ax)$ $n=0,1,\ldots$	$(-1)^n\frac{1}{2}\pi(2n)!(a^2+y^2)^{-n-\frac{1}{2}}P_{2n}\{y(a^2+y^2)^{-\frac{1}{2}}\}$
17.9	$x^{2n+1}K_o(ax)$ $n=0,1,\ldots$	$(-1)^{n+1}(2n+1)!(a^2+y^2)^{-n-1}Q_{2n+1}\{y(a^2+y^2)^{-\frac{1}{2}}\}$
17.10	$x^{-\frac{1}{2}}\log xK_o(x)$	$-(\frac{1}{2}\pi z/y)^{\frac{1}{2}}\left[\gamma+\frac{1}{2}\pi+\log 4+\frac{1}{2}\log(1+y^2)\right]$ $\cdot\left[K\{(\frac{1}{2}+\frac{1}{2}z)^{\frac{1}{2}}\}+K\{(\frac{1}{2}-\frac{1}{2}z)^{\frac{1}{2}}\}\right],\ z=y(1+y^2)^{-\frac{1}{2}}$
17.11	$\sinh(\frac{1}{2}ax)K_1(\frac{1}{2}ax)$	$\pi a^2(2y)^{-\frac{1}{2}}(a^2+y^2)^{-\frac{1}{2}}\{y+(a^2+y^2)^{\frac{1}{2}}\}^{-3/2}$
17.12	$K_o(ax)I_o(bx)$ $a\geq b$	$\{y^2+(a+b)^2\}^{-\frac{1}{2}}K\left[2(ab)^{\frac{1}{2}}/\{y^2+(a+b)^2\}^{\frac{1}{2}}\right]$

	$f(x)$	$g_c(y)$
17.13	$K_0(ax)K_0(bx)$	$\pi\{y^2+(a+b)^2\}^{-\frac{1}{2}}K\left(\{y^2+(a-b)^2\}^{\frac{1}{2}}/\{y^2+(a+b^2\}^{\frac{1}{2}}\right)$
17.14	$K_\nu(ax)$ $-1<\mathrm{Re}\nu<1$	$\frac{1}{4}\pi\sec(\frac{1}{2}\pi\nu)(a^2+y^2)^{-\frac{1}{2}}\left((\frac{z}{a})^\nu+(\frac{a}{z})^\nu\right),$ $z=y+(y^2+a^2)^{\frac{1}{2}}$
17.15	$x^\nu K_\nu(ax)$ $\mathrm{Re}\nu>-\frac{1}{2}$	$\frac{1}{2}\pi^{\frac{1}{2}}(2a)^\nu\Gamma(\frac{1}{2}+\nu)(y^2+a^2)^{-\nu-\frac{1}{2}}$
17.16	$x^\mu K_\nu(ax)$ $\mathrm{Re}(\mu\pm\nu)>-1$	$\frac{1}{4}\pi\sec(\frac{1}{2}\pi\mu-\frac{1}{2}\pi\nu)\Gamma(1+\mu+\nu)(a^2+y^2)^{-\frac{1}{2}\mu-\frac{1}{2}}$ $\cdot\left(P_\mu^{-\nu}(z)+P_\mu^{-\nu}(-z)\right),z=y(a^2+y^2)^{-\frac{1}{2}}$
17.17	$x^{-\frac{1}{2}}\log x\, K_\nu(ax)$ $-\frac{1}{2}<\mathrm{Re}\nu<\frac{1}{2}$ $z=y(a^2+y^2)^{-\frac{1}{2}}$	$\frac{1}{4}\pi\sec(\frac{1}{2}\pi\nu+\frac{\pi}{4})\Gamma(\frac{1}{2}+\nu)(a^2+y^2)^{-\frac{1}{4}}\{P_{-\frac{1}{2}}^{-\nu}(z)+P_{-\frac{1}{2}}^{-\nu}(-z)\}$ $\cdot\{\Psi(\frac{1}{2}+\nu)-\frac{1}{2}\log(1+y^2)-\frac{1}{2}\pi\tan(\frac{1}{2}\pi\nu+\frac{1}{4}\pi)\}$
17.18	$x^{-\nu}\cosh(\frac{1}{2}ax)$ $K_\nu(\frac{1}{2}ax),-\frac{1}{2}<\mathrm{Re}\nu<\frac{1}{2}$	$\frac{1}{2}\pi^{3/2}\{\Gamma(\frac{1}{2}+\nu)\}^{-1}a^{-\nu}\sec(\pi\nu)y^{\nu-\frac{1}{2}}(a^2+y^2)^{\frac{1}{2}\nu-\frac{1}{4}}$ $\cdot\cos\{(\nu-\frac{1}{2})\mathrm{arccot}(y/a)\}$
17.19	$x^{\frac{1}{2}}I_{\nu-\frac{1}{4}}(ax)K_{\nu+\frac{1}{4}}(ax)$ $\mathrm{Re}\nu>-3/4$	$a^{-2\nu}(\frac{1}{2}\pi/y)^{\frac{1}{2}}(y^2+4a^2)^{-\frac{1}{2}}\{(4a^2+y^2)^{\frac{1}{2}}-y\}^{2\nu}$
17.20	$x^{-2\nu}I_\nu(ax)K_\nu(ax)$ $-\frac{1}{2}<\mathrm{Re}\nu<\frac{1}{2}$	$\frac{1}{2}\pi^{\frac{1}{2}}(\frac{1}{2}y)^{2\nu}y^{-1}\Gamma(\frac{1}{2}-\nu)\{\Gamma(1+\nu)\}^{-1}$ $_2F_1(\frac{1}{2},\frac{1}{2}-\nu;1+\nu;-\frac{1}{4}a^2/y^2)$
17.21	$x^{2\nu+1}I_\nu(ax)K_\nu(ax)$ $-\frac{1}{2}<\mathrm{Re}\nu<0$	$-\frac{1}{2}a^{-1}\sin(\pi\nu)y^{-2\nu-1}{}_2F_1(\frac{1}{2},\frac{1}{2}+\nu;-\nu;-\frac{1}{4}y^2/a^2)$
17.22	$I_\nu(bx)K_\nu(ax)$ $a>b\ ,\ \mathrm{Re}\nu>-\frac{1}{2}$	$\frac{1}{2}(ab)^{-\frac{1}{2}}Q_{\nu-\frac{1}{2}}\{(a^2+b^2+y^2)/(2ab)\}$
17.23	$K_\nu(ax)K_\nu(bx)$ $-\frac{1}{2}<\mathrm{Re}\nu<\frac{1}{2}$	$\frac{1}{4}\pi^2\sec(\pi\nu)(ab)^{-\frac{1}{2}}P_{\nu-\frac{1}{2}}\{(a^2+b^2+y^2)/(2ab)\}$

	$f(x)$	$g_c(y)$
17.24	$x^{\lambda-1}K_\nu(ax)K_\mu(ax)$ $\quad$ $\operatorname{Re}\lambda < \lvert\operatorname{Re}\nu\rvert + \lvert\operatorname{Re}\mu\rvert$	$\frac{1}{8}(2/a)^\lambda \{\Gamma(\lambda)\}^{-1}\Gamma(\tfrac{1}{2}\mu+\tfrac{1}{2}\nu+\tfrac{1}{2}\lambda)\,\Gamma(\tfrac{1}{2}\lambda+\tfrac{1}{2}\mu-\tfrac{1}{2}\nu)$ $\cdot\Gamma(\tfrac{1}{2}\lambda-\tfrac{1}{2}\mu+\tfrac{1}{2}\nu)\,\Gamma(\tfrac{1}{2}\lambda-\tfrac{1}{2}\mu-\tfrac{1}{2}\nu)$ ${}_4F_3(\tfrac{1}{2}\lambda+\tfrac{1}{2}\mu+\tfrac{1}{2}\nu,\tfrac{1}{2}\lambda+\tfrac{1}{2}\mu-\tfrac{1}{2}\nu,\tfrac{1}{2}\lambda-\tfrac{1}{2}\mu+\tfrac{1}{2}\nu,\tfrac{1}{2}\lambda-\tfrac{1}{2}\mu-\tfrac{1}{2}\nu;$ $\tfrac{1}{2},\tfrac{1}{2}\lambda,\tfrac{1}{2}+\tfrac{1}{2}\lambda;-\tfrac{1}{4}y^2/a^2)$
17.25	$x^{\nu-\mu}I_\mu(bx)K_\nu(ax)$ $\quad$ $a>b,\ \operatorname{Re}\nu>-\tfrac{1}{2}$	$\tfrac{1}{2}\pi\displaystyle\int_0^\infty t^{\nu-\mu}J_\nu(at)J_\mu(bt)e^{-ty}dt$
17.26	$x^{\nu+\mu}K_\nu(ax)K_\mu(bx)$ $\quad$ $\operatorname{Re}(\nu,\mu)>-\tfrac{1}{2}$	$-\tfrac{1}{4}\pi^2\displaystyle\int_0^\infty t^{\nu+\mu}\{J_\nu(at)Y_\mu(bt)+Y_\nu(at)J_\mu(bt)\}e^{-ty}dt$
17.27	$x^{-\nu}(b^2+x^2)^{-1}K_\nu(ax)$ $\quad$ $\operatorname{Re}\nu<\tfrac{1}{2}$	$\tfrac{1}{4}\pi^{\frac{1}{2}}(2a)^{-\nu}\Gamma(\tfrac{1}{2}-\nu)b^{-1}\Big(e^{by}\displaystyle\int_y^\infty e^{-bt}(a^2+t^2)^{\nu-\frac{1}{2}}dt$ $+e^{-by}\displaystyle\int_{-y}^\infty e^{-bt}(a^2+t^2)^{\nu-\frac{1}{2}}dt\Big)$
17.28	$x^{\frac{1}{2}}e^{-ax^2}I_{-\frac{1}{4}}(ax^2)$	$\tfrac{1}{2}(ay)^{-\frac{1}{2}}\exp(-\tfrac{1}{8}y^2/a)$
17.29	$x^{\frac{1}{2}}e^{-ax^2}I_{-\frac{1}{4}}(bx^2)$ $\quad$ $a>b$	$\tfrac{1}{2}(\tfrac{1}{2}\pi y)^{\frac{1}{2}}z^{-\frac{1}{2}}\exp(-\tfrac{a}{4}y^2/z)I_{-\frac{1}{4}}(\tfrac{1}{4}by^2/z);z=(a^2-b^2)$
17.30	$x^{1/3}e^{-ax^2}I_{-\frac{1}{3}}(ax^2)$	$\tfrac{1}{4}2^{-1/6}\pi^{\frac{1}{2}}a^{-5/6}y^{1/3}\exp(-\tfrac{y^2}{16a})I_{\frac{1}{3}}(\tfrac{1}{16}y^2/a)$
17.31	$e^{-ax^2}I_0(ax^2)$	$\tfrac{1}{2}(2\pi a)^{-\frac{1}{2}}\exp(-\tfrac{y^2}{16a})K_0(\tfrac{y^2}{16a})$
17.32	$K_0(ax^2)$	$\tfrac{1}{8}\pi a^{-1}yK_{\frac{1}{4}}(z)\{I_{\frac{1}{4}}(z)+I_{-\frac{1}{4}}(z)\},\ z=\tfrac{1}{8}y^2/a$
17.33	$\exp(ax^2)K_0(ax^2)$	$\tfrac{1}{2}(\tfrac{1}{2}\pi/a)^{\frac{1}{2}}\exp(\tfrac{y^2}{16a})K_0(\tfrac{y^2}{16a})$
17.34	$\exp(-ax^2)K_0(ax^2)$	$\tfrac{1}{2}\pi(\tfrac{1}{2}\pi/a)^{\frac{1}{2}}\exp(-\tfrac{y^2}{16a})I_0(\tfrac{y^2}{16a})$
17.35	$x^{\frac{1}{2}}K_{\frac{1}{4}}(ax^2)$	$\tfrac{1}{4}\pi a^{-1}(\tfrac{1}{2}\pi y)^{\frac{1}{2}}\{I_{-\frac{1}{4}}(z)-\mathbf{L}_{-\frac{1}{4}}(z)\},\ z=\tfrac{1}{4}y^2/a$

	$f(x)$	$g_c(y)$
17.36	$x^{3/2} K_{\frac{1}{4}}(ax^2)$	$\frac{1}{4}(\frac{1}{2}\pi)^{3/2} a^{-2} y^{3/2} \{I_{-\frac{3}{4}}(z) - \mathbf{L}_{-\frac{3}{4}}(z)\}, z=\frac{1}{4}y^2/a$
17.37	$x^{-2\nu} e^{-x^2} I_{\nu}(x^2)$ $\mathrm{Re}\,\nu > -\frac{1}{2}$	$2^{-\frac{1}{2}\nu} y^{\nu-1} \exp(\frac{y^2}{16}) W_{-\frac{3\nu}{2}, \frac{1}{2}\nu}(y^2/8)$
17.38	$x^{2\nu} e^{-x^2} I_{\nu}(x^2)$ $-\frac{1}{4} < \mathrm{Re}\,\nu < \frac{1}{2}$	$\pi^{-1} 2^{-\nu-3/2} \Gamma(-\nu)\Gamma(\frac{1}{2}+2\nu) \exp(-y^2/8)$ $\cdot {}_1F_1(\frac{1}{2}-\nu;1+\nu;y^2/8) + 2^{2\nu-3/2}\Gamma(\nu)\{\Gamma(\frac{1}{2}-\nu)\}^{-1}$ $\cdot\exp(-y^2/8){}_1F_1(\frac{1}{2}-2\nu;1-\nu;y^2/a)$
17.39	$x^{2\nu+1} e^{-ax^2} I_{\nu}(ax^2)$ $-1 < \mathrm{Re}\,\nu < 0$	$2^{2\nu-\frac{1}{2}} a^{-\frac{1}{2}} \Gamma(\frac{1}{2}+\nu)\{\Gamma(-\nu)\}^{-1} y^{-2\nu-1}$ $\cdot\exp(-\frac{1}{4}y^2/a({}_1F_1(-\frac{1}{2}-2\nu;-\nu;\frac{1}{8}y^2/a)$
17.40	$x^{2\nu} \exp(-ax^2) K_{\nu}(ax^2)$ $\mathrm{Re}\,\nu > -\frac{1}{4}$	$\frac{1}{2}\pi a^{-\nu-\frac{1}{2}} \Gamma(\frac{1}{2}+2\nu)\{\Gamma(1+\nu)\}^{-1}$ $\cdot\exp(-\frac{1}{8}y^2/a){}_1F_1(\frac{1}{2}-\nu;1+\nu;\frac{1}{8}y^2/a)$
17.41	$x^{2\nu} K_{\nu}(ax^2)$ $\mathrm{Re}\,\nu > -\frac{1}{4}$	$2^{\nu-2}\pi a^{-\nu-\frac{1}{2}}\Big(\Gamma(\frac{1}{4}+\nu)\{\Gamma(\frac{3}{4})\}^{-1} {}_1F_2(\frac{1}{4}+\nu;\frac{1}{2},\frac{3}{4};\frac{y^2}{64a^2})$ $\quad - \frac{y^2}{4a}\{\Gamma(\frac{5}{4})\}^{-1}\Gamma(\frac{3}{4}+\nu){}_1F_2(\frac{3}{4}+\nu;\frac{3}{2},\frac{5}{4};\frac{y^2}{64a^2})\Big)$
17.42	$x^{-2\nu} e^{ax^2} K_{\nu}(ax^2)$ $-\frac{1}{2} < \mathrm{Re}\,\nu < \frac{1}{2}$	$\pi(\frac{1}{2}a)^{\frac{1}{2}\nu}\Gamma(\frac{1}{2}-2\nu)\{\Gamma(1+\nu)\}^{-1} y^{\nu-1}$ $\exp(\frac{y^2}{16a}) W_{\frac{3}{2}\nu,-\frac{1}{2}\nu}(\frac{y^2}{8a})$
17.43	$x^{2\lambda-1} e^{-ax^2} K_{\nu}(ax^2)$ $\mathrm{Re}\,\lambda < \|\mathrm{Re}\,\nu\|$	$\frac{1}{2}\pi^{\frac{1}{2}} (2a)^{-\lambda} \Gamma(\lambda+\nu)\Gamma(\lambda-\nu)\{\Gamma(\frac{1}{2}+\lambda)\}^{-1}$ $\cdot {}_2F_2(\lambda+\nu,\lambda-\nu;\frac{1}{2},\frac{1}{2}+\lambda;-\frac{1}{8}y^2/a)$
17.44	$x^{\frac{1}{2}} I_{-\frac{1}{8}}(ax^2) K_{\frac{1}{8}}(ax^2)$	$\frac{1}{4}a^{-1}(\frac{1}{2}\pi y)^{\frac{1}{2}} I_{-\frac{1}{8}}(z) K_{\frac{1}{8}}(z) \; ; \; z=y^2/(16a)$
17.45	$x\{K_{\frac{1}{4}}(ax^2)\}^2$	$\frac{1}{2}2^{-\frac{1}{2}}\pi^2 a^{-1} \{I_0(z) - \mathbf{L}_0(z)\}, \; z=\frac{1}{8}y^2/a$

	$f(x)$	$g_c(y)$
17.46	$x^{\frac{1}{2}} I_{-\frac{1}{8}-\nu}(u) K_{\frac{1}{8}-\nu}(u)$ $\mathrm{Re}\,\nu < \frac{3}{8}$, $u = ax^2$	$\{\Gamma(\frac{3}{4})\}^{-1} \Gamma(\frac{3}{8}-\nu) y^{-3/2} (2\pi)^{\frac{1}{2}} W_{\nu,-\frac{1}{8}}(z) M_{-\nu,-\frac{1}{8}}(z)$ $z = \frac{1}{8} y^2 / a$
17.47	$x^{-1} K_0(a/x)$	$-\pi K_0(z) Y_0(z)$, $z = (2ay)^{\frac{1}{2}}$
17.48	$x^{-3} K_0(a/x)$	$-\pi a^{-1} y K_1(z) Y_1(z)$, $z = (2ay)^{\frac{1}{2}}$
17.49	$x^{-1} K_\nu(a/x)$ $-1 < \mathrm{Re}\,\nu < 1$	$-\pi K_\nu(z) \{J_\nu(z) \sin(\frac{1}{2}\pi\nu) + Y_\nu(z) \cos(\frac{1}{2}\pi\nu)\}$, $z = (2ay)^{\frac{1}{2}}$
17.50	$x^{-1} e^{-a/x} K_\nu(b/x)$ $-1 < \mathrm{Re}\,\nu < 1$	$K_\nu(ui^{\frac{1}{2}}) K_\nu(vi^{\frac{1}{2}}) + K_\nu\{u(-i)^{\frac{1}{2}}\} K_\nu\{v(-i)^{\frac{1}{2}}\}$ $\begin{matrix}u\\v\end{matrix} = (ay+by)^{\frac{1}{2}} \pm (ay-by)^{\frac{1}{2}}$

1.18 Modified Bessel Functions of Argument $(ax^2+bx+c)^{\frac{1}{2}}$

	$f(x)$	$g_c(y)$
18.1	$e^{-bx} I_0(ax^{\frac{1}{2}})$	$4a^{-2} z e^{bz} \{b \cos(yz) - y \sin(yz)\}$ $z = \frac{1}{4} a^2 / (b^2 + y^2)$
18.2	$K_0(ax^{\frac{1}{2}})$	$(2y)^{-1} \{\mathrm{Ci}(z) \sin z - \mathrm{si}(z) \cos z\}, z = \frac{1}{4} a^2 / y$
18.3	$\{K_0(ax^{\frac{1}{2}})$ $\quad -\frac{\pi}{2} Y_0(ax^{\frac{1}{2}})\} \log(bx)$	$\pi y^{-1} \cos(\frac{1}{4} a^2 / y) \log(\frac{1}{2} ab^{\frac{1}{2}} / y)$
18.4	$x^{-\frac{1}{2}} \{K_1(ax^{\frac{1}{2}}) + \frac{1}{2}\pi Y_1(ax^{\frac{1}{2}})\}$	$-\pi a^{-1} \sin(\frac{1}{4} a^2 / y)$
18.5	$J_0(ax^{\frac{1}{2}}) K_0(ax^{\frac{1}{2}})$	$\frac{1}{4}\pi y^{-1} \{I_0(\frac{1}{4} a^2 / y) - \mathbf{L}_0(\frac{1}{4} a^2 / y)\}$
18.6	$Y_0(ax^{\frac{1}{2}}) K_0(ax^{\frac{1}{2}})$	$-\frac{1}{2} y^{-1} K_0(\frac{1}{4} a^2 / y)$
18.7	$x Y_1(ax^{\frac{1}{2}}) K_1(ax^{\frac{1}{2}})$	$-\frac{1}{4} a^2 / y^{-3} K_0(\frac{1}{4} a^2 / y)$
18.8	$I_0(ax^{\frac{1}{2}}) K_0(ax^{\frac{1}{2}})$	$\frac{1}{4}\pi y^{-1} \{J_0(z) \cos z + Y_0(z) \sin z\}, z = \frac{1}{2} a^2 / y$
18.9	$K_0\{a(ix)^{\frac{1}{2}}\} K_0\{a(-ix)^{\frac{1}{2}}\}$	$\frac{1}{8}\pi^2 / y \{\mathbf{H}_0(\frac{1}{2} a^2 / y) - Y_0(\frac{1}{2} a^2 / y)\}$

	$f(x)$	$g_c(y)$
18.10	$x^{-\frac{1}{2}}K_\nu(ax^{\frac{1}{2}})$ $-1<\mathrm{Re}\,\nu<1$	$-\frac{1}{4}\pi\sec(\frac{1}{2}\pi\nu)(\pi/y)^{\frac{1}{2}}\Big[J_{\frac{1}{2}\nu}(z)\sin(\frac{1}{4}\pi\nu-\frac{1}{4}\pi-z)$ $+Y_{\frac{1}{2}\nu}(z)\cos(\frac{1}{4}\pi\nu-\frac{1}{4}\pi-z)\Big]$, $z=\frac{1}{8}a^2/y$
18.11	$J_\nu(ax^{\frac{1}{2}})K_\nu(ax^{\frac{1}{2}})$ $\mathrm{Re}\,\nu>-1$	$\frac{1}{2}\pi\mathrm{cosec}(\pi\nu)y^{-1}\Big[\sin(\frac{1}{2}\pi\nu)I_\nu(\frac{1}{2}a^2/y)$ $+\frac{1}{2}iJ_\nu(\frac{1}{2}ia^2/y)-\frac{1}{2}iJ_\nu(-\frac{1}{2}ia^2/y)\Big]$
18.12	$x^{-\frac{1}{2}}J_\nu(ax^{\frac{1}{2}})K_\nu(ax^{\frac{1}{2}})$ $\mathrm{Re}\,\nu>-\frac{1}{2}$	$a^{-2}\Gamma(\frac{1}{4}+\frac{1}{2}\nu)\{\Gamma(1+\nu)\}^{-1}$ $(\frac{1}{2}\pi y)^{\frac{1}{2}}W_{\frac{1}{4},\frac{1}{2}\nu}(\frac{1}{2}a^2/y)M_{-\frac{1}{4},\frac{1}{2}\nu}(\frac{1}{2}a^2/y)$
18.13	$\{I_\nu(ax^{\frac{1}{2}})+I_{-\nu}(ax^{\frac{1}{2}})\}$ $\cdot K_\nu(ax^{\frac{1}{2}})$, $-1<\mathrm{Re}\,\nu<1$	$\frac{1}{2}\pi y^{-1}\Big[\cos(\frac{1}{2}\pi\nu-z)J_\nu(z)-\sin(\frac{1}{2}\pi\nu-z)Y_\nu(z)\Big]$ $z=\frac{1}{2}a^2/y$
18.14	$x^{-\frac{1}{2}}K_\nu(ax^{\frac{1}{2}})$ $\cdot\{\cos(\frac{1}{2}\pi\nu-\frac{1}{4}\pi)J_\nu(ax^{\frac{1}{2}})$ $+\cos(\frac{1}{2}\pi\nu+\frac{1}{4}\pi)Y_\nu(ax^{\frac{1}{2}})$ $-\frac{1}{2}<\mathrm{Re}\,\nu<\frac{1}{2}$	$-a^{-2}(\frac{1}{2}\pi y)^{\frac{1}{2}}W_{-\frac{1}{4},\frac{1}{2}\nu}(\frac{1}{2}a^2/y)W_{\frac{1}{4},\frac{1}{2}\nu}(\frac{1}{2}a^2/y)$
18.15	$K_\nu(ax^{\frac{1}{2}})$ $\cdot\{\sin(\frac{1}{2}\pi\nu)J_\nu(ax^{\frac{1}{2}})$ $+\cos(\frac{1}{2}\pi\nu)Y_\nu(ax^{\frac{1}{2}})\}$ $-1<\mathrm{Re}\,\nu<1$	$-\frac{1}{2}y^{-1}K_\nu(\frac{1}{2}a^2/y)$
18.16	$x^{-\frac{1}{2}\nu}\{K_\nu(ax^{\frac{1}{2}})\cos(\pi\nu)$ $-\frac{1}{2}\pi Y_\nu(ax^{\frac{1}{2}})\}$, $-1<\mathrm{Re}\,\nu<1$	$\frac{1}{2}\pi(\frac{1}{2}a)^{-\nu}y^{\nu-1}\cos(\frac{1}{4}a^2/y-\frac{1}{2}\pi\nu)$
18.17	$x^{-\frac{1}{2}\nu}\{K_\nu(ax^{\frac{1}{2}})\sin(\pi\nu)$ $-\frac{1}{2}\pi J_\nu(ax^{\frac{1}{2}})\}$, $-1<\mathrm{Re}\,\nu<1$	$-\frac{1}{2}\pi(\frac{1}{2}a)^{-\nu}y^{\nu-1}\sin(\frac{1}{4}a^2/y-\frac{1}{2}\pi\nu)$
18.18	$K_\nu\{a(ix)^{\frac{1}{2}}\}K_\nu\{a(-ix)^{\frac{1}{2}}\}$ $-1<\mathrm{Re}\,\nu<1$	$\frac{1}{4}\pi y^{-1}\sec(\frac{1}{2}\pi\nu)S_{0,\nu}(\frac{1}{2}a^2/y)$
18.19	$x^{-\frac{1}{2}}K_\nu\{a(ix)^{\frac{1}{2}}\}$ $\cdot K_\nu\{a(-ix)^{\frac{1}{2}}\}$, $-\frac{1}{2}<\mathrm{Re}\,\nu<\frac{1}{2}$	$\frac{1}{2}a^{-2}(\frac{1}{2}\pi y)^{\frac{1}{2}}\Gamma(\frac{1}{4}+\frac{1}{2}\nu)\Gamma(\frac{1}{4}-\frac{1}{2}\nu)$ $\cdot W_{\frac{1}{4},\frac{1}{2}\nu}(\frac{1}{2}ia^2/y)W_{\frac{1}{4},\frac{1}{2}\nu}(-\frac{1}{2}ia^2/y)$

	$f(x)$	$g_c(y)$
18.20	$x^{\frac{1}{2}\nu}\{e^{\frac{1}{4}i\pi\nu}K_\nu\{a(ix)^{\frac{1}{2}}\}$ $+e^{-\frac{1}{4}i\pi\nu}K_\nu\{a(-ix)^{\frac{1}{2}}\}$ $\mathrm{Re}\,\nu>-1$	$\pi a^\nu(2y)^{-\nu-1}\exp(-\tfrac{1}{4}a^2/y)$
18.21	$K_0(br)$	$\tfrac{1}{2}\pi q^{-1}e^{-aq}$
18.22	$r^{-1}K_1(br)$	$\tfrac{1}{2}\pi(ab)^{-1}e^{-aq}$
18.23	$r^{-2}K_0(br)$	$-\tfrac{1}{4}\pi a^{-1}\{e^{ay}\mathrm{Ei}(-au)+e^{-ay}\mathrm{Ei}(-av)\}$, $\begin{smallmatrix}u\\v\end{smallmatrix}=q\pm y$
18.24	$\log r K_0(br)$	$-\tfrac{1}{2}\pi q^{-1}\left(e^{-aq}\log\{q/(ab)\}+e^{aq}\mathrm{Ei}(-2aq)\right]$
18.25	$K_0\{b(x^2+2ax)^{\frac{1}{2}}\}$	$\tfrac{1}{2}q^{-1}\{\sin u\,\mathrm{Ci}(u)+\sin v\,\mathrm{Ci}(v)$ $\qquad-\cos u\,\mathrm{si}(u)-\cos v\,\mathrm{si}(v)\}$, $\begin{smallmatrix}u\\v\end{smallmatrix}=a(q\pm y)$
18.26	$K_{2n}(br)$ $n=1,2,\ldots$	$(\tfrac{1}{2}\pi a)^{\frac{1}{2}}n\sum_{k=0}^{n}(n+k-1)!\{k!(n-k)!\}^{-1}(\tfrac{1}{2}b^2a)^{-k}$ $\cdot q^{k-\frac{1}{2}}K_{k-\frac{1}{2}}(aq)$
18.27	$r^{-1}K_{2\nu}(br)$	$\tfrac{1}{2}K_\nu(u)K_\nu(v)$, $\begin{smallmatrix}u\\v\end{smallmatrix}=\tfrac{1}{2}a(q\pm y)$
18.28	$r^{-2}K_{2\nu}(br)$	$\dfrac{b}{8\nu}\{K_{\nu+\frac{1}{2}}(u)K_{\nu+\frac{1}{2}}(v)-K_{\nu-\frac{1}{2}}(u)K_{\nu-\frac{1}{2}}(v)\}$, $\begin{smallmatrix}u\\v\end{smallmatrix}=\tfrac{1}{2}a(q\pm y)$
18.29	$r^{-\nu}K_\nu(br)$	$(\tfrac{1}{2}\pi a)^{\frac{1}{2}}(ab)^{-\nu}q^{\nu-\frac{1}{2}}K_{\nu-\frac{1}{2}}(aq)$
18.30	$(x^2+2ax)^{-\frac{1}{2}}$ $\cdot K_{2\nu}\{b(x^2+2ax)^{\frac{1}{2}}\}$ $-\tfrac{1}{2}<\mathrm{Re}\,\nu<\tfrac{1}{2}$	$\tfrac{1}{8}\pi^2\sec(\pi\nu)\bigg(\cos(ay)\{J_\nu(u)J_\nu(v)+Y_\nu(u)Y_\nu(v)\}$ $-\sin(ay)\{J_\nu(u)Y_\nu(v)-J_\nu(v)Y_\nu(u)\}\bigg]$ $\begin{smallmatrix}u\\v\end{smallmatrix}=\tfrac{1}{2}a\{(b^2+y^2)^{\frac{1}{2}}\pm y\}$
18.31	$I_0(bu)\quad x<a$ $\quad 0\qquad x>a$	$s^{-1}\sinh(as)\qquad\qquad y<b$ $s^{-1}\sin(aS)\qquad\qquad y>b$
18.32	$K_0(bu)\quad x<a$ $\quad 0\qquad x>a$	$-\tfrac{1}{2}S^{-1}\bigg(\sin(aS)\{\mathrm{Ci}(u)+\mathrm{Ci}(v)\}$ $-\cos(aS)\{\mathrm{si}(u)-\mathrm{si}(v)\}\bigg]$, $\begin{smallmatrix}u\\v\end{smallmatrix}=a(y\pm S),y>b$

$r=(a^2+x^2)^{\frac{1}{2}}, u=(a^2-x^2)^{\frac{1}{2}}, w=(2ax-x^2)^{\frac{1}{2}}, s=(b^2-y^2)^{\frac{1}{2}}, S=(y^2-b^2)^{\frac{1}{2}}, q=(b^2+y^2)^{\frac{1}{2}}$

	$f(x)$	$g_c(y)$
18.33	$u^{-1}I_{2\nu}(bu)\quad x<a$ $0\quad x>a$ $\text{Re}\,\nu>-\frac{1}{2}$	$\frac{1}{2}\pi J_\nu(u)J_\nu(v)\qquad \frac{u}{v}=\frac{1}{2}a(y\pm S)$
18.34	$u^{-1}K_{2\nu}(bu)\quad x<a$ $0\quad x>a$ $-\frac{1}{2}<\text{Re}\,\nu<\frac{1}{2}$	$\frac{1}{4}\pi^2\operatorname{cosec}(2\pi\nu)\{J_{-\nu}(u)J_{-\nu}(v)-J_\nu(u)J_\nu(v)\}$ $=-\frac{1}{8}\pi^2\Big\{J_\nu(u)Y_\nu(v)+J_\nu(v)Y_\nu(u)-\tan(\pi\nu)$ $\cdot\{J_\nu(u)J_\nu(v)-Y_\nu(u)Y_\nu(v)\}\Big\}\Big|_{v=\frac{1}{2}a(y\pm S)}, y>b$
18.35	$u^\nu I_\nu(bu)\quad x<a$ $0\quad x>a$ $\text{Re}\,\nu>-1$	$(\frac{1}{2}\pi a)^{\frac{1}{2}}s^{-\nu-\frac{1}{2}}(ab)^\nu I_{\nu+\frac{1}{2}}(as)\qquad y<b$ $(\frac{1}{2}\pi a)^{\frac{1}{2}}S^{-\nu-\frac{1}{2}}(ab)^\nu J_{\nu+\frac{1}{2}}(aS)\qquad y>b$
18.36	$I_0(bw)\quad x<2a$ $0\quad x>2a$	$2\cos(ay)\begin{array}{ll}s^{-1}\sinh(as) & y<b\\ s^{-1}\sin(as) & y>b\end{array}$
18.37	$w^{-1}I_\nu(bw)\quad x<2a$ $0\quad x>2a$ $\text{Re}\,\nu>-1$	$\pi\cos(ay)J_{\frac{1}{2}\nu}(u)J_{\frac{1}{2}\nu}(v)\ ,\ \frac{u}{v}=\frac{1}{2}a(y\pm S)\quad y>b$
18.38	$w^\nu I_\nu(bw)\quad x<2a$ $0\quad x>2a$ $\text{Re}\,\nu>-1$	$(2\pi a)^{\frac{1}{2}}(ab)^\nu\cos(ay)s^{-\nu-\frac{1}{2}}I_{\nu+\frac{1}{2}}(as)\qquad y<b$ $(2\pi a)^{\frac{1}{2}}(ab)^\nu\cos(ay)S^{-\nu-\frac{1}{2}}J_{\nu+\frac{1}{2}}(aS)\qquad y>b$
18.39	$w^{-1}K_{2\nu}(bw)\quad x<2a$ $0\quad x>2a$ $-\frac{1}{2}<\text{Re}\,\nu<\frac{1}{2}$	$\frac{1}{2}\pi^2\operatorname{cosec}(2\pi\nu)\cos(ay)\{J_{-\nu}(u)J_{-\nu}(v)$ $-J_\nu(u)J_\nu(v)\}\ ;\ \frac{u}{v}=\frac{1}{2}a(y\pm S)\ ,\qquad y>b$
18.40	$0\quad x<a$ $K_0(bU)\quad x>a$	$\frac{1}{2}r^{-1}\Big(\sin(aq)\{Ci(u)+Ci(v)\}$ $-\cos(aq)\{si(u)+si(v)\}\quad \frac{u}{v}=a(q\pm y)$
18.41	$0\quad x<a$ $U^{-1}K_{2\nu}(bU)\quad x>a$ $-\frac{1}{2}<\text{Re}\,\nu<\frac{1}{2}$	$\frac{1}{8}\pi^2\sec(\pi\nu)\{J_\nu(u)J_\nu(v)+Y_\nu(u)Y_\nu(v)\}$ $\frac{u}{v}=\frac{1}{2}a(q\pm y)$

$u=(a^2-x^2)^{\frac{1}{2}}, U=(x^2-a^2)^{\frac{1}{2}}, w=(2ax-x^2)^{\frac{1}{2}}, s=(b^2-y^2)^{\frac{1}{2}}, S=(y^2-b^2)^{\frac{1}{2}}, q=(b^2+y^2)^{\frac{1}{2}}$

	$f(x)$	$g_c(y)$
18.42	$u^\nu Y_\nu(bu)$ $x<a$ $-2\pi^{-1}\cos(\pi\nu)$ $U^\nu K_\nu(bU)$ $x>a$ $\mathrm{Re}\,\nu>-1$	$(\tfrac{1}{2}\pi a)^{\frac{1}{2}}(ab)^\nu q^{-\nu-\frac{1}{2}}Y_{\nu+\frac{1}{2}}(aq)$
18 43	$I_0(\tfrac{1}{2}br-\tfrac{1}{2}bx)I_0(\tfrac{1}{2}br+\tfrac{1}{2}bx)$	$\tfrac{1}{2}\pi q^{-1}\{I_0(\tfrac{1}{2}aq)-\mathbf{L}_0(\tfrac{1}{2}aq)\}$
18.44	$x^{\frac{1}{2}}I_{-\frac{1}{4}}(\tfrac{1}{2}br-\tfrac{1}{2}ab)K_{\frac{1}{4}}(\tfrac{1}{2}br+\tfrac{1}{2}ab)$	$(\tfrac{1}{2}\pi/y)^{\frac{1}{2}}q^{-1}e^{-aq}$
18.45	$K_0(\tfrac{1}{2}bx+\tfrac{1}{2}bU)K_0(\tfrac{1}{2}bx-\tfrac{1}{2}bU)$	$\tfrac{1}{2}\pi^2 q^{-1}\{\mathbf{H}_0(aq)-Y_0(aq)\}$
18.46	$K_\nu(\tfrac{1}{2}bx+\tfrac{1}{2}bU)K_\nu(\tfrac{1}{2}bx-\tfrac{1}{2}bU)$	$\pi q^{-1}S_{0,2\nu}(aq)$
18.47	$K_\nu(\tfrac{1}{2}br-\tfrac{1}{2}bx)K_\nu(\tfrac{1}{2}br+\tfrac{1}{2}bx)$	$\pi q^{-1}K_{2\nu}(aq)$
18.48	$I_\nu(kr)K_\mu(br)$ $k\leq b$	$\tfrac{1}{2}\pi\displaystyle\int_0^\infty t^{\mu-\nu+1}J_\nu(kt)J_\mu(bt)$ $\cdot(a^2+t^2)^{-\frac{1}{2}}\exp\{-y(a^2+t^2)^{\frac{1}{2}}\}dt$

1.19 Modified Bessel Functions of Trigonometric and Hyperbolic Arguments

	$f(x)$	$g_c(y)$
19.1	$K_0\{(a^2+b^2-2ab\cos x)^{\frac{1}{2}}\}$, $x<\pi$ 0 , $x>\pi$	$y\sin(\pi y)\displaystyle\sum_{n=0}^\infty(-1)^n\varepsilon_n(y^2-n^2)^{-1}I_n(b)K_n(a)$ $a\leq b$
19.2	$I_{2\nu}\{2a\cos(\tfrac{1}{2}x)\}$ $x<\pi$ 0 $x>\pi$ $\mathrm{Re}\,\nu>-\tfrac{1}{2}$	$\pi I_{\nu-y}(a)I_{\nu+y}(a)$
19.3	$I_{2\nu}(2a\sin x)$ $x<\pi$ 0 $x>\pi$ $\mathrm{Re}\,\nu>-\tfrac{1}{2}$	$\pi\cos(\tfrac{1}{2}\pi y)I_{\nu-\frac{1}{2}y}(a)I_{\nu+\frac{1}{2}y}(a)$

$r=(a^2+x^2)^{\frac{1}{2}},\quad u=(a^2-x^2)^{\frac{1}{2}}$
$U=(x^2-a^2)^{\frac{1}{2}}$

$s=(b^2-y^2)^{\frac{1}{2}},\quad S=(y^2-b^2)^{\frac{1}{2}}$
$q=(b^2+y^2)^{\frac{1}{2}}$

	$f(x)$	$g_c(y)$
19.4	$K_{2\nu}\{2a\cos(\tfrac{1}{2}x)\}$ $\quad x\ \pi$ $0\qquad\qquad\qquad x>\pi$ $-\tfrac{1}{2}<\mathrm{Re}\,\nu<\tfrac{1}{2}$	$\pi\,\mathrm{cosec}(2\pi\nu)\Big[I_{y-\nu}(a)K_{y+\nu}(a)\sin(\pi y+\pi\nu)$ $-I_{y+\nu}(a)K_{y-\nu}(a)\sin(\pi y-\pi\nu)\Big]$
19.5	$K_{2\nu}(2a\sin x)\qquad x<\pi$ $0\qquad\qquad\qquad x>\pi$ $-\tfrac{1}{2}<\mathrm{Re}\,\nu<\tfrac{1}{2}$	$\tfrac{1}{2}\pi^2\mathrm{cosec}(2\pi\nu)\cos(\tfrac{1}{2}\pi y)$ $\cdot\Big[(I_{-\nu-\frac{1}{2}y}(a)I_{-\nu+\frac{1}{2}y}(a)-I_{\nu-\frac{1}{2}y}(a)I_{\nu+\frac{1}{2}y}(a)\Big]$
19.6	$\sec(\tfrac{1}{2}x)I_{2\nu}\{2a\cos(\tfrac{1}{2}x)\}$ $\qquad\qquad\qquad\qquad x<\pi$ $0\qquad\qquad\qquad x>\pi$ $\mathrm{Re}\,\nu>0$	$\tfrac{1}{2}\pi a\nu^{-1}\Big[I_{\nu-\frac{1}{2}-y}(a)I_{\nu-\frac{1}{2}+y}(a)$ $-I_{\nu+\frac{1}{2}-y}(a)I_{\nu+\frac{1}{2}+y}(a)\Big]$
19.7	$\mathrm{cosec}\,x\,I_{2\nu}(2a\sin x)$ $\qquad\qquad\qquad\qquad x<\pi$ $0\qquad\qquad\qquad x>\pi$ $\mathrm{Re}\,\nu>0$	$\tfrac{1}{2}\pi a\nu^{-1}\cos(\tfrac{1}{2}\pi y)\Big[I_{\nu-\frac{1}{2}+\frac{1}{2}y}(a)I_{\nu-\frac{1}{2}-\frac{1}{2}y}(a)$ $-I_{\nu+\frac{1}{2}+\frac{1}{2}y}(a)I_{\nu+\frac{1}{2}-\frac{1}{2}y}(a)\Big]$
19.8	$\sec^m(\tfrac{1}{2}x)I_m(2a\cos\tfrac{x}{2})$ $\qquad\qquad\qquad\qquad ,x<\pi$ $0\qquad\qquad\qquad ,x>\pi$ $m=1,2,\ldots$	$\pi a^m m!\sum_{n=0}^{m}(-1)^n\varepsilon_n\{(m+n)!(m-n)!\}^{-1}$ $\cdot I_{n-y}(a)I_{n+y}(a)$
19.9	$\mathrm{cosec}^m x\,I_m(2a\sin x)$ $\qquad\qquad\qquad\qquad ,x<\pi$ $0\qquad\qquad\qquad x>\pi$ $m=1,2,\ldots$	$\pi a^m m!\cos(\tfrac{1}{2}\pi y)\sum_{n=0}^{m}(-1)^n\varepsilon_n\{(m+n)!(m-n)!\}^{-1}$ $\cdot I_{n-\frac{1}{2}y}(a)I_{n+\frac{1}{2}y}(a)$
19.10	$\{\sec x\cos(\tfrac{1}{2}x)\}^{\frac{1}{2}}$ $\cdot\exp(-a^2\sec x)$ $\cdot K_{\frac{1}{4}}[a^2(1+\sec x)]\quad x<\tfrac{1}{2}\pi$ $0\qquad\qquad\qquad x>\tfrac{1}{2}\pi$	$\pi(2a)^{-\frac{1}{2}}\exp(-a^2)D_{-\frac{3}{4}+y}(2a)D_{-\frac{3}{4}-y}(2a)$
19.11	$K_0\{2a\sinh(\tfrac{1}{2}x)\}$	$\tfrac{1}{4}\pi^2\Big[\{J_{iy}(a)\}^2+\{Y_{iy}(a)\}^2\Big]$
19.12	$K_0\{a(2\cosh x)^{\frac{1}{2}}\}$	$K_{iy}(ai^{\frac{1}{2}})K_{iy}\{a(-i)^{\frac{1}{2}}\}$

	$f(x)$	$g_c(y)$
19.13	$2\pi^{-1}K_0\{a(2\sinh x)^{\frac{1}{2}}\}$ $-Y_0\{a(2\sinh x)^{\frac{1}{2}}\}$	$-K_{iy}(a)\{Y_{iy}(a)+Y_{-iy}(a)\}$
19.14	$K_0\{(a^2+b^2+2ab\cosh x)^{\frac{1}{2}}\}$	$K_{iy}(a)K_{iy}(b)$
19.15	$K_0\{(b^2+a^2\sinh^2 x)^{\frac{1}{2}}\}$	$\frac{1}{2}K_{\frac{1}{2}iy}(u)K_{\frac{1}{2}iy}(v)$, ${u \atop v} = \frac{1}{2}\{b\pm(b^2-a^2)^{\frac{1}{2}}\}$
19.16	$K_0\{(b^2+a^2\cosh^2 x)^{\frac{1}{2}}\}$	$\frac{1}{2}K_{\frac{1}{2}iy}(u)K_{\frac{1}{2}iy}(v)$, ${u \atop v} = \frac{1}{2}\{(a^2+b^2)^{\frac{1}{2}}\pm b\}$
19.17	$K_0\{a(i\cosh x)^{\frac{1}{2}}\}$ $\cdot K_0\{a(-i\cosh x)^{\frac{1}{2}}\}$	$-\frac{1}{2}\pi\,\mathrm{sech}(\frac{1}{2}\pi y)K_{iy}(z)\{Y_{iy}(z)+Y_{-iy}(z)\}$ $z=2^{-\frac{1}{2}}a$
19.18	$(\cosh x)^{\frac{1}{2}}\exp(-a\sinh^2 x)$ $\cdot K_{\frac{1}{4}}(a\cosh^2 x)$	$2^{-9/4}a^{-3/4}W_{-\frac{1}{4},\frac{1}{2}iy}(2a)$
19.19	$(\cosh x)^{\frac{1}{2}}\exp(a\sinh^2 x)$ $\cdot K_{\frac{1}{4}}(a\cosh^2 x)$	$(2a)^{-3/4}\Gamma(\frac{1}{4}+\frac{1}{2}iy)\Gamma(\frac{1}{4}-\frac{1}{2}iy)W_{\frac{1}{4},\frac{1}{2}iy}(2a)$
19.20	$K_{2\nu}\{2a\cosh(\frac{1}{2}x)\}$	$K_{\nu+iy}(a)K_{\nu-iy}(a)$
19.21	$K_{2\nu}\{2a\sinh(\frac{1}{2}x)\}$ $-\frac{1}{2}<\mathrm{Re}\,\nu<\frac{1}{2}$	$\frac{1}{4}\pi^2\Big[J_{iy-\nu}(a)J_{iy+\nu}(a)+Y_{iy-\nu}(a)Y_{iy+\nu}(a)$ $+\tan(\pi\nu)\{J_{iy+\nu}(a)Y_{iy-\nu}(a)$ $-J_{iy-\nu}(a)Y_{iy+\nu}(a)\}\Big]$
19.22	$\mathrm{sech}\,x\,K_\nu(2a\cosh x)$	$-\frac{1}{2}a\nu^{-1}\Big(K_{\frac{1}{2}-\frac{1}{2}\nu+\frac{1}{2}iy}(a)K_{\frac{1}{2}-\frac{1}{2}\nu-\frac{1}{2}iy}(a)$ $-K_{\frac{1}{2}+\frac{1}{2}\nu+\frac{1}{2}iy}(a)K_{\frac{1}{2}+\frac{1}{2}\nu-\frac{1}{2}iy}(a)\Big]$
19.23	$(\mathrm{sech}\,x)^m K_m(2a\cosh x)$ $m=1,2,\ldots$	$(-1)^m\frac{1}{2}a^m m!\sum_{n=o}^{m}(-1)^n\varepsilon_n\{(m+n)!(m-n)!\}^{-1}$ $\cdot K_{n+\frac{1}{2}iy}(a)K_{n-\frac{1}{2}iy}(a)$
19.24	$\Big[\{(a+be^x)/(b+ae^x)\}^{\frac{1}{2}\nu}$ $+\{(b+ae^x)/(a+be^x)\}^{\frac{1}{2}\nu}\Big]$ $\cdot K_\nu\{(a^2+b^2+2ab\cosh x)^{\frac{1}{2}}\}$	$K_{\frac{1}{2}\nu+iy}(a)K_{\frac{1}{2}\nu-iy}(b)+K_{\frac{1}{2}\nu+iy}(b)K_{\frac{1}{2}\nu-iy}(a)$

1.20   Modified Bessel Functions of Variable Order

	$f(x)$	$g_c(y)$
20.1	$I_x(a)$	MacRobert,T.M.Proc.Roy.Soc.Edinburgh Vol.55,P.87,1934
20.2	$I_{\nu+x}(a)I_{\nu-x}(a)$ $\text{Re}\nu>-\tfrac{1}{2}$	$\tfrac{1}{2}I_{2\nu}\{2a\cos(\tfrac{1}{2}y)\}$ $y<\pi$ 0 $y>\pi$
20.3	$\cosh(bx)\operatorname{sech}(\pi x)$ $\cdot\{I_{ix}(a)+I_{-ix}(a)\}$ $b<\tfrac{1}{2}\pi$	$-\tfrac{1}{2}i\left(e^{-a\cosh u}\text{Erf}\{i(2a)^{\frac{1}{2}}\cosh(\tfrac{1}{2}u)\}\right.$ $\left.+e^{-a\cosh v}\text{Erf}\{i(2a)^{\frac{1}{2}}\cosh(\tfrac{1}{2}v)\}\right)\;\substack{u\\v}=y\pm ib$
20.4	$\sinh(bx)\operatorname{sech}(\pi x)$ $\cdot\{I_{ix}(a)-I_{-ix}(a)\}$ $=2i\pi^{-1}\tanh(\pi x)$ $\cdot\sinh(bx)K_{ix}(a)$ $b<\tfrac{1}{2}$	$\tfrac{1}{2}\left(e^{-a\cosh v}\text{Erf}\{i(2a)^{\frac{1}{2}}\sinh(\tfrac{1}{2}v)\}\right.$ $\left.-e^{-a\cosh u}\text{Erf}\{i(2a)^{\frac{1}{2}}\sinh(\tfrac{1}{2}u)\}\right),\;\substack{u\\v}=y\pm ib$
20.5	$K_{ix}(a)$	$\tfrac{1}{2}\pi\exp(-a\cosh y)$
20.6	$\cosh(bx)K_{ix}(a),b\leq\tfrac{1}{2}\pi$	$\tfrac{1}{2}\pi\cos(a\sin b\sinh y)\exp(-a\cos b\cosh y)$
20.7	$\operatorname{sech}(\pi x)K_{ix}(a)$	$\tfrac{1}{2}\pi\exp(a\cosh y)\text{Erfc}\{(2a)^{\frac{1}{2}}\cosh(\tfrac{1}{2}y)\}$
20.8	$\operatorname{sech}(\tfrac{1}{2}\pi x)K_{ix}(a)$	$a\int_0^\infty(a^2+t^2)^{-\frac{1}{2}}\exp(-t\cosh y)K_1\{(a^2+t^2)^{\frac{1}{2}}\}dt$
20.9	$\sinh(\pi x)\operatorname{cosech}(bx)K_{ix}(a)$ $b\geq\tfrac{1}{2}\pi$	$\tfrac{1}{2}\pi^2 b^{-1}\sum_0^\infty(-1)^n\varepsilon_n I_{n\pi/b}(a)\cosh(n\pi/b)$
20.10	$\operatorname{sech}(\tfrac{\pi}{2}x)\{I_{ix}(a)$ $\quad +I_{-ix}(a)\}$	$-2i\exp\{-2a\sinh^2(\tfrac{y}{2})\}\text{Erfc}\{i(2a)^{\frac{1}{2}}\cosh y\}$
20.11	$\operatorname{sech}(\pi x)\cosh(bx)K_{ix}(a)$	$\tfrac{1}{4}\pi\left(\exp(a\cosh u)\text{Erfc}\{(2a)^{\frac{1}{2}}\cosh(\tfrac{1}{2}u)\}\right.$ $\left.+\exp(a\cosh v)\text{Erfc}\{(2a)^{\frac{1}{2}}\cosh(\tfrac{1}{2}v)\}\right),\substack{u\\v}=y\pm ib$
20.12	$\{J_{ix}(a)+J_{-ix}(a)\}K_{ix}(a)$	$\tfrac{1}{2}\pi J_0\{a(2\sinh y)^{\frac{1}{2}}\}$
20.13	$\{Y_{ix}(a)+Y_{-ix}(a)\}K_{ix}(a)$	$\tfrac{1}{2}\pi Y_0(z)-K_0(z)$, $z=a(2\sinh y)^{\frac{1}{2}}$

	$f(x)$	$g_c(y)$
20.14	$\{I_{ix}(a)+I_{-ix}(a)\}K_{ix}(a)$	$\frac{1}{2}\pi J_0\{2a\sinh(\frac{1}{2}y)\}$
20.15	$\mathrm{sech}(\pi x)\{I_{ix}(a)+I_{-ix}(a)\}$ $\cdot K_{ix}(a)$	$\frac{1}{2}\pi\{I_0(z)-\mathbf{L}_0(z)\};\ z=2a\cosh(\frac{1}{2}y)$
20.16	$\{I_{ix}(a)+I_{-ix}(a)\}K_{ix}(b)$	$0\quad,\quad 2ab\cosh y<a^2+b^2$ $\frac{1}{2}\pi J_0\{(2ab\cosh y-a^2-b^2)^{\frac{1}{2}}\},$ $,\quad 2ab\cosh y>a^2+b^2$
20.17	$\mathrm{sech}(\frac{\pi}{2}x)\{J_{ix}(b)+J_{ix}(b)\}$ $\cdot K_{ix}(a),\quad a\underline{\le}b$	$iK_0(u)-iK_0(v),$ $\begin{matrix}u\\v\end{matrix}=(a^2-b^2\pm i2ab\cosh y)^{\frac{1}{2}}$
20.18	$\mathrm{cosec}(\frac{1}{2}\Pi x)$ $\cdot\{J_{ix}(b)-J_{-ix}(b)\}K_{ix}(a)$	$-iK_0(u)-iK_0(v),\begin{matrix}u\\v\end{matrix}=(a^2-b^2\pm 2ab\cosh y$
20.19	$\{K_{ix}(a)\}^2$	$\frac{1}{2}\pi K_0\{2a\cosh(\frac{1}{2}y)\}$
20.20	$\cosh(\pi x)\{K_{ix}(a)\}^2$	$-\frac{1}{4}\pi^2 Y_0\{2a\sinh(\frac{1}{2}y)\}$
20.21	$\cosh(\pi x)K_{ix}(a)K_{ix}(b)$	$\frac{1}{2}\pi K_0\{(a^2+b^2-2ab\cosh y)^{\frac{1}{2}}\},$ $2ab\cosh y<a^2+b^2$ $-\frac{1}{4}\pi^2 Y_0\{(2ab\cosh y-a^2-b^2)^{\frac{1}{2}}\}$ $2ab\cosh y>a^2+b^2$
20.22	$K_{ix}(a)K_{ix}(b)$	$\frac{1}{2}\pi K_0\{(a^2+b^2+2ab\cosh y)^{\frac{1}{2}}\}$
20.23	$\mathrm{sech}(\pi x)K_{ix}(a)K_{ix}(b)$	$\frac{1}{2}\pi\int_0^\infty\exp\{-(a+b)\cosh t-2(ab)^{\frac{1}{2}}\cosh(\frac{1}{2}y)$ $\cdot\sinh t\}dt$
20.24	$K_{\nu+ix}(a)K_{\nu-ix}(a)$	$\frac{1}{2}\pi K_{2\nu}\{2a\cosh(\frac{1}{2}y)\}$
20.25	$K_{\nu+ix}(a)K_{\nu-ix}(b)$ $+K_{\nu+ix}(b)K_{\nu-ix}(a)$	$\frac{1}{2}\pi(z^\nu+z^{-\nu})K_0\{(a^2+b^2+2ab\cosh y)^{\frac{1}{2}}\}$ $z=(a+be^y)/(b+ae^y)$

	$f(x)$	$g_c(y)$
20.26	$\text{sech}(\pi x)$ $\cdot \left[I_{\nu+ix}(a)K_{\nu-ix}(a) \right.$ $\left. +I_{\nu-ix}(a)K_{\nu+ix}(a) \right]$	$\frac{1}{2}i\pi\cosec(2\nu)\left[J_{2\nu}(iz)-J_{2\nu}(-iz) \right.$ $\left. -2i\,\sin(\pi\nu)I_{2\nu}(z) \right]$, $z=2a\cosh(\frac{1}{2}y)$
20.27	$\text{cosech}(\pi x)$ $\cdot \left[I_{\nu+ix}(a)K_{\nu-ix}(a) \right.$ $\left. -I_{\nu-ix}(a)K_{\nu+ix}(a) \right]$	$-\frac{1}{2}i\pi\cosec(2\nu)\left[J_{2\nu}(iz)+J_{2\nu}(-iz) \right.$ $\left. -2\cos(\pi\nu)I_{2\nu}(z) \right]$, $z=2a\cosh(\frac{1}{2}y)$

1.21 Functions Related to Bessel Functions

	$f(x)$	$g_c(y)$
21.1	$\mathbf{J}_\nu(ax)+\mathbf{J}_{-\nu}(ax)$ $=\cot(\frac{\pi}{2}\nu)\{\mathbf{E}_\nu(ax)-\mathbf{E}_{-\nu}(ax)\}$	$2\cos(\frac{1}{2}\pi\nu)(a^2-y^2)^{-\frac{1}{2}}\cos\{\nu\arccos(y/a)\}$, $y<a$ $\qquad\qquad 0 \qquad\qquad$, $y>a$
21.2	$\mathbf{J}_\nu(ax^2)-\mathbf{J}_\nu(ax^2)$ $\operatorname{Re}\nu>-\frac{1}{2}$	$\sin(\pi\nu)(2\pi a)^{-\frac{1}{2}}\Gamma(\frac{1}{2}+\nu)$ $\cdot D_{-\nu-\frac{1}{2}}(u)D_{-\nu-\frac{1}{2}}(v)$; $\begin{smallmatrix}u\\v\end{smallmatrix}=y(\pm 2ia)^{-\frac{1}{2}}$
21.3	$Y_\nu(ax^2)+\mathbf{E}_\nu(ax^2)$ $-\frac{1}{2}<\operatorname{Re}<\frac{1}{2}$	$(2\pi a)^{-\frac{1}{2}}\left(\Gamma(\frac{1}{2}-\nu)D_{\nu-\frac{1}{2}}(v)D_{\nu-\frac{1}{2}}(u) \right.$ $\left. +\cos(\pi\nu)\Gamma(\frac{1}{2}+\nu)D_{-\nu-\frac{1}{2}}(u)D_{-\nu-\frac{1}{2}}(v) \right)$ $\begin{smallmatrix}u\\v\end{smallmatrix}=y(\pm 2ia)^{-\frac{1}{2}}$
21.4	$r^{-1}\{\mathbf{J}_\nu(br)-\mathbf{J}_{-\nu}(br)\}$ $r=(a^2+x^2)^{\frac{1}{2}}$	$\qquad\qquad 0 \qquad\qquad y>b$
21.5	$\mathbf{J}_\nu(2a\cosh x)$	$\frac{1}{2}\pi i\,\text{cosech}(\pi y)\left[\cos(\frac{1}{2}\pi\nu+\frac{1}{2}i\pi y) \right.$ $\cdot J_{\frac{1}{2}\nu+\frac{1}{2}iy}(a)J_{-\frac{1}{2}\nu+\frac{1}{2}iy}(a)-\cos(\frac{1}{2}\pi\nu-\frac{1}{2}i\pi y)$ $\left. \cdot J_{\frac{1}{2}\nu-\frac{1}{2}iy}(a)J_{-\frac{1}{2}\nu-\frac{1}{2}iy}(a) \right]$
21.6	$\mathbf{E}_\nu(2a\cosh x)$	$\frac{1}{2}i\pi\text{cosech}(\pi y)\left[\sin(\frac{1}{2}\pi\nu+\frac{1}{2}i\pi y) \right.$ $\cdot J_{-\frac{1}{2}\nu+\frac{1}{2}iy}(a)J_{\frac{1}{2}\nu+\frac{1}{2}iy}(a)-\sin(\frac{1}{2}\pi\nu-\frac{1}{2}i\pi y)$ $\left. \cdot J_{\frac{1}{2}\nu-\frac{1}{2}iy}(a)J_{-\frac{1}{2}\nu-\frac{1}{2}iy}(a) \right]$
21.7	$\mathbf{J}_{2\nu}(z)+\mathbf{J}_{-2\nu}(z)$ $=\cot(\pi\nu)\{\mathbf{E}_{2\nu}(Z)-\mathbf{E}_{-2\nu}(z)\}$ $z=2a\sinh x$	$-\frac{1}{2}i\pi\cos(\pi\nu)\text{cosech}(\frac{1}{2}\pi y)$ $\left[I_{-\nu-\frac{1}{2}iy}(a)I_{\nu-\frac{1}{2}iy}(a) \right.$ $\left. -I_{-\nu+\frac{1}{2}iy}(a)I_{\nu+\frac{1}{2}iy}(a) \right]$

	$f(x)$	$g_c(y)$	
21.8	$\cot(\tfrac{1}{2}\pi x)\{\mathbf{J}_x(a)-\mathbf{J}_{-x}(a)\}$ $=-\mathbf{E}_x(a)-\mathbf{E}_{-x}(a)$	$\sin(a\sin y)$ 0	$y<\pi$ $y>\pi$
21.9	$\operatorname{cosec}(\tfrac{\pi}{2}x)\{\mathbf{J}_x(a)-\mathbf{J}_{-x}(a)\}$ $=-\sec(\tfrac{\pi}{2}x)\{E_x(a)+E_{-x}(a)\}$	$2\sin(a\cos y)$ 0	$y<\tfrac{1}{2}\pi$ $y>\tfrac{1}{2}\pi$
21.10	$\mathbf{J}_x(a)+\mathbf{J}_{-x}(a)$ $=\cot(\tfrac{\pi}{2}x)\{\mathbf{E}_x(a)-\mathbf{E}_{-x}(a)\}$	$\cos(a\sin y)$ 0	$y<\pi$ $y>\pi$
21.11	$\sec(\tfrac{\pi}{2}x)\{\mathbf{J}_x(a)+\mathbf{J}_{-x}(a)\}$ $=\operatorname{cosec}(\tfrac{\pi}{2}x)\{\mathbf{E}_x(a)-\mathbf{E}_{-x}(a)\}$	$2\cos(a\cos y)$ 0	$y<\tfrac{1}{2}\pi$ $y>\tfrac{1}{2}\pi$
21.12	$\operatorname{cosech}(\pi x)\left(\mathbf{J}_{ix}(a)-J_{ix}(a)\right.$ $\left.+J_{-ix}(a)-\mathbf{J}_{-ix}(a)\right)$	$2i\pi^{-1}\exp(-a\sinh y)$	
21.13	$\mathbf{J}_{2\nu}(z)-J_{2\nu}(z)$ $z=2a\cosh(\tfrac{1}{2}y)$	$-\tfrac{1}{2}i\pi\sin(2\pi\nu)\operatorname{cosech}(2\pi y)$ $\cdot\left(J_{\nu+iy}(a)Y_{\nu-iy}(a)-J_{\nu-iy}(a)Y_{\nu+iy}(a)\right)$	
21.14	$\mathbf{H}_0(ax)$	$2\pi^{-1}(a^2-y^2)^{-\tfrac{1}{2}}\log\left[\tfrac{a}{y}+\{(\tfrac{a}{y})^2-1\}^{\tfrac{1}{2}}\right]$ $-2\pi^{-1}(y^2-a^2)^{-\tfrac{1}{2}}\operatorname{arccot}\left[\{(\tfrac{y}{a})^2-1\}^{\tfrac{1}{2}}\right]$	$y<a$ $y>a$
21.15	$x^{-1}\mathbf{H}_0(ax)$	$\arccos(y/a)$ 0	$y<a$ $y>a$
21.16	$x^{-1}\{\sin x-\mathbf{H}_0(x)\}$	$\arcsin y$ 0	$y<1$ $y>1$
21.17	$\mathbf{H}_{-1}(ax)$	$a^{-1}y(a^2-y^2)^{-\tfrac{1}{2}}$ 0	$y<a$ $y>a$
21.18	$x^{-\nu-1}\mathbf{H}_\nu(x)$ $\operatorname{Re}\nu>-\tfrac{1}{2}$	$(2\pi)^{\tfrac{1}{2}}(1-y^2)^{\tfrac{1}{2}\nu+\tfrac{1}{4}}P_{\nu-\tfrac{1}{2}}^{-\nu-\tfrac{1}{2}}(y)$ 0	$y<1$ $y>1$
21.19	$x^{-\nu}\{\mathbf{H}_\nu(ax)-Y_\nu(ax)\}$ $\operatorname{Re}\nu<\tfrac{1}{2}$	$2^{-\nu}\pi^{-\tfrac{1}{2}}a^{\nu-1}\{(\tfrac{1}{2}-\nu)\Gamma(\tfrac{1}{2}+\nu)\}^{-1}$ $\cdot{}_2F_1(\tfrac{1}{2}-\nu,1;3/2-\nu;1-y^2/a^2)$	

	$f(x)$	$g_c(y)$
21.20	$x^\nu\{\mathbf{H}_\nu(ax)-Y_\nu(ax)\}$ $-\tfrac{1}{2}<\mathrm{Re}\,\nu<\tfrac{1}{2}$	$(\tfrac{1}{2}\pi)^{-\frac{1}{2}}\cos(\pi\nu)\,\Gamma(1+2\nu)\,a^\nu y^{-\nu-\frac{1}{2}}$ $(a^2-y^2)^{-\frac{1}{2}\nu-\frac{1}{4}}P_{-\nu-\frac{1}{2}}^{-\nu-\frac{1}{2}}(a/y)$
21.21	$I_0(ax)-\mathbf{L}_0(ax)$	$2\pi^{-1}(a^2+y^2)^{-\frac{1}{2}}\log\left\{a/y+(1+a^2/y^2)^{\frac{1}{2}}\right\}$
21.22	$x^{-1}[1-I_0(ax)+\mathbf{L}_0(ax)]$	$\log[\tfrac{1}{2}+\tfrac{1}{2}(1+a^2/y^2)^{\frac{1}{2}}]$
21.23	$x^{-\nu}\{I_\nu(ax)-\mathbf{L}_\nu(ax)\}$	$\pi^{-\frac{1}{2}}a(\tfrac{1}{2}a)^\nu\{\Gamma(3/2+\nu)\}^{-1}y^{-1}$ $\cdot\,_2F_1(1,1;3/2+\nu;-a^2/y^2)$
21.24	$x^{\nu+1}\{I_\nu(ax)-\mathbf{L}_\nu(ax)\}$ $-1<\mathrm{Re}\,\nu<1$	$4(2a)^{\nu-1}\{\Gamma(-\nu)\}^{-1}y^{-2\nu-1}$ $_2F_1(1,\tfrac{1}{2};-\nu;-y^2/a^2)$
21.25	$\mathbf{H}_0(ax^2)$	$\tfrac{1}{8}\pi a^{-1}y\{J_{\frac{1}{4}}^2(z)-Y_{-\frac{1}{4}}^2(z)\}$; $z=\tfrac{1}{8}y^2/a$
21.26	$\mathbf{H}_0(ax^2)-Y_0(ax^2)$	$\tfrac{1}{8}\pi a^{-1}y\{J_{\frac{1}{4}}^2(z)+Y_{\frac{1}{4}}^2(z)\}$; $z=\tfrac{1}{8}y^2/a$
21.27	$x^{\frac{1}{2}}\mathbf{H}_{-\frac{1}{4}}(ax^2)$	$-\tfrac{1}{2}a^{-1}(\tfrac{1}{2}\pi y)^{\frac{1}{2}}Y_{-\frac{1}{4}}(\tfrac{1}{4}y^2/a)$
21.28	$x^{3/2}\mathbf{H}_{-3/4}(ax^2)$	$-\tfrac{1}{4}a^{-2}(\tfrac{1}{2}\pi)^{\frac{1}{2}}y^{3/2}Y_{\frac{1}{4}}(\tfrac{1}{4}y^2/a)$
21.29	$I_0(ax^2)-\mathbf{L}_0(ax^2)$	$2^{-\frac{1}{2}}(2\pi a)^{-1}y\{K_{\frac{1}{4}}(\tfrac{1}{8}y^2/a)\}^2$
21.30	$x^{-1}\{\mathbf{H}_0(a/x)-Y_0(a/x)\}$	$4\pi^{-1}K_0\{(2iay)^{\frac{1}{2}}\}K_0\{(-2iay)^{\frac{1}{2}}\}$
21.31	$x^{-1}\{I_0(a/x)-\mathbf{L}_0(a/x)\}$	$2J_0\{(2ay)^{\frac{1}{2}}\}K_0\{(2ay)^{\frac{1}{2}}\}$
21.32	$x^{\frac{1}{2}\nu}\mathbf{H}_\nu(ax^{\frac{1}{2}})$ $-3/2<\mathrm{Re}\,\nu<\tfrac{1}{2}$	$-2^{\frac{1}{2}}y^{-1}(2y/a)^{-\nu}\left[\cos(z-\tfrac{1}{2}\pi\nu+\tfrac{1}{4}\pi)C(z)\right.$ $\left.+\sin(z-\tfrac{1}{2}\pi\nu+\tfrac{1}{4}\pi)S(z)\right]$, $z=\tfrac{1}{4}a^2/y$
21.33	$x^{\frac{1}{2}\nu}\{I_\nu(ax^{\frac{1}{2}})-\mathbf{L}_\nu(ax^{\frac{1}{2}})\}$ $-1<\mathrm{Re}\,\nu<\tfrac{1}{2}$	$(\tfrac{1}{2}a)^\nu y^{-\nu-1}\left[\cos(z+\tfrac{1}{2}\pi\nu)\{C(z)-S(z)\}\right.$ $\left.-\sin(z+\tfrac{1}{2}\pi\nu)\{1-C(z)-S(z)\}\right]$, $z=\tfrac{1}{4}a^2/y$
21.34	$r^{-\nu-1}\mathbf{H}_\nu(br)$, $r=(a^2+x^2)^{\frac{1}{2}}$	0 $y>b$

	$f(x)$	$g_c(y)$
21.35	$r^{-1}\{\mathbf{H}_0(br)-Y_0(br)\}$	$\pi^{-1}K_0(u)K_0(v)$, $\dfrac{u}{v}=\tfrac12 a\{y\pm(y^2-b^2)^{\frac12}\}$
21.36	$r^{-1}\{I_0(br)-\mathbf{L}_0(br)\}$ $r=(a^2+x^2)^{\frac12}$	$K_0(u)I_0(v)$; , $\dfrac{u}{v}=\tfrac12 a\{(b^2+y^2)^{\frac12}\pm y\}$
21.37	$\mathbf{H}_0(2a\cosh\tfrac12 x)$	$\tfrac12\pi\,\mathrm{sech}(\pi y)\left(\{J_{iy}(a)\}^2+\{J_{-iy}(a)\}^2\right)$
21.38	$\mathbf{H}_0(2a\cosh\tfrac12 x)$ $-Y_0(2a\cosh\tfrac12 x)$	$\tfrac12\pi\,\mathrm{sech}(\pi y)\left(\{J_{iy}(a)\}^2+\{Y_{iy}(a)\}^2\right)$
21.39	$I_0(2a\cosh\tfrac12 x)$ $-\mathbf{L}_0(2a\cosh\tfrac12 x)$	$\mathrm{sech}(\pi y)K_{iy}(a)\{I_{iy}(a)+I_{-iy}(a)\}$
21.40	$\tfrac12\pi r^{-1}\,\mathrm{sech}(\tfrac12\pi\nu)\Big[I_\nu(br)$ $-\tfrac12 i\,\sec(\tfrac12\pi\nu)\{\mathbf{E}_\nu(ibr)$ $+\mathbf{E}_{-\nu}(ibr)\}\Big];r=(a^2+x^2)^{\frac12}$	$\tfrac12\pi K_{\frac12\nu}(u)I_{\frac12\nu}(v)$, $\dfrac{u}{v}=\tfrac12 a\{(b^2+y^2)^{\frac12}\pm y\}$
21.41	$x^{-\nu-1}\mathbf{s}_{\nu,\nu+2}(x)$ $\mathrm{Re}\,\nu>-3/2$	$-\tfrac12\pi y(1-y^2)^{\nu+\frac12}$ $y<1$ 0 $y>1$
21.42	$x^{-\mu-1}\mathbf{s}_{\mu,\nu}(x)$ $\mathrm{Re}\,\mu>-3/2$	$(\tfrac12\pi)^{\frac12}2^\mu\Gamma(\tfrac12+\tfrac12\mu+\tfrac12\nu)\,\Gamma(\tfrac12+\tfrac12\mu-\tfrac12\nu)$ $\cdot(1-y^2)^{\frac12\mu+\frac14}P_{\nu-\frac12}^{-\mu-\frac12}(y)$ $y>1$ 0 $y<1$
21.43	$x^{-\mu-1}\mathbf{s}_{\mu,\nu}(ax)$ $\mathrm{Re}(\mu\pm\nu)<0$	$\tfrac14(\tfrac12\pi/a)^{\frac12}2^{-\mu}\Gamma(-\tfrac12\nu-\tfrac12\mu)\,\Gamma(\tfrac12\nu-\tfrac12\mu)$ $\cdot(y^2-a^2)^{\frac12\mu+\frac14}P_{\nu-\frac12}^{\frac12+\mu}(y/a)$
21.44	$x^\nu \mathbf{s}_{\mu,\nu}(ax)$ $\mathrm{Re}\,\nu>-\tfrac12$; $-2<\mathrm{Re}(\mu+\nu)<1$	$\pi^{\frac12}(2\nu+1)^{-1}2^{\nu+\mu}a^\nu\Gamma(1+\tfrac12\mu+\tfrac12\nu)\{\Gamma(\tfrac12-\tfrac12\mu-\tfrac12\nu)\}^{-1}$ $\cdot{}_2F_1(\tfrac12+\nu,\tfrac12-\tfrac12\mu+\tfrac12\nu;\tfrac32+\nu;1-a^2/y^2)$
21.45	$r^{-1}\mathbf{s}_{0,\nu}(br),r=(a^2+x^2)^{\frac12}$	$\tfrac12 K_{\frac12\nu}(u)K_{\frac12\nu}(v)$, $\dfrac{u}{v}=\tfrac12 a\{y\pm(y^2-b^2)^{\frac12}\}$
21.46	$x\mathbf{s}_{\mu,\frac12}(ax^2)$ $-2<\mathrm{Re}\,\mu<\tfrac12$	$\tfrac14\pi^{\frac12}a^{-3/2}\Gamma(2+\mu)\{\Gamma(\tfrac12-\mu)\}^{-1}$ $\cdot y\mathbf{s}_{-\mu-3/2,\frac12}(\tfrac14 y^2/a)$

	$f(x)$	$g_c(y)$
21.47	$S_{\mu,\nu}(a\cosh x)$ $\operatorname{Re}\mu<1$	$\frac{1}{2}(2a)^{\mu+1}\{\Gamma(\frac{1}{2}-\frac{1}{2}\mu-\frac{1}{2}\nu)\Gamma(\frac{1}{2}-\frac{1}{2}\mu+\frac{1}{2}\nu)\}^{-1}$ $\cdot\int_0^\infty t^{-\mu}(a^2+b^2)^{-1}K_{\frac{1}{2}\nu+\frac{1}{2}iy}(\frac{1}{2}t)K_{\frac{1}{2}\nu-\frac{1}{2}iy}(\frac{1}{2}t)\,dt$
21.48	$(\cosh x)^{\frac{1}{2}}S_{\mu,\frac{1}{2}}(a\cosh x)$ $\operatorname{Re}\mu<\frac{1}{2}$	$2^{-\mu-1}(2a)^{-\frac{1}{2}}\{\Gamma(\frac{1}{4}-\mu)\}^{-1}\Gamma(\frac{1}{4}-\frac{1}{2}\mu-\frac{1}{2}iy)$ $\cdot\Gamma(\frac{1}{4}-\frac{1}{2}\mu+\frac{1}{2}iy)S_{\mu+\frac{1}{2},iy}(a)$
21.49	$S_{o,ix}(a)$	$\frac{1}{2}\pi\exp(-a\sinh y)$
21.50	$x\operatorname{cosech}(\frac{1}{2}\pi x)S_{-1,ix}(a)$	$-\cos z\;\mathrm{Ci}(z)-\sin z\;\mathrm{si}(z),\quad z=a\cosh y$
21.51	$\Gamma(\frac{1}{2}-\frac{1}{2}\mu-\frac{1}{2}ix)\Gamma(\frac{1}{2}-\frac{1}{2}\mu+\frac{1}{2}ix)$ $\cdot S_{\mu,ix}(a)$ $\operatorname{Re}\mu<1$	$\pi 2^{\mu}a^{\frac{1}{2}}\Gamma(1-\mu)(\cosh y)^{\frac{1}{2}}S_{\mu-\frac{1}{2},\frac{1}{2}}(a\cosh y)$
21.52	$\operatorname{sech}(\frac{1}{2}\pi x)S_{o,ix}(a)$	$\sin z\;\mathrm{Ci}(z)-\cos z\;\mathrm{si}(z),\quad z=a\cosh y$

1.22 Parabolic Cylinder- and Whittaker Functions

	$f(x)$	$g_c(y)$
22.1	$e^{\frac{1}{4}x^2}D_{-2}(x)$	$(\frac{1}{2}\pi)^{\frac{1}{2}}e^{\frac{1}{4}y^2}D_{-2}(y)$
22.2	$e^{-\frac{1}{4}x^2}D_{2n}(x)$ $n=0,1,\ldots$	$(-1)^n(\frac{1}{2}\pi)^{\frac{1}{2}}y^{2n}e^{-\frac{1}{4}y^2}$
22.3	$\exp(\frac{1}{4}a^2x^2)D_{\nu}(ax)$ $\operatorname{Re}\nu<0$	$\pi^{\frac{1}{2}}2^{\frac{1}{4}+3/4\nu}\{\Gamma(-\frac{1}{2}\nu)\}^{-1}a^{\frac{1}{2}+\frac{1}{2}\nu}y^{-\frac{1}{2}\nu-3/2}$ $\cdot\exp(\frac{1}{4}y^2/a^2)W_{\frac{1}{4}\nu-\frac{1}{4},\frac{1}{4}\nu+\frac{1}{4}}(\frac{1}{2}y^2/a^2)$
22.4	$\exp(-\frac{1}{4}a^2x^2)D_{\nu}(ax)$	$\pi^{\frac{1}{2}}2^{\frac{1}{2}\nu-\frac{1}{2}}\{\Gamma(1-\frac{1}{2}\nu)\}^{-1}a^{-1}{}_1F_1(1;1-\frac{1}{2}\nu;-\frac{1}{2}y^2/a^2)$
22.5	$x^{\mu}\exp(-\frac{1}{4}x^2)D_{\nu}(x)$ $\operatorname{Re}\mu>-1$	$2^{\frac{1}{2}(\nu-\mu-1)}\pi^{\frac{1}{2}}\Gamma(1+\mu)\{\Gamma(1+\frac{1}{2}\mu-\frac{1}{2}\nu)\}^{-1}$ ${}_2F_2(\frac{1}{2}+\frac{1}{2}\mu,1+\frac{1}{2}\mu;\frac{1}{2},1+\frac{1}{2}\mu-\frac{1}{2}\nu;-\frac{1}{2}y^2)$

	$f(x)$	$g_c(y)$
22.6	$D_\nu(axi^{\frac{1}{2}})D_\nu\{ax(-1)^{\frac{1}{2}}\}$ $\mathrm{Re}\,\nu<0$	$\frac{1}{2}\pi^{3/2}\{a\Gamma(-\nu)\}^{-1}\sec(\pi\nu)$ $\cdot\{J_{-\nu-\frac{1}{2}}(\frac{1}{2}y^2/a^2)-J_{-\nu-\frac{1}{2}}(\frac{1}{2}y^2/a^2)\}$
22.7	$D_{2\nu-\frac{1}{2}}(z)\{D_{-2\nu-\frac{1}{2}}(z)$ $+D_{-2\nu-\frac{1}{2}}(-z)\},z=(2x)^{\frac{1}{2}}$	$\pi^{-1}\sin(\frac{1}{4}\pi-\pi\nu)y^{-2\nu-\frac{1}{2}}(1+y^2)^{-\frac{1}{2}}\{1+(1+y^2)^{\frac{1}{2}}\}^{2\nu}$
22.8	$\exp(-\frac{1}{2}x^2)\{D_{2\nu-\frac{1}{2}}(x)$ $+D_{2\nu-\frac{1}{2}}(-x)\}$	$2^{1-2\nu}\pi^{\frac{1}{2}}\sin(\frac{1}{4}\pi+\pi\nu)y^{2\nu-\frac{1}{2}}e^{-\frac{1}{4}y^2}$
22.9	$x^{-\nu}\exp(-\frac{1}{4}a^2/x)$ $\cdot D_{2\nu-1}(ax^{-\frac{1}{2}})$	$2(\frac{1}{2}\pi)^{\frac{1}{2}}(2y)^{\nu-1}\sin(\frac{1}{2}\pi\nu-ay^{\frac{1}{2}})\exp(-ay^{\frac{1}{2}})$
22.10	$x^{-\nu}\exp(\frac{1}{4}a^2x)D_{2\nu-1}(ax^{\frac{1}{2}})$ $\mathrm{Re}\,\nu<1$	$(\pi/y)^{\frac{1}{2}}\{y+(a+y^{\frac{1}{2}})^2\}^{\nu-\frac{1}{2}}$ $\cdot\cos\left[(2\nu-1)\arctan\{y^{\frac{1}{2}}/(a+y^{\frac{1}{2}})\}-\frac{1}{4}\pi\right]$
22.11	$x^{-\nu-1}\exp(\frac{1}{4}a^2x)D_{2\nu-1}(ax^{\frac{1}{2}})$ $\mathrm{Re}\,\nu<0$	$-(\frac{1}{2}\pi)^{\frac{1}{2}}\nu^{-1}\{y+(a+y^{\frac{1}{2}})^2\}^\nu$ $\cdot\cos\left[2\nu\arctan\{y^{\frac{1}{2}}/(a+y^{\frac{1}{2}})\}\right]$
22.12	$(\cos x)^{-\frac{1}{2}\nu-\frac{1}{2}}\exp(-a^2\sec x)$ $\cdot D_\nu\{2a(1+\sec x)^{\frac{1}{2}}\},\ x<\frac{1}{2}\pi$ $\qquad\qquad 0\qquad\quad x>\frac{1}{2}\pi$	$\pi^{\frac{1}{2}}2^{\frac{1}{2}\nu}e^{-a^2}D_{\frac{1}{2}\nu-\frac{1}{2}+y}(2a)D_{\frac{1}{2}\nu-\frac{1}{2}-y}(2a)$
22.13	$\exp\{(a\sinh x)^2\}$ $\cdot D_\nu(2a\cosh x);\ \mathrm{Re}\,\nu<0$	$\frac{1}{2}2^{-\frac{1}{2}\nu-\frac{1}{2}}a^{-1}\Gamma(\frac{1}{2}iy-\frac{1}{2}\nu)\Gamma(-\frac{1}{2}iy-\frac{1}{2}\nu)$ $\cdot\{\Gamma(-\nu)\}^{-1}W_{\frac{1}{2}\nu+\frac{1}{2},\frac{1}{2}iy}(2a^2)$
22.14	$\exp\{-(a\sinh x)^2\}$ $\cdot D_\nu(2a\cosh x)$	$2^{\frac{1}{2}\nu-3/2}\pi^{\frac{1}{2}}a^{-1}W_{\frac{1}{2}\nu,\frac{1}{2}iy}(2a^2)$
22.15	$\exp(-z^2)\{D_\nu(2z)+D_\nu(-2z)\}$ $z=a\sinh x,\ \mathrm{Re}\,\nu>-1$	$2^{\frac{1}{2}\nu}(2\pi a^2)^{-\frac{1}{2}}\cos(\frac{1}{2}\pi\nu)\exp(a^2)\cosh(\frac{1}{2}\pi y)$ $\cdot\Gamma(\frac{1}{2}+\frac{1}{2}\nu+\frac{1}{2}iy)\Gamma(\frac{1}{2}+\frac{1}{2}\nu-\frac{1}{2}iy)W_{-\frac{1}{2}\nu,\frac{1}{2}iy}(2a^2)$

	$f(x)$	$g_c(y)$
22.16	$D_{x-\frac{1}{2}}(a)D_{-x-\frac{1}{2}}(a)$	$\frac{1}{2}(\pi\sec y)^{\frac{1}{2}}\exp(-\frac{1}{2}a^2\sec y)$ $\qquad y<\frac{1}{2}\pi$ $0 \qquad\qquad\qquad\qquad\qquad\qquad y>\frac{1}{2}\pi$
22.17	$D_{-1+x}(a)D_{-1-x}(a)$	$\frac{1}{2}\pi\exp(\frac{1}{2}a^2)\mathrm{Erfc}\{2^{-\frac{1}{2}}a(1+\sec y)^{\frac{1}{2}}\}\ y<\frac{1}{2}\pi$ $0 \qquad\qquad\qquad\qquad\qquad\qquad y>\frac{1}{2}\pi$
22.18	$D_{\nu+x}(a)D_{\nu-x}(a)$	$(\frac{1}{2}\pi)^{\frac{1}{2}}2^{-\nu-1}(\cos y)^{-\nu-1}\exp[\frac{1}{4}a^2(1-\sec y)]$ $\cdot D_{2\nu+1}\{a(1+\sec y)^{\frac{1}{2}}\} \qquad\qquad\qquad y<\frac{1}{2}\pi$ $0 \qquad\qquad\qquad\qquad\qquad\qquad\qquad y>\frac{1}{2}\pi$
22.19	$\mathrm{sech}(\pi x)D_{-\frac{1}{2}+ix}(a)$ $\cdot D_{-\frac{1}{2}-ix}(a)$	$\frac{1}{2}(\pi\mathrm{sech} y)^{\frac{1}{2}}\exp(\frac{1}{2}a^2\mathrm{sech} y)$ $\cdot\mathrm{Erfc}\{a(\frac{1}{2}+\frac{1}{2}\mathrm{sech} y)^{\frac{1}{2}}\}$
22.20	$x^{2\nu}W_{\mu,\nu}(ax)M_{-\mu,\nu}(ax)$ $\mathrm{Re}\,\nu>-\frac{1}{2}\ ,\ \mathrm{Re}(\nu+\mu)<0$	$\pi^{\frac{1}{2}}2^{2\nu}(2a)^{2\mu}\Gamma(1+2\nu)\{\Gamma(1-\nu-\mu)\}^{-1}y^{-2\nu-2\mu}$ $\cdot y^{-1}{}_3F_2(\frac{1}{2}-\mu,1-\mu,\frac{1}{2}-\mu+\nu;1-2\mu,-\nu-\mu;-y^2/a^2)$
22.21	$x^{2\nu}\exp(-\frac{1}{4}x^2)M_{\mu,\nu}(\frac{1}{2}x^2)$ $-\frac{1}{2}<\mathrm{Re}\,\nu<-\frac{1}{2}+\mathrm{Re}\,\mu$	$\pi^{\frac{1}{2}}2^{-\frac{1}{2}\nu+3/2\mu}\Gamma(1+2\nu)\{\Gamma(\mu-\nu)\}^{-1}y^{\mu-\nu-1}$ $\exp(-\frac{1}{4}y^2)M_{\alpha,\beta}(\frac{1}{2}y^2);2\alpha=1+\mu+3\nu,2\beta=\mu-\nu-1$
22.22	$x^{-2\nu-1}\exp(-\frac{1}{4}x^2)W_{\mu,\nu}(\frac{1}{2}x^2)$ $\mathrm{Re}\,\nu<\frac{1}{4}$	$\pi^{\frac{1}{2}}2^{-\frac{1}{2}-\frac{1}{2}\mu-3/2\nu}\Gamma(\frac{1}{2}-2\nu)\{\Gamma(1-\mu-\nu)\}^{-1}$ $\cdot y^{\mu+\nu-1}\exp(-\frac{1}{4}y^2)M_{\alpha,\beta}(\frac{1}{2}y^2)$ $2\alpha=\mu-3\nu;\ 2\beta=-\mu-\nu$
22.23	$x^{-2\nu-1}\exp(-\frac{1}{2}x^2)M_{-\mu,\nu}(x^2)$ $\mathrm{Re}(\mu-\nu)<\frac{1}{2},\mathrm{Re}\,\nu>-\frac{1}{2}$	$\pi^{\frac{1}{2}}2^{\mu-\nu}\Gamma(1+2\nu)\{\Gamma(\frac{1}{2}-\mu+\nu)\}^{-1}y^{\nu-\mu-1}$ $\cdot\exp(-\frac{1}{8}y^2)W_{\alpha,\beta}(\frac{1}{4}y^2);2\alpha=-\mu-\frac{3}{2}\nu,2\beta=\nu-\mu$
22.24	$x^{-2\nu-1}\exp(\frac{1}{4}x^2)W_{\mu,\nu}(\frac{1}{2}x^2)$ $\mathrm{Re}\,\nu<\frac{1}{4},\ \mathrm{Re}(\mu-\nu)<0$	$\pi^{\frac{1}{2}}2^{\frac{1}{2}(\mu-3\nu-1)}\Gamma(\frac{1}{2}-2\nu)\{\Gamma(\frac{1}{2}+\nu-\mu)\}^{-1}y^{\nu-\mu-1}$ $\cdot\exp(\frac{1}{4}y^2)W_{\alpha,\beta}(\frac{1}{2}y^2);\ 2\alpha=\mu+3\nu,\ 2\beta=\mu-\nu$
22.25	$x^{-\frac{1}{2}}K_{\nu+\frac{1}{2}}(\frac{1}{2}x^2)M_{\mu,\nu}(x^2)$ $\mathrm{Re}\,\mu>-\frac{3}{4}\ ,\ \mathrm{Re}\,\nu>-\frac{1}{2}$	$\frac{1}{2}(\frac{1}{2}\pi)^{\frac{1}{2}}\Gamma(1+2\nu)\{\Gamma(\frac{1}{2}+\mu)\}^{-1}y^{-\frac{1}{2}}$ $W_{\alpha,\beta}(\frac{1}{2}y^2)M_{\alpha,\gamma}(\frac{1}{2}y^2);\ 2\alpha=\mu-\nu,$ $2\beta=\frac{1}{4}+\mu,\ 2\gamma=\mu-\frac{1}{4}$

	$f(x)$	$g_c(y)$
22.26	$x^{-3/2} W_{\nu+\mu,k-\frac{1}{2}}(\frac{1}{2}x^2)$ $\cdot M_{\nu-\mu,-\frac{1}{2}-k}(\frac{1}{2}x^2)$ $\operatorname{Re}\mu<\frac{1}{2},\ \operatorname{Re} k<1/8$	$y^{-1}(\frac{1}{2}\pi y)^{\frac{1}{2}}\Gamma(\frac{3}{4}-2\mu)\{\Gamma(\frac{3}{4}-2k)\}^{-1}$ $W_{\alpha,\beta}(\frac{1}{2}y^2)M_{\gamma,\delta}(\frac{1}{2}y^2)$ $\alpha=\nu+k,\ \beta=\mu-\frac{1}{8},\ \gamma=\nu-k,\ \delta=-\nu-\frac{3}{8}$
22.27	$W_{-\frac{1}{2},ix}(a)$	$\frac{1}{2}\pi a^{\frac{1}{2}}\exp(\frac{1}{2}a\cosh y)\operatorname{Erfc}\{a^{\frac{1}{2}}\cosh(\frac{1}{2}y)\}$
22.28	$W_{k,ix}(a)$	$\frac{1}{2}(\pi a)^{\frac{1}{2}}2^{-k}\exp\{-\frac{1}{2}a\sinh^2(\frac{1}{2}y)\}$ $\cdot D_{2k}\{(2a)^{\frac{1}{2}}\cosh(\frac{1}{2}y)\}$
22.29	$\Gamma(\frac{1}{2}-k+ix)\Gamma(\frac{1}{2}-k-ix)$ $\cdot W_{k,ix}(a),\ \operatorname{Re}k<\frac{1}{2}$	$\pi 2^k(\frac{1}{2}a)^{\frac{1}{2}}\Gamma(1-2k)\exp\{\frac{1}{2}a\sinh^2(\frac{1}{2}y)\}$ $\cdot D_{2k-1}\{(2a)^{\frac{1}{2}}\cosh(\frac{1}{2}y)\}$
22.30	$\cosh(\pi x)\Gamma(\frac{1}{2}-k+ix)$ $\Gamma(\frac{1}{2}-k-ix)W_{k,ix}(a)$	$2^{k-2}\pi^{3/2}\sec(\pi\nu)\exp\{-\frac{1}{2}a\cosh^2(\frac{1}{2}y)\}a^{\frac{1}{2}}$ $\cdot\{D_{-2k}(z)+D_{-2k}(-z)\},\ z=(2a)^{\frac{1}{2}}\sinh(\frac{1}{2}y)$

1.23 Elliptic Integrals*

	$f(x)$	$g_c(y)$
23.1	$a^{-1}K(x/a)\qquad\qquad x<a$ $x^{-1}K(a/x)\qquad\qquad x>a$	$-\frac{1}{4}\pi^2 J_o(\frac{1}{2}ay)Y_o(\frac{1}{2}ay)$
23.2	$(a+x)^{-\frac{1}{2}}K\{x^{\frac{1}{2}}(a+x)^{-\frac{1}{2}}\}$	$-\frac{1}{4}\pi^{3/2}y^{-\frac{1}{2}}\{J_o(\frac{1}{2}ay)\cos(\frac{1}{4}\pi+\frac{1}{2}ay)$ $+Y_o(\frac{1}{2}ay)\sin(\frac{1}{4}\pi+\frac{1}{2}ay)\}$
23.3	$(a^2+x^2)^{-\frac{1}{2}}K\{b(a^2+x^2)^{-\frac{1}{2}}\}$ $a>b$	$\frac{1}{2}\pi K_o(u)K_o(v)\ ;\quad \frac{u}{v}=\frac{1}{2}y\{a\pm(a^2-b^2)^{\frac{1}{2}}\}$
23.4	$K\{(\frac{1}{2}-\frac{1}{2}x)^{\frac{1}{2}}\}\qquad x<1$ $0\qquad\qquad\qquad x>1$	$\frac{1}{16}\pi^{3/2}\{\Gamma(\frac{5}{4})\}^{-2}y^{-\frac{1}{2}}s_{-\frac{1}{2},o}(y)$
23.5	$K\{(1-x^2/a^2)^{\frac{1}{2}}\}\qquad x<a$ $0\qquad\qquad\qquad x>a$	$\frac{1}{4}\pi^2 a\{J_o(\frac{1}{2}ay)\}^2$

*Note that the usual notations for the elliptic integrals K(k), E(k)
(in bold letters) are here replaced by the ordinary letters K and E.

	$f(x)$	$g_c(y)$
23.6	$(1+x^2)^{-\frac{1}{2}}$ $\cdot K\{(x^2-1)^{\frac{1}{2}}(x^2+1)^{-\frac{1}{2}}\}$ $x<1$ $\quad\quad\quad 0 \quad\quad\quad\quad x>1$	$\frac{1}{2}K_0\{y(\frac{1}{2}i)^{\frac{1}{2}}\}K_0\{y(-\frac{1}{2}i)^{\frac{1}{2}}\}$
23.7	$\quad\quad 0 \quad\quad\quad\quad x<1$ $(1+x)^{-\frac{1}{2}}K(\frac{x-1}{x+1})^{\frac{1}{2}} \quad x>1$	$-\frac{1}{2}\pi(\frac{1}{2}\pi/y)^{\frac{1}{2}}\{J_0(y)+Y_0(y)\}$
23.8	$\quad\quad 0 \quad\quad\quad\quad x<a$ $x^{-1}K\{(1-a^2/x^2)^{\frac{1}{2}}\} \quad x>a$	$\frac{1}{8}\pi^2\left[\{Y_0(\frac{1}{2}ay)\}^2-\{J_0(\frac{1}{2}ay)\}^2\right]$
23.9	$a^{-1}K\{(1-x^2/a^2)^{\frac{1}{2}}\} \quad x<a$ $2x^{-1}K\{(1-a^2/x^2)^{\frac{1}{2}}\} \quad x>a$	$\frac{1}{4}\pi^2\{Y_0(\frac{1}{2}ay)\}^2$
23.10	$\quad\quad 0 \quad\quad\quad\quad x<a$ $(x^2-b^2)^{-\frac{1}{2}}$ $K\{(x^2-a^2)^{\frac{1}{2}}(x^2-b^2)^{-\frac{1}{2}}\}x>a$	$\frac{1}{8}\pi^2\{Y_0(u)Y_0(v)-J_0(u)J_0(v)\}$ $a>b$ $\quad \begin{matrix} u \\ v \end{matrix} = \frac{1}{2}y(a\pm b)$
23.11	$(a^2-x^2)^{-\frac{1}{2}}$ $K\{(a^2-b^2)^{\frac{1}{2}}(a^2-x^2)^{-\frac{1}{2}}\}x<a$ $\quad(a^2-b^2)^{-\frac{1}{2}}$ $K\{(a^2-x^2)^{\frac{1}{2}}(a^2-b^2)^{-\frac{1}{2}}\}$ $\quad\quad a>b \quad\quad b<x<a$	$\frac{1}{2}\pi^2 J_0(u)J_0(v) \quad\quad \begin{matrix} u \\ v \end{matrix} = \frac{1}{2}y(a\pm b)$
23.12	$K\{\cos(\frac{1}{2}x)\} \quad\quad x<\pi$ $\quad\quad 0 \quad\quad\quad\quad x>\pi$	$\frac{1}{4}\pi\cos(\pi y)\Gamma(\frac{1}{4}+\frac{1}{2}y)\Gamma(\frac{1}{4}-\frac{1}{2}y)$ $\cdot\{\Gamma(\frac{3}{4}+\frac{1}{2}y)\Gamma(\frac{3}{4}-\frac{1}{2}y)\}^{-1}$
23.13	$(1+a^2\cos^2 x)^{-\frac{1}{2}}$ $\cdot K\{a\cos x(1+a^2\cos^2 x)^{-\frac{1}{2}}\}$ $\quad\quad\quad\quad x<\frac{1}{2}\pi$ $\quad\quad 0 \quad\quad\quad\quad x>\frac{1}{2}\pi$	$\frac{1}{2}a^{-1}\cos(\frac{1}{2}\pi y)q_{-\frac{1}{2}+\frac{1}{2}y}(z)q_{-\frac{1}{2}-\frac{1}{2}y}(z)$ $\quad\quad z = (1+a^{-2})^{\frac{1}{2}}$
23.14	$\operatorname{sech} x\ K(\operatorname{sech} x)$	$\frac{1}{8}\pi\Gamma(\frac{1}{4}+\frac{1}{4}iy)\Gamma(\frac{1}{4}-\frac{1}{4}iy)\{\Gamma(\frac{3}{4}+\frac{1}{4}iy)\Gamma(\frac{3}{4}-\frac{1}{4}iy)\}^{-1}$

	$f(x)$	$g_c(y)$
23.15	$\operatorname{sech}(ax)\,K\{\tanh(ax)\}$	$\frac{1}{16}(\pi a)^{-1}\left\|\Gamma(\tfrac{1}{4}+\tfrac{1}{4}iy/a)\right\|^4$
23.16	$(z+1)^{-\frac{1}{2}}K\left(\left(\dfrac{z-1}{z+1}\right)^{\frac{1}{2}}\right)$ $z=a\cosh x,\qquad a>1$	$2^{-7/2}(\pi/a)^{\frac{1}{2}}\operatorname{sech}(\pi y)\,\Gamma(\tfrac{1}{4}+\tfrac{1}{2}iy)\,\Gamma(\tfrac{1}{4}-\tfrac{1}{2}iy)$ $\left[P_{-\frac{1}{2}+iy}(u)+P_{-\frac{1}{2}+iy}(-u)\right],\ u=(1-a^{-2})^{\frac{1}{2}}$
23.17	$(1+z)^{-\frac{1}{2}}K\left(\{2(1+z)^{-1}\}^{\frac{1}{2}}\right)$ $z=a\cosh x,\qquad a>1$	$\tfrac{1}{2}(\tfrac{1}{2}\pi/a)^{\frac{1}{2}}\Gamma(\tfrac{1}{4}+\tfrac{1}{2}iy)\,\Gamma(\tfrac{1}{4}-\tfrac{1}{2}iy)\,P_{-\frac{1}{2}+iy}(u)$ $u=(1-a^{-2})^{\frac{1}{2}}$
23.18	$z^{-\frac{1}{2}}K(z^{-\frac{1}{2}})$ $z=1+a^2\sinh^2 x,\qquad a>1$	$\tfrac{1}{4}\pi^2 a^{-1}\operatorname{sech}(\tfrac{1}{2}\pi y)\,P_{-\frac{1}{2}+\frac{1}{2}iy}(u)\,P_{-\frac{1}{2}+\frac{1}{2}iy}(-u)$ $u=(1-a^{-2})^{\frac{1}{2}}$
23.19	$z^{-\frac{1}{2}}K(z^{-\frac{1}{2}}a\cosh x)$ $z=1+a^2\cosh^2 x$	$\tfrac{1}{4}\pi a^{-1}\operatorname{sech}(\tfrac{1}{2}\pi y)\,p_{-\frac{1}{2}+\frac{1}{2}iy}(u)$ $\cdot\left[q_{-\frac{1}{2}+\frac{1}{2}iy}(u)+q_{-\frac{1}{2}-\frac{1}{2}iy}(u)\right],\ u=(1+a^{-2})^{\frac{1}{2}}$
23.20	$\operatorname{sech} x\,K\{(1-a^2\operatorname{sech}^2 x)^{\frac{1}{2}}\}$ $\qquad\qquad a<1$	$\tfrac{1}{4}\pi^2\operatorname{sech}^2(\tfrac{1}{2}\pi y)\,P_{-\frac{1}{2}+\frac{1}{2}iy}(u)\,P_{-\frac{1}{2}+\frac{1}{2}iy}(-u)$ $u=(1-a^2)^{\frac{1}{2}}$
23.21	$\qquad\qquad 0\qquad \sinh x<a$ $\operatorname{cosech} x$ $\cdot K\{(1-a^2\operatorname{cosech}^2 x)^{\frac{1}{2}}\}$ $\qquad\qquad\qquad \sinh x>a$	$\tfrac{1}{4}\left[\{q_{-\frac{1}{2}+\frac{1}{2}iy}(u)\}^2+\{q_{-\frac{1}{2}-\frac{1}{2}iy}(u)\}^2\right]$ $u=(1+a^2)^{\frac{1}{2}}$
23.22	$\operatorname{sech} x\,K(a\operatorname{sech} x)\quad a\underline{\le}1$	$\tfrac{1}{4}\pi^2\operatorname{sech}(\tfrac{1}{2}\pi y)\left[P_{-\frac{1}{2}+\frac{1}{2}iy}(u)\right]^2,\ u=(1-a^2)^{\frac{1}{2}}$

Part II
Fourier Sine Transforms
(Tables II)

2.1 Algebraic Functions

	$f(x)$		$g_s(y)$
1.1	1	$x<a$	$2y^{-1}\sin^2(\tfrac{1}{2}ay)$
	0	$x>a$	
1.2	x	$a<x<b$	$y^{-1}\Big(a\,\cos(ay)-b\,\cos(by)\Big)+y^{-2}\Big(\sin(by)-\sin(ay)\Big)$
	0	$x>b$	
1.3	x	$x<a$	$4y^{-2}\sin(ay)\sin^2(\tfrac{1}{2}ay)$
	$2a-x$	$a<x<2a$	
	0	$x>2a$	
1.4	0	$x<a$	$-\mathrm{si}(ay)$
	x^{-1}	$x>a$	
1.5	x^{-1}	$x<a$	$\mathrm{Si}(ay)$
	0	$x>a$	
1.6	$x^{-\frac{1}{2}}$		$(\tfrac{1}{2}\pi/y)^{\frac{1}{2}}$
1.7	$x^{-\frac{1}{2}}$	$x<a$	$(2\pi/y)^{\frac{1}{2}}S(ay)$
	0	$x>a$	
1.8	0	$x<a$	$(2\pi/y)^{\frac{1}{2}}\{\tfrac{1}{2}-S(ay)\}$
	$x^{-\frac{1}{2}}$	$x>a$	
1.9	$x^{-3/2}$		$(2\pi y)^{\frac{1}{2}}$
1.10	$(a+x)^{-1}$	$x<b$	$\cos(ay)\Big(\mathrm{si}(ay+by)-\mathrm{si}(ay)\Big)$
	0	$x>b$	$-\sin(ay)\Big(\mathrm{Ci}(ay+by)-\mathrm{Ci}(ay)\Big)$
1.11	0	$x<b$	$\sin(ay)\mathrm{Ci}(ay+by)-\cos(ay)\mathrm{si}(ay+by)$
	$(a+x)^{-1}$	$x>b$	
1.12	$(a-x)^{-1}$	$x<b$	$\sin(ay)\Big(\mathrm{Ci}(ay)-\mathrm{Ci}(ay-by)\Big)$
	0	$x>b$	$-\cos(ay)\Big(\mathrm{si}(ay)-\mathrm{si}(ay-by)\Big)$
	$a>b$		

	$f(x)$	$g_s(y)$
1.13	$0 \quad x<b$ $(x-a)^{-1} \quad x>b$ $b>a$	$-\sin(ay)\,Ci(by-ay)-\cos(ay)\,si(by-ay)$
1.14	$0 \quad x<b$ $(a+x)^{-1} \quad x>b$	$\sin(ay)\,Ci(ay+by)-\cos(ay)\,si(ay+by)$
1.15	$x^{-1}(a+x)^{-1}$	$a^{-1}\left(\tfrac{1}{2}\pi+\cos(ay)\,si(ay)-\sin(ay)\,Ci(ay)\right]$
1.16	$0 \quad x<b$ $x^{-1}(a+x)^{-1} \quad x>b$	$a^{-1}\{\cos(ay)\,si(ay+by)$ $-\sin(ay)\,Ci(ay+by)-si(ay)\}$
1.17	$x^{\frac{1}{2}}(a+x)^{-1}$	$(2y/\pi)^{-\frac{1}{2}}+\pi a^{\frac{1}{2}}\sin(ay)\{1-C(ay)-S(by)\}$ $-\pi a^{\frac{1}{2}}\cos(ay)\{C(ay)-S(ay)\}$
1.18	$x^{-\frac{1}{2}}(a+x)^{-1}$	$\pi a^{-\frac{1}{2}}\left(\cos(ay)\{C(ay)-S(ay)\}\right.$ $\left.-\sin(ay)\{1-C(ay)-S(ay)\}\right]$
1.19	$(a+x)^{-\frac{1}{2}} \quad x<b$ $0 \quad x>b$	$\{2\pi/y\}^{\frac{1}{2}}\left[\cos(ay)\{S(ay+by)-S(ay)\}\right.$ $\left.-\sin(ay)\{C(ay+by)-C(ay)\}\right]$
1.20	$0 \quad x<b$ $(a+x)^{-\frac{1}{2}} \quad x>b$	$(2\pi/y)^{\frac{1}{2}}\left[\cos(ay)\{\tfrac{1}{2}-S(ay+by)\}\right.$ $\left.-\sin(ay)\{\tfrac{1}{2}-C)ay+by)\}\right]$
1.21	$(a+x)^{-3/2}$	$(2\pi y)^{\frac{1}{2}}\left(\sin(ay)\{1-2S(ay)\}+\cos(ay)\{1-2C(ay)\}\right]$
1.22	$(a-x)^{-1}$	$\sin(ay)\,Ci(ay)-\cos(ay)\{\tfrac{1}{2}\pi+Si(ay)\}$ (Cauchy principal value)
1.23	$0 \quad x<a$ $(x-a)^{-\frac{1}{2}} \quad x>a$	$(2y/\pi)^{-\frac{1}{2}}\{\sin(ay)+\cos(ay)\}$
1.24	$(a-x)^{-\frac{1}{2}} \quad x<a$ $0 \quad x>a$	$(2\pi/y)^{\frac{1}{2}}\{\sin(ay)\,C(ay)-\cos(ay)\,S(ay)\}$

	$(f(x)$	$g_s(y)$
1.25	$\begin{array}{ll}0 & x<a \\ x^{-1}(x-a)^{\frac{1}{2}} & x>a\end{array}$	$(\pi/y)^{\frac{1}{2}}\sin(\frac{1}{4}\pi+ay)-\pi a^{\frac{1}{2}}\{C(ay)-S(ay)\}$
1.26	$\begin{array}{ll}x^{-1}(x-a)^{-\frac{1}{2}} & x>a \\ 0 & x<a\end{array}$	$\pi a^{-\frac{1}{2}}\{C(ay)-S(ay)\}$
1.27	$\begin{array}{ll}0 & x<a \\ (x-a)^{-\frac{1}{2}}(a+x)^{-1} & x>a\end{array}$	$\pi(2a)^{-\frac{1}{2}}\Big(\cos(ay)\{C(2ay)-S(2ay)\} \\ \quad -\{1-C(2ay)-S(2ay)\}\sin(ay)\Big]$
1.28	$\begin{array}{ll}0 & x<b \\ (a+x)^{-1}(x-b)^{-\frac{1}{2}} & x>b\end{array}$	$\pi(a+b)^{-\frac{1}{2}}\Big(\cos(ay)\{C(ay+by)-S(ay-by)\} \\ \quad -\sin(ay)\{1-C(ay+by)-S(ay+by)\}\Big]$
1.29	$x^{-3/2}\{a+x+(2ax)^{\frac{1}{2}}\}^{-1}$	$2^{-\frac{1}{2}}a^{-3/2}\pi\Big(2(ay/\pi)^{\frac{1}{2}}-1+e^{ay}\text{Erfc}\{(ay)^{\frac{1}{2}}\}\Big]$
1.30	$x^{-1}(a+x)^{-\frac{1}{2}}\{(a+x)^{\frac{1}{2}}-a^{\frac{1}{2}}\}$	$\frac{1}{2}\pi\text{Erfc}\{(iay)^{\frac{1}{2}}\}\text{Erfc}\{(-iay)^{\frac{1}{2}}\} \\ =\frac{1}{2}\pi\Big(1-2C(ay)\{1-C(ay)\}-2S(ay)\{1-S(ay)\}\Big]$
1.31	$(a^2+x^2)^{-1}$	$(2a)^{-1}\{e^{-ay}\overline{\text{Ei}}(ay)-e^{ay}\text{Ei}(-ay)\}$
1.32	$x(a^2+x^2)^{-1}$	$\frac{1}{2}\pi e^{-ay}$
1.33	$x^{-1}(a^2+x^2)^{-1}$	$\frac{1}{2}\pi a^{-2}(1-e^{-ay})$
1.34	$\begin{array}{ll}(a^2-x^2)^{-1} & x<b \\ 0 & x>b \\ b<a\end{array}$	$(2a)^{-1}\Big(\cos(ay)\{\text{si}(ay+by)+\text{si}(ay-by) \\ \quad -2\text{si}(ay)\}-\sin(ay)\{\text{Ci}(ay+by)+\text{Ci}(ay-by) \\ \quad -2\text{Ci}(ay)\}\Big]$
1.35	$\begin{array}{ll}0 & x<b \\ (x^2-a^2)^{-1} & x>b \\ b>a\end{array}$	$-(2a)^{-1}\Big(\sin(ay)\{\text{Ci}(by-ay)+\text{Ci}(by+ay)\} \\ \quad +\cos(ay)\{\text{si}(by-ay)-\text{si}(by+ay)\}\Big]$
1.36	$(a^2-x^2)^{-1}$	$a^{-1}\{\sin(ay)\text{Ci}(ay)-\cos(ay)\text{Si}(ay)\}$ (Cauchy principal value)

	$f(x)$	$g_s(y)$
1.37	$x(a^2-x^2)^{-1}$	$-\frac{1}{2}\pi\cos(ay)$ (Cauchy principal value)
1.38	$x^{-1}(a^2-x^2)^{-1}$	$\frac{1}{2}\pi a^{-2}(1-\cos(ay))$ (Cauchy principal value)
1.39	$x^{-\frac{1}{2}}(a^2-x^2)^{-1}$	$(\frac{1}{2}\pi y)^{\frac{1}{2}}a^{-1}S_{0,\frac{1}{2}}(ay)-\frac{1}{2}a^{-3/2}\pi\cos(ay)$ (Cauchy principal value)
1.40	$(a^2+x^2)^{-\frac{1}{2}}$	$\frac{1}{2}\pi\{I_0(ay)-\mathbf{L}_0(ay)\}$
1.41	$x(a^2+x^2)^{-\frac{1}{2}}-1$	$aK_1(ay)-\frac{1}{2}$
1.42	$\{x+(a^2+x^2)^{\frac{1}{2}}\}^{-1}$	$\frac{1}{2}\pi(ay)^{-1}\{I_1(ay)-\mathbf{L}_1(ay)\}$
1.43	$(a^2-x^2)^{-\frac{1}{2}} \quad x<a$ $0 \qquad\qquad x>a$	$\frac{1}{2}\pi\mathbf{H}_0(ay)$
1.44	$x(a^2-x^2)^{-\frac{1}{2}} \quad x<a$ $0 \qquad\qquad x>a$	$\frac{1}{2}a\pi J_1(ay)$
1.45	$0 \qquad\qquad x<a$ $x^{-1}(x^2-a^2)^{-\frac{1}{2}} \quad x>a$	$-\frac{1}{4}\pi^2 y\{\mathbf{H}_0(ay)Y_1(ay)+Y_0(ay)\mathbf{H}_{-1}(ay)\}$
1.46	$x^{-\frac{1}{2}}(a^2+x^2)-\frac{1}{2}$	$(\frac{1}{2}\pi y)^{\frac{1}{2}}I_{\frac{1}{4}}(\frac{1}{2}ay)K_{\frac{1}{4}}(\frac{1}{2}ay)$
1.47	$x^{-\frac{1}{2}}(a^2-x^2)^{-\frac{1}{2}} \quad x<a$ $0 \qquad\qquad x>a$	$\frac{1}{2}\pi(\frac{1}{2}\pi y)^{\frac{1}{2}}\{J_{\frac{1}{4}}(\frac{1}{2}ay)\}^2$
1.48	$0 \qquad\qquad x<a$ $x^{-\frac{1}{2}}(x^2-a^2)^{-\frac{1}{2}} \quad x>a$	$-\frac{1}{2}\pi(\frac{1}{2}\pi y)^{\frac{1}{2}}J_{\frac{1}{4}}(\frac{1}{2}ay)Y_{\frac{1}{4}}(\frac{1}{2}ay)$
1.49	$(a^2+x^2)^{-\frac{1}{2}}\{a+(a^2+x^2)^{\frac{1}{2}}\}^{-\frac{1}{2}}$	$-i\pi(2a)^{-\frac{1}{2}}\text{Erf}\{i(ay)^{\frac{1}{2}}\}\text{Erfc}\{(ay)^{\frac{1}{2}}\}$
1.50	$(a^2+x^2)^{-\frac{1}{2}}\{(a^2+x^2)^{\frac{1}{2}}-a\}^{-\frac{1}{2}}$	$\pi(2a)^{-\frac{1}{2}}\text{Erf}(ay)$

	$f(x)$	$g_s(y)$
1.51	$(a^2x+x^3)^{-\frac{1}{2}}\{x+(a^2+x^2)^{\frac{1}{2}}\}^{\frac{1}{2}}$	$2^{-\frac{1}{2}}e^{-\frac{1}{2}ay}I_o(\frac{1}{2}ay)$
1.52	$(a^2x+x^3)^{-\frac{1}{2}}\{x+(a^2+x^2)^{\frac{1}{2}}\}^{-\frac{1}{2}}$	$2^{\frac{1}{2}}a^{-1}\sinh(\frac{1}{2}ay)K_o(\frac{1}{2}ay)$
1.53	$(a^2x+x^3)^{-\frac{1}{2}}\{x+(a^2+x^2)^{\frac{1}{2}}\}^{-\frac{3}{2}}$	$2^{-\frac{1}{2}}\pi a^{-2}e^{-\frac{1}{2}ay}I_1(\frac{1}{2}ay)$
1.54	$(a^2+x^2)^{-\frac{1}{2}}\{(a^2+x^2)^{\frac{1}{2}}-a\}^{\frac{1}{2}}$	$(\frac{1}{2}\pi/y)^{\frac{1}{2}}e^{-ay}$
1.55	$(a^2x^2+x^4)^{-\frac{1}{2}}\{(a^2+x^2)^{\frac{1}{2}}+a\}^{\frac{1}{2}}$	$\pi(2a)^{-\frac{1}{2}}\text{Erf}\{(ay)^{\frac{1}{2}}\}$
1.56	$x\{(a^2+x^2)(b^2+x^2)\}^{-1}$	$\frac{1}{2}\pi(a^2-b^2)^{-1}(e^{-by}-e^{-ay})$
1.57	$x(x^4+2a^2x^2\cos u+a^4)^{-1}$ $u<\pi$	$\frac{1}{2}\pi a^{-2}\text{cosec }u\,\sin(ay\sin\frac{u}{2})\exp(-ay\cos\frac{u}{2})$
1.58	$x^3(x^4+2a^2x^2\cos u+a^4)^{-1}$ $u<\pi$	$\frac{1}{2}\pi\text{cosec }u\,\sin(u-ay\sin\frac{u}{2})\exp(-ay\cos\frac{u}{2})$
1.59	$x(a^4+x^4)^{-1}$	$\frac{1}{2}\pi a^{-2}\exp(-2^{-\frac{1}{2}}ay)\sin(2^{-\frac{1}{2}}ay)$
1.60	$x(a^2+x^2)^{-n}$ $n=2,3,\ldots$	$4a^3(2a)^{-2n}\pi\{(n-1)!\}^{-1}ye^{-ay}$ $\cdot\displaystyle\sum_{m=o}^{n-1}(2n-m-4)!\{m!(n-m-2)!\}^{-1}(2ay)^m$
1.61	$x^{-1}(a^2+x^2)^{-n}$ $n=1,2,\ldots$	$\frac{1}{2}\pi a^{-2n}\left(1-2^{1-n}e^{-ay}F_{n-1}(ay)\{(n-1)!\}^{-1}\right)$ $F_o(z)=1\;;\;F_n(z)=(z+2n)F_{n-1}(z)$ $-zF'_{n-1}(z)$
1.62	$x^{2m+1}(a^2+x^2)^{-n-\frac{1}{2}}$ $-1\leq m<n$	$(-1)^{m+1}(2a)^{-n}\pi^{\frac{1}{2}}\{\Gamma(\frac{1}{2}+m)\}^{-1}$ $\cdot\dfrac{d^{2m+1}}{dy^{2m+1}}y^nK_n(ay)$
1.63	$x^{2m-1}(a^{2n}+x^{2n})^{-1}$ $m\leq n,\;n,m=1,2,\ldots$	$-\frac{1}{2}\pi a^{2m-2n}n^{-1}\displaystyle\sum_{k=1}^{n}\exp\{-ay\sin(k-\frac{1}{2})\pi/n\}$ $\cdot\cos\{2m\pi(k-\frac{1}{2})/n+ay\cos(k-\frac{1}{2})\pi/n\}$

	$f(x)$	$g_s(y)$
1.64	$x^{-1}(a^{2n}+x^{2n})^{-1}$ $n=1,2,\ldots$	$\frac{1}{2}\pi a^{-2n}\left\{1-n^{-1}\sum_{m=1}^{n}\exp\{-ay\,\sin(m-\frac{1}{2})\,\pi/n\}\right.$ $\left. \cdot\cos\{ay\,\cos m-\frac{1}{2})\,\pi/n)\}\right\}$
1.65	$x^{2m+1}(a^2+x^2)^{-n-1}$ $m\leq n;\; n,m=1,2,\ldots$	$\frac{1}{2}\pi\frac{(-1)^{n+m}}{n!}\frac{d^n}{da^n}(a^m e^{-a^{\frac{1}{2}}}y)$
1.66	$0 \qquad\qquad x<a$ $(x^2-a^2)^{-\frac{1}{2}}$ $\cdot\{x+(x^2-a^2)^{\frac{1}{2}}\}^{-n} \quad x>a$ $n=1,2,\ldots$	$\frac{1}{2}\pi a^{-n-2}\{\cos(\frac{1}{2}\pi n)J_n(ay)-\sin(\frac{1}{2}\pi n)Y_n(ay)\}$ $-\sum_{m=1}^{n}m!\,(n+m-1)!\,\{(2m)!\,(n-m)!\}^{-1}(\frac{1}{2}ay)^{-m}$ $\cdot\sin(ay+\frac{1}{2}\pi m)$
1.67	$0 \qquad\qquad x<b$ $(a+x)^{-n} \qquad x>b$ $n=1,2,\ldots$	$\sum_{m=1}^{n-1}\frac{(m-1)!}{(n-1)!}\cos\{\frac{1}{2}\pi(n-m)-by\}(a+b)^{-m}(-y)^{n-m-1}$ $-\frac{(-y)^{n-1}}{(n-1)!}\{\cos(ay+\frac{1}{2}\pi n)\,\mathrm{Ci}(ay+by)$ $+\sin(ay+\frac{1}{2}\pi n)\,\mathrm{si}(ay+by)\}$
1.68	$0 \qquad\qquad x<a$ $(x^2-a^2)^{-\frac{1}{2}} \quad a<x<b$ $0 \qquad\qquad x>b$	$\pi\sum_{n=0}^{\infty}J_{n+\frac{1}{2}}(\frac{1}{2}by+\frac{1}{2}ay)J_{n+\frac{1}{2}}(\frac{1}{2}by-\frac{1}{2}ay)$

2.2 Arbitrary Powers

	$f(x)$	$g_s(y)$
2.1	$(a+x)^{\nu}$ $\mathrm{Re}\,\nu<0$	$\frac{1}{2}y^{-\nu-1}\left(\exp(\frac{1}{2}i\pi\nu-iay)\,\Gamma(1+\nu,-iay)\right.$ $\left.+\exp(iay-\frac{1}{2}i\pi\nu)\,\Gamma(1+\nu,iay)\right)$
2.2	$(a+x)^{\nu} \qquad x<b$ $0 \qquad\qquad x>b$	$\frac{1}{2}y^{-\nu-1}\left(\exp(\frac{1}{2}i\pi\nu-iay)\{\gamma(1+\nu,-iay-iby)\right.$ $-\gamma(1+\nu,-iay)\}+\exp(iay-\frac{1}{2}i\pi\nu)\{\Gamma(1+\nu,iay+iby)$ $\left.-\gamma(1+\nu,iay)\}\right)$
2.3	$x^{\nu}(a+x)^{-1},\,-2<\mathrm{Re}\,\nu<1$	$\frac{1}{2}ia^{\nu}\Gamma(1+\nu)\{e^{iay}\Gamma(-\nu,iay)-e^{-iay}\Gamma(-\nu,-iay)\}$
2.4	$(a-x)^{\nu} \qquad x<a$ $0 \qquad\qquad x>a$ $\mathrm{Re}\,\nu>-1$	$-\frac{1}{2}y^{-\nu-1}\left(\exp(iay-\frac{1}{2}i\pi\nu)\,\gamma(1+\nu,iay)\right.$ $\left.+\exp(\frac{1}{2}i\pi\nu-iay)\,\gamma(1+\nu,-iay)\right)$

	$f(x)$	$g_s(y)$
2.5	$x^\nu (a-x)^\mu$ $\quad x<a$ $0 \qquad x>a$ $\mathrm{Re}\,\nu>-2,\ \mathrm{Re}\,\mu>-1$	$\tfrac{1}{2} i B(\nu+1,\mu+1)a^{\nu+\mu+1}$ $\cdot \left[{}_1F_1(1+\nu;2+\nu+\mu;-iay)-{}_1F_1(1+\nu;2+\nu+\mu;iay)\right]$
2.6	$0 \qquad x<a$ $x^{-1}(x-a)^\nu \quad x>a$ $-1<\mathrm{Re}\,\nu<1$	$\tfrac{1}{2}ia^\nu \Gamma(1+\nu)\{\Gamma(-\nu,iay)-\Gamma(-\nu,-iay)\}$
2.7	$x^\nu (a^2+x^2)^{-1}$ $-2<\mathrm{Re}\,\nu<2$	$-\tfrac{1}{2}i\pi a^{\nu-1}\mathrm{cosec}(\tfrac{1}{2}\pi\nu)\sinh(ay)-2^\nu(\pi y)^{\frac{1}{2}}$ $\cdot (iay)^{\nu-\frac{1}{2}}\Gamma(1+\tfrac{1}{2}\nu)\{\Gamma(\tfrac{1}{2}-\tfrac{1}{2}\nu)\}^{-1}s_{-\nu-\frac{1}{2},\frac{1}{2}}(iay)$
2.8	$(a^2+x^2)^{-\nu-\frac{1}{2}},\ \mathrm{Re}\,\nu>-\frac{1}{2}$	$\pi^{\frac{1}{2}}2^{-\nu-1}\Gamma(\tfrac{1}{2}-\nu)(y/a)^\nu\{I_\nu(ay)-\mathbf{L}_{-\nu}(ay)\}$
2.9	$x^{-1}(a^2+x^2)^{-\nu-\frac{1}{2}}$ $\mathrm{Re}\,\nu>-1$	$\tfrac{1}{2}\pi a^{-2\nu}y\left(K_\nu(ay)\mathbf{L}_{\nu-1}(ay)+\mathbf{L}_\nu(ay)K_{\nu-1}(ay)\right)$
2.10	$x(a^2+x^2)^{\nu-3/2}$ $\mathrm{Re}\,\nu<1$	$\tfrac{1}{2}\pi^{\frac{1}{2}}(2a)^\nu\{\Gamma(\tfrac{3}{2}-\nu)\}^{-1}y^{1-\nu}K_\nu(ay)$
2.11	$(x^2+2ax)^{-\nu-\frac{1}{2}}$ $-\frac{1}{2}<\mathrm{Re}\,\nu<\frac{1}{2}$	$\tfrac{1}{2}\pi^{\frac{1}{2}}\Gamma(\tfrac{1}{2}-\nu)(\tfrac{1}{2}y/a)^\nu\{J_\nu(ay)\cos(ay)$ $+\ Y_\nu(ay)\sin(ay)\}$
2.12	$\{(a^2+x^2)^{\frac{1}{2}}+x\}^{-\nu}$ $\mathrm{Re}\,\nu>0$	$a^{-\nu}y^{-1}+a^{-\nu}\pi\nu\,\mathrm{cosec}(\pi\nu)y^{-1}$ $\cdot\{\cos(\tfrac{1}{2}\pi\nu)I_\nu(ay)-\tfrac{1}{2}\mathbf{J}_\nu(iay)-\tfrac{1}{2}\mathbf{J}_\nu(-iay)\}$
2.13	$(a^2+x^2)^{-\frac{1}{2}}\{(a^2+x^2)^{\frac{1}{2}}+x\}^{-\nu}$ $\mathrm{Re}\,\nu>-1$	$\mathrm{cosec}(\pi\nu)a^{-\nu}\{\sin(\tfrac{1}{2}\pi\nu)I_\nu(ay)$ $+\ \tfrac{1}{2}i\mathbf{J}_\nu(iay)-\tfrac{1}{2}i\mathbf{J}_\nu(-iay)\}$
2.14	$(a^2x+x^3)^{-\frac{1}{2}}\{(a^2+x^2)^{\frac{1}{2}}+x\}^\nu$ $\mathrm{Re}\,\nu<3/2$	$a^\nu(\tfrac{1}{2}\pi y)^{\frac{1}{2}}I_{\frac{1}{4}-\frac{1}{2}\nu}(\tfrac{1}{2}ay)K_{\frac{1}{4}+\frac{1}{2}\nu}(\tfrac{1}{2}ay)$
2.15	$x^{-\nu-\frac{1}{2}}(a^2+x^2)^{-\frac{1}{2}}$ $\cdot\{(a^2+x^2)^{\frac{1}{2}}+a\}^\nu$ $\mathrm{Re}\,\nu<3/2$	$a^{-1}\Gamma(\tfrac{3}{4}-\tfrac{1}{2}\nu)(\tfrac{1}{2}y)^{-\frac{1}{2}}W_{\frac{1}{2}\nu,\frac{1}{4}}(ay)M_{-\frac{1}{2}\nu,\frac{1}{4}}(ay)$ $=-\tfrac{1}{2}(\pi/a)^{\frac{1}{2}}\mathrm{cosec}(\tfrac{1}{4}\pi+\tfrac{1}{2}\pi\nu)D_{\nu-\frac{1}{2}}\{(2ay)^{\frac{1}{2}}\}$ $\cdot\left(D_{-\nu-\frac{1}{2}}\{(2ay)^{\frac{1}{2}}\}-D_{-\nu-\frac{1}{2}}\{-(2ay)^{\frac{1}{2}}\}\right)$

	$f(x)$	$g_s(y)$
2.16	$(a+ix)^\nu-(a-ix)^\nu$ $\mathrm{Re}\,\nu<0$	$i\pi\{\Gamma(-\nu)\}^{-1}y^{-\nu-1}e^{-ay}$
2.17	$x^{-1}\{(a+ix)^{-\nu}+(a-ix)^{-\nu}\}$ $\mathrm{Re}\,\nu>-1$	$\pi\{\Gamma(1+\nu)\}^{-1}y^\nu {}_1F_1(\nu;1+\nu;-ay)$
2.18	$x\{(a+ix)^{-\nu}+(a-ix)^{-\nu}\}$ $\mathrm{Re}\,\nu>-1$	$\pi a^{-\nu}\{\Gamma(\nu)\}^{-1}y^{\nu-2}(1-ay)e^{-ay}$
2.19	$x^\nu(a^2-x^2)^{-1}$ $-2<\mathrm{Re}\,\nu<2$	$-\tfrac{1}{2}\pi a^{\nu-1}\cos(ay)+(\pi/a)^{\frac{1}{2}}(2a)^\nu$ $\cdot\Gamma(1+\tfrac{1}{2}\nu)\{\Gamma(\tfrac{1}{2}-\tfrac{1}{2}\nu)\}^{-1}y^{\frac{1}{2}}S_{-\nu-\frac{1}{2},\frac{1}{2}}(ay)$ Cauchy principal value
2.20	$(a^2+x^2)^{-\frac{1}{2}}\Big\{\{(a^2+x^2)^{\frac{1}{2}}+x\}^\nu$ $-\{(a^2+x^2)^{\frac{1}{2}}-x\}^\nu\Big\},\,\mid\mathrm{Re}\,\nu\mid<1$	$2a^\nu\sin(\tfrac{1}{2}\pi\nu)K_\nu(ay)$
2.21	$x^\nu(a^2+x^2)^{-\mu-1}$ $\mathrm{Re}\,\mu>-2,\ \mathrm{Re}\,(\nu-2\mu)<2$	$\tfrac{1}{2}a^{\nu-2\mu}B(1+\tfrac{1}{2}\nu,\mu-\tfrac{1}{2}\nu)y{}_1F_2(1+\tfrac{\nu}{2};1+\tfrac{\nu}{2}-\mu,\tfrac{3}{2};\tfrac{a^2y^2}{4})$ $+2^{\nu-2\mu-2}\pi^{\frac{1}{2}}\Gamma(\tfrac{1}{2}\nu-\mu)\{\Gamma(\mu-\tfrac{1}{2}\nu+3/2)\}^{-1}$ $\cdot y^{2\mu-\nu+1}{}_1F_2(1+\mu;\mu-\tfrac{1}{2}\nu+3/2,1+\mu-\tfrac{1}{2}\nu;\tfrac{1}{4}a^2y^2)$
2.22	$(a^2-x^2)^{\nu-\frac{1}{2}}\quad\quad x<a$ $\quad\quad\quad 0\quad\quad\quad\quad x>a$ $\mathrm{Re}\,\nu>-\tfrac{1}{2}$	$\tfrac{1}{2}\pi^{\frac{1}{2}}(2a/y)^\nu\Gamma(\tfrac{1}{2}+\nu)\mathbf{H}_\nu(ay)$
2.23	$x(a^2-x^2)^{\nu-\frac{1}{2}}\quad\quad x<a$ $\quad\quad\quad 0\quad\quad\quad\quad x>a$ $\mathrm{Re}\,\nu>-\tfrac{1}{2}$	$\tfrac{1}{4}\pi^{\frac{1}{2}}(2a)^{\nu+1}y^{-\nu}\Gamma(\tfrac{1}{2}+\nu)J_{\nu+1}(ay)$
2.24	$(2ax-x^2)^{\nu-\frac{1}{2}}\quad\quad x<2a$ $\quad\quad\quad 0\quad\quad\quad\quad x>2a$ $\mathrm{Re}\,\nu>-\tfrac{1}{2}$	$\pi^{\frac{1}{2}}\Gamma(\tfrac{1}{2}+\nu)(2a/y)^\nu\sin(ay)J_\nu(ay)$
2.25	$x^{-1}(a^2-x^2)^{\nu-\frac{1}{2}}\quad\quad x<a$ $\quad\quad\quad 0\quad\quad\quad\quad x>a$ $\mathrm{Re}\,\nu>-\tfrac{1}{2}$	$\tfrac{1}{4}\pi^2\sec(\pi\nu)a^{2\nu}y$ $\cdot\Big(J_\nu(ay)\mathbf{H}_{-\nu-1}(ay)+\mathbf{H}_{-\nu}(ay)J_{\nu+1}(ay)\Big)$

	$f(x)$		$g_s(y)$
2.26	$(2ax-x^2)^{\nu-\frac{1}{2}}$	$x<a$	$\frac{1}{2}\pi^{\frac{1}{2}}\Gamma(\frac{1}{2}+\nu)(2a/y)^{\nu}$
	0	$x>a$	$\cdot\left[\sin(ay)J_{\nu}(ay)-\cos(ay)\mathbf{H}_{\nu}(ay)\right]$
	$\mathrm{Re}\,\nu>-3/2$		
2.27	$x^{\nu}(a^2-x^2)^{\mu}$	$x<a$	$\frac{1}{2}a^{\nu+\mu+2}B(1+\mu,1+\frac{1}{2}\nu)$
	0	$x>a$	$\cdot y_1F_2(1+\frac{1}{2}\nu;3/2,2+\mu+\frac{1}{2}\nu;-\frac{1}{4}a^2y^2)$
	$\mathrm{Re}\,\nu>-2,\ \mathrm{Re}\,\mu>-1$		
2.28	$(a^2-x^2)^{-\frac{1}{2}}\Big[\{x+i(a^2-x^2)^{\frac{1}{2}}\}^{\nu}$		$\frac{1}{2}\pi a^{\nu}\mathrm{cosec}(\frac{1}{2}\pi\nu)\{J_{\nu}(ay)-\mathbf{J}_{-\nu}(ay)\}$
	$+\{x-i(a^2-x^2)^{\frac{1}{2}}\}^{\nu}\Big]$	$x<a$	$=-\frac{1}{2}\pi a^{\nu}\sec(\frac{1}{2}\pi\nu)\{\mathbf{E}_{\nu}(ay)+\mathbf{E}_{-\nu}(ay)\}$
	0	$x>a$	
2.29	$(a^2x-x^3)^{-\frac{1}{2}}\Big[\{x+i(a^2-x^2)^{\frac{1}{2}}\}^{\nu}$		$\pi(\frac{1}{2}\pi y)^{\frac{1}{2}}a^{\nu}J_{\frac{1}{4}+\frac{1}{2}\nu}(\frac{1}{2}ay)J_{\frac{1}{4}-\frac{1}{2}\nu}(\frac{1}{2}ay)$
	$+\{x-i(a^2-x^2)^{\frac{1}{2}}\}^{\nu}\Big]$	$x<a$	
	0	$x>a$	
2.30	$x^{-\nu-\frac{1}{2}}(a^2-x^2)^{-\frac{1}{2}}$		$(2a)^{\frac{1}{2}}B(\frac{3}{4}+\frac{1}{2}\nu,\frac{3}{4}-\frac{1}{2}\nu)$
	$\cdot\Big[\{a+(a^2-x^2)^{\frac{1}{2}}\}^{\nu}$		$\cdot y_1F_1(\frac{3}{4}-\frac{1}{2}\nu;\frac{3}{2};-iay)_1F_1(\frac{3}{4}-\frac{1}{2}\nu;\frac{3}{2};iay)$
	$+\{a-(a^2-x^2)^{\frac{1}{2}}\}^{\nu}\Big]$	$x<a$	
	0	$x>a$	
	$-3/2<\mathrm{Re}\,\nu<3/2$		
2.31	0	$x<a$	$\frac{1}{2}\pi^{\frac{1}{2}}\Gamma(\frac{1}{2}-\nu)(\frac{1}{2}y/a)^{\nu}J_{\nu}(ay)$
	$(x^2-a^2)^{-\nu-\frac{1}{2}}$	$x>a$	
	$-\frac{1}{2}<\mathrm{Re}\,\nu<\frac{1}{2}$		
2.32	0	$x<a$	$\frac{1}{4}\pi^{\frac{1}{2}}(2a)^{\nu+1}\Gamma(\frac{1}{2}+\nu)y^{-\nu}\mathbf{Y}_{-\nu-1}$
	$x(x^2-a^2)^{\nu-\frac{1}{2}}$	$x>a$	
	$-\frac{1}{2}<\mathrm{Re}\,\nu<0$		
2.33	0	$x<a$	$\frac{1}{4}\pi^2\sec(\pi\nu)a^{-2\nu}y$
	$x^{-1}(x^2-a^2)^{-\nu-\frac{1}{2}}$	$x>a$	$\{\mathbf{H}_{\nu}(ay)Y_{\nu-1}(ay)-Y_{\nu}(ay)\mathbf{H}_{\nu-1}(ay)\}$
	$-1<\mathrm{Re}\,\nu<\frac{1}{2}$		

	$f(x)$	$g_s(y)$
2.34	$0 \qquad\qquad x<2a$ $(x^2-2ax)^{-\nu-\frac{1}{2}} \qquad x>2a$ $-\frac{1}{2}<\mathrm{Re}\,\nu<\frac{1}{2}$	$\frac{1}{2}\pi^{\frac{1}{2}}\Gamma(\frac{1}{2}-\nu)(\frac{1}{2}y/a)^{\nu}$ $\cdot\Big(J_{\nu}(ay)\cos(ay)-Y_{\nu}(ay)\sin(ay)\Big]$
2.35	$0 \qquad\qquad\qquad x<a$ $(x^2-a^2)^{-\frac{1}{2}}\Big({\{x+(x^2-a^2)^{\frac{1}{2}}\}}^{\nu}$ $+{\{x-(x^2-a^2)^{\frac{1}{2}}\}}^{\nu}\Big] \quad x>a$ $-1<\mathrm{Re}\,\nu<1$	$\pi a^{\nu}\Big(\cos(\frac{1}{2}\pi\nu)J_{\nu}(ay)-\sin(\frac{1}{2}\pi\nu)Y_{\nu}(ay)\Big]$
2.36	$0 \qquad\qquad\qquad x<a$ $x^{-\frac{1}{2}}(x^2-a^2)^{-\frac{1}{2}}$ $\cdot\Big({\{x+(x^2-a^2)^{\frac{1}{2}}\}}^{\nu}$ $+{\{x-(x^2-a^2)^{\frac{1}{2}}\}}^{\nu}\Big] \quad x>a$ $-3/2<\mathrm{Re}\,\nu<3/2$	$-\frac{1}{2}\pi a^{\nu}(\frac{1}{2}\pi y)^{\frac{1}{2}}$ $\cdot\Big(J_{\frac{1}{4}+\frac{1}{2}\nu}(\frac{1}{2}ay)Y_{\frac{1}{4}-\frac{1}{2}\nu}(\frac{1}{2}ay)$ $+J_{\frac{1}{4}-\frac{1}{2}\nu}(\frac{1}{2}ay)Y_{\frac{1}{4}+\frac{1}{2}\nu}(\frac{1}{2}ay)\Big]$

2.3 Exponential Functions

	$f(x)$	$g_s(y)$
3.1	e^{-ax}	$y(a^2+y^2)^{-1}$
3.2	$x^{-1}e^{-ax}$	$\arctan(y/a)$
3.3	$x^{-2}(e^{-ax}-e^{-bx})$	$\frac{1}{2}y\,\log\{(b^2+y^2)/(a^2+y^2)\}$ $+b\,\arctan(y/b)-a\,\arctan(y/a)$
3.4	$x^{\frac{1}{2}}e^{-ax}$	$\frac{1}{2}\pi^{\frac{1}{2}}(a^2+y^2)^{-3/4}\sin\{\frac{3}{2}\arctan(y/a)\}$
3.5	$x^{-\frac{1}{2}}e^{-ax}$	$(\frac{1}{2}\pi)^{\frac{1}{2}}(a^2+y^2)^{-\frac{1}{2}}\{(a^2+y^2)^{\frac{1}{2}}-a\}^{\frac{1}{2}}$
3.6	$x^{-3/2}e^{-ax}$	$(2\pi)^{\frac{1}{2}}\{(a^2+y^2)^{\frac{1}{2}}-a\}^{\frac{1}{2}}$
3.7	$x^{n}e^{-ax}$ $n=0,1,2,\ldots$	$n!a^{n+1}(a^2+y^2)^{-n-1}$ $\cdot\sum_{m=0}^{[\frac{1}{2}n]}(-1)^{m}\binom{n+1}{2m+1}(y/a)^{2m+1}$

	$f(x)$	$g_s(y)$
3.8	$x^{n-\frac{1}{2}}e^{-ax}$ $n=0,1,2,\ldots$	$(-1)^n (\tfrac{1}{2}\pi)^{\frac{1}{2}} \dfrac{d^n}{da^n}\left[(a^2+y^2)^{-\frac{1}{2}}\{(a^2+y^2)^{\frac{1}{2}}-a\}^{\frac{1}{2}} \right]$
3.9	$x^{\nu-1}e^{-ax}$ $\mathrm{Re}\,\nu>-1$	$\Gamma(\nu)\,(a^2+y^2)^{-\frac{1}{2}\nu}\sin\{\nu\arctan(y/a)\}$
3.10	$\begin{array}{ll} 0 & x<b \\ (x-b)^{\nu}e^{-bx} & x>b \end{array}$ $\mathrm{Re}\,\nu>-1$	$\Gamma(1+\nu)\,(a^2+y^2)^{-\frac{1}{2}\nu-\frac{1}{2}}e^{-ab}$ $\cdot\sin\{by+(\nu+1)\arctan(y/a)\}$
3.11	$(e^{ax}+1)^{-1}$	$\tfrac{1}{2}y^{-1}-\tfrac{1}{2}\pi a^{-1}\mathrm{cosech}(\pi y/a)$
3.12	$(e^{ax}-1)^{-1}$	$\tfrac{1}{2}\pi a^{-1}\coth(\pi y/a)-\tfrac{1}{2}y^{-1}$
3.13	$x^{\nu-1}(e^{ax}-1)^{-1}$ $\mathrm{Re}\,\nu>0$	$\tfrac{1}{2}ia^{-\nu}\Gamma(\nu)\{\zeta(\nu,1+iy/a)-\zeta(\nu,1-iy/a)\}$
3.14	$x^{\nu-1}(e^{ax}+1)^{-1}$ $\mathrm{Re}\,\nu>-1$	$\Gamma(\nu)\left[y^{-\nu}\sin(\tfrac{1}{2}\pi\nu)+\tfrac{1}{2}i\,(2\pi)^{-\nu}\right.$ $\cdot\{\zeta(\nu,\tfrac{1}{2}+\tfrac{1}{2}iy/a)-\zeta(\nu,\tfrac{1}{2}-\tfrac{1}{2}iy/a)-\zeta(\nu,\tfrac{1}{2}iy/a)$ $\left. +\zeta(\nu,-\tfrac{1}{2}iy/a)\}\right]$
3.15	$x^{-1}(e^{ax}+1)^{-1}$	$\tfrac{1}{2}\pi+\tfrac{1}{2}i\left[\log\{\Gamma(\tfrac{1}{2}+\tfrac{1}{2}iy/a)\}+\log\{\Gamma(-\tfrac{1}{2}iy/a)\}\right.$ $\left. -\log\{\Gamma(\tfrac{1}{2}-\tfrac{1}{2}iy/a)\}-\log\{\Gamma(\tfrac{1}{2}iy/a)\}\right]$
3.16	$e^{-\frac{1}{2}x}(1-e^{-x})^{-1}$	$-\tfrac{1}{2}\tanh(\pi y)$
3.17	$e^{-ax}(1-e^{-x})^{-1}$	$\tfrac{1}{2}i\{\Psi(a-iy)-\Psi(a+iy)\}$
3.18	$(e^{ax}-e^{bx})^{-1}$	$\tfrac{1}{2}i(a-b)^{-1}\left[\Psi\{(a-iy)/(a-b)\}-\Psi\{(a+iy)/(a-b)\}\right]$
3.19	$e^{-nx}(1-e^{-x})^{-1}$ $n=1,2,\ldots$	$\tfrac{1}{2}\pi-\tfrac{1}{2}y^{-1}+\pi(e^{2\pi y}-1)^{-1}-y\displaystyle\sum_{m=1}^{n-1}(y^2+m^2)^{-1}$
3.20	$x^{-1}(1-e^{-cx})^{-1}$ $\cdot(e^{-ax}-e^{-bx})$	$\tfrac{1}{2}i\,\log\left(\dfrac{\Gamma\{(b-iy)/c\}\Gamma\{(a+iy)/c\}}{\Gamma\{(b+iy)/c\}\Gamma\{(a-iy)/c\}}\right)$

	$f(x)$	$g_s(y)$
3.21	$e^{-ax}(1-e^{-bx})^{\nu-1}$ $\operatorname{Re}\nu>-1$	$\tfrac{1}{2}ib^{-1}\left(B(\nu,\tfrac{a+iy}{b})-B(\nu,\tfrac{a-iy}{b})\right)$
3.22	$x^{-2}(1-e^{-ax})^2$	$2a\,\arctan\{ay/(y^2+2a^2)\}$ $-\tfrac{1}{2}y\,\log\{y^2(y^2+4a^2)/(a^2+b^2)\}$
3.23	e^{-ax^2}	$-\tfrac{1}{2}i(\pi/a)^{\frac{1}{2}}e^{-\frac{1}{4}y^2/a}\operatorname{Erf}(\tfrac{1}{2}iya^{-\frac{1}{2}})$
3.24	xe^{-ax^2}	$\tfrac{1}{4}a^{-1}(\pi/a)^{\frac{1}{2}}ye^{-\frac{1}{4}y^2/a}$
3.25	$x^{-1}e^{-ax^2}$	$\tfrac{1}{2}\pi\operatorname{Erf}(\tfrac{1}{2}ya^{-\frac{1}{2}})$
3.26	$x^{\frac{1}{2}}e^{-ax^2}$	$\tfrac{1}{4}\pi(\tfrac{1}{2}y/a)^{3/2}e^{-u}\{I_{-\frac{1}{4}}(u)-I_{3/4}(u)\},\ \ u=\tfrac{1}{8}y^2/a$
3.27	$x^{-\frac{1}{2}}e^{-ax^2}$	$\tfrac{1}{2}\pi^{\frac{1}{2}}(y/a)^{\frac{1}{2}}e^{-u}I_{\frac{1}{4}}(u),\ \ u=\tfrac{1}{8}y^2/a$
3.28	$x^{2n+1}e^{-ax^2}$ $n=0,1,2,\ldots$	$(-1)^n\pi^{\frac{1}{2}}2^{-n-3/2}a^{-n-1}e^{-\frac{1}{4}y^2/a}$ $\cdot He_{2n+1}\{(2a)^{-\frac{1}{2}}y\}$
3.29	$x(b^2+x^2)^{-1}e^{-a^2x^2}$	$\tfrac{1}{4}\pi e^{a^2b^2}\left(e^{-by}\operatorname{Erfc}(ab-\tfrac{1}{2}y/a)\right.$ $\left.-e^{by}\operatorname{Erfc}(ab+\tfrac{1}{2}y/a)\right]$
3.30	e^{-ax-bx^2}	$-\tfrac{1}{4}i(\pi/b)^{\frac{1}{2}}\left(e^{u^2}\operatorname{Erfc}(u)-e^{v^2}\operatorname{Erfc}(v)\right]$ $\begin{smallmatrix}v\\u\end{smallmatrix}=\tfrac{1}{2}b^{-\frac{1}{2}}(a\pm iy)$
3.31	$x^{\nu-1}e^{-ax^2}$ $\operatorname{Re}\nu>-1$	$-\tfrac{1}{2}(\tfrac{1}{2}\pi)^{\frac{1}{2}}(2a)^{-\frac{1}{2}\nu}\sec(\tfrac{1}{2}\pi\nu)\exp(-\tfrac{1}{8}y^2/a)$ $\cdot\left(D_{\nu-1}\{(2a)^{-\frac{1}{2}}y\}-D_{\nu-1}\{-(2a)^{-\frac{1}{2}}y\}\right]$
3.32	$x^{\nu-1}\exp(-ax-bx^2)$ $\operatorname{Re}\nu>-1$	$\tfrac{1}{2}i\Gamma(\nu)(2b)^{-\frac{1}{2}\nu}\exp\{(a^2-y^2)/(8b)\}$ $\cdot\left(e^{\frac{1}{4}iay/b}D_{-\nu}(u)-e^{-\frac{1}{4}iay/b}D_{-\nu}(v)\right]$ $\begin{smallmatrix}u\\v\end{smallmatrix}=(2b)^{-\frac{1}{2}}(a\pm iy)$

	$f(x)$	$g_s(y)$
3.33	$x^{\mu}\exp(-ax^c)$ $0<c<1,\ \mathrm{Re}\,\mu>-2$	$\sum\limits_{n=0}^{\infty}(-a)^n\Gamma(1+\mu+nc)\cos\{\tfrac{\pi}{2}(\mu+nc)\}y^{-\mu-1-nc}/n!$
3.34	$x^{\mu}\exp(-ax^c)$ $c>1,\ \mathrm{Re}\,\mu>-2$	$c^{-1}\sum\limits_{n=0}^{\infty}(-1)^n\{(2n+1)!\}^{-1}\Gamma\{(2+\mu+2n)/c\}$ $\cdot a^{-(\mu+2+2n)/c}y^{2n+1}$
3.35	$\exp(-a^3x^3)$	$\tfrac{1}{3}izy^{-1}\left(e^{\frac{1}{6}i\pi}S_{0,1/3}(ze^{i3\pi/4})\right.$ $\left.-e^{-\frac{1}{6}i\pi}S_{0,1/3}(ze^{-i3\pi/4})\right),\quad z=2(\tfrac{1}{3}y/a)^{3/2}$
3.36	$x^{-\frac{1}{2}}e^{-a/x}$	$(\tfrac{1}{2}\pi/y)^{\frac{1}{2}}e^{-z}(\cos z+\sin z)\ ,\quad z=(2ay)^{\frac{1}{2}}$
3.37	$x^{-3/2}e^{-a/x}$	$(\pi/a)^{\frac{1}{2}}e^{-z}\sin z\ ,\qquad\qquad z=(2ay)^{\frac{1}{2}}$
3.38	$x^{\nu-1}e^{-a/x}$	$i(y/a)^{\frac{1}{2}\nu}\left(e^{-\frac{1}{4}i\pi\nu}K_{\nu}(u)-e^{\frac{1}{4}i\pi\nu}K_{\nu}(v)\right),\ {}^u_v=2(\pm iay)^{\frac{1}{2}}$
3.39	$x^{-\frac{1}{2}}\exp(-ax-b/x)$	$\tfrac{1}{2}(\pi/b)^{\frac{1}{2}}(a^2+y^2)^{-\frac{1}{2}}e^{-u}(u\sin v+v\cos v)$ ${}^u_v=(2b)^{\frac{1}{2}}\{(a^2+y^2)^{\frac{1}{2}}\pm a\}^{\frac{1}{2}}$
3.40	$x^{-3/2}\exp(-ax-b/x)$	$(\pi/b)^{\frac{1}{2}}e^{-u}\sin v\ ,\quad {}^u_v=(2b)^{\frac{1}{2}}\{(a^2+y^2)^{\frac{1}{2}}\pm a\}^{\frac{1}{2}}$
3.41	$x^{\nu-1}\exp(-ax-b/x)$	$ib^{\frac{1}{2}\nu}\{u^{-\frac{1}{2}\nu}K_{\nu}(2b^{\frac{1}{2}}u)-v^{-\frac{1}{2}\nu}K_{\nu}(2b^{\frac{1}{2}}v)\},\ {}^u_v=(a\pm iy)^{\frac{1}{2}}$
3.42	$x^{-2}\exp(-a/x^2)$	$y\sum\limits_{n=0}^{\infty}\{n!(2n+1)!\}^{-1}(y^2/a)^n\left[\psi(2n+2)+\tfrac{1}{2}\psi(n+1)\right.$ $\left.-\log(ya^{\frac{1}{2}})\right]$
3.43	$\exp(-ax^{\frac{1}{2}})$	$y^{-1}+y(\tfrac{1}{2}\pi/y)^{\frac{1}{2}}\left[\{\tfrac{1}{2}-C(z)\}\sin z-\{\tfrac{1}{2}-S(z)\}\cos z\right],$ $z=\dfrac{a^2}{4y}$
3.44	$x^{-1}\exp(-ax^{\frac{1}{2}})$	$\pi\{\tfrac{1}{2}-C(z)\}^2+\pi\{\tfrac{1}{2}-S(z)\}^2\ ,\quad z=\tfrac{1}{4}a^2/y$
3.45	$x^{-\frac{1}{2}}\exp(-ax^{\frac{1}{2}})$	$(2\pi/y)^{\frac{1}{2}}\left[\{\tfrac{1}{2}-C(z)\}\cos z+\{(\tfrac{1}{2}-S(z)\}\sin z\right],z=\tfrac{1}{4}a^2/y$
3.46	$x^{-3/4}\exp(-ax^{\frac{1}{2}})$	$-\tfrac{1}{2}\pi(y/a)^{-\frac{1}{2}}\left[J_{\frac{1}{4}}(z)\cos(\tfrac{1}{8}\pi+z)+Y_{\frac{1}{4}}(z)\sin(\tfrac{1}{8}\pi+z)\right]$ $z=\tfrac{1}{8}a^2/y$

	$f(x)$	$g_s(y)$
3.47	$x^{\nu-1}\exp(-ax^{\frac{1}{2}})$ $\mathrm{Re}\,\nu > -1$	$i\Gamma(2\nu)(3y)^{-\nu}\Big(\exp(-\tfrac{1}{2}i\pi\nu-\tfrac{1}{2}iz)D_{-2\nu}\{(-2iz)^{\frac{1}{2}}\}$ $-\exp(\tfrac{1}{2}i\pi\nu+\tfrac{1}{2}iz)D_{-2\nu}\{(2iz)^{\frac{1}{2}}\}\Big)$, $z=\tfrac{1}{4}a^2/y$
3.48	$r^{-1}(r-a)^{\frac{1}{2}}e^{-br}$	$(\tfrac{1}{2}\pi)^{\frac{1}{2}}y(s+b)^{-\frac{1}{2}}s^{-1}e^{-bs}$
3.49	$r^{-1}(r+a)^{-\frac{1}{2}}e^{-br}$	$-i\pi(2a)^{-\frac{1}{2}}e^{-ab}\mathrm{Erfc}(u)\mathrm{Erf}(iv)$, $\begin{matrix}u\\v\end{matrix} = (as\pm ab)^{\frac{1}{2}}$
3.50	$(rx)^{-1}(r+a)^{\frac{1}{2}}e^{-br}$	$\pi(2a)^{-\frac{1}{2}}e^{-ab}\mathrm{Erf}(v2^{-\frac{1}{2}})$, $v=(2as-2ab)^{\frac{1}{2}}$
3.51	$r^{-1}\{(r+x)^{\nu}-(r-x)^{\nu}\}$ $\cdot\exp(-br)$	$2a^{\nu}\sin\{\nu\arctan(y/b)\}K_{\nu}(as)$
3.52	$r^{-1}(r+x)^{\nu}e^{-br}$	$a^{\nu}\mathrm{cosec}(\pi\nu)\Big(\pi\sin\{\nu\arctan(y/b)\}I_{-\nu}(as)$ $-\int_{o}^{\pi}\exp(ab\cos t)\sinh(ay\sin t)\sin(\nu t)\,dt\Big)$
3.53	$x^{\nu-\frac{1}{2}}r^{-1}(r+a)^{-\nu}e^{-br}$ $\mathrm{Re}\,\nu>-3/2$	$\tfrac{1}{2}(\tfrac{\pi}{a})^{\frac{1}{2}}\mathrm{cosec}(\tfrac{1}{2}\pi\nu-\tfrac{1}{4}\pi)D_{-\nu-\frac{1}{2}}(u)\{D_{\nu-\frac{1}{2}}(v)$ $\qquad\qquad\qquad\qquad -D_{\nu-\frac{1}{2}}(-v)\}$ $=(2a)^{-\frac{1}{2}}i\Gamma(\tfrac{1}{2}+\nu)D_{-\nu-\frac{1}{2}}(u)\{D_{-\nu-\frac{1}{2}}(iv)$ $-D_{-\nu-\frac{1}{2}}(-iv)\}$ $\qquad\qquad \begin{matrix}u\\v\end{matrix} = (2as\pm 2ab)$
3.54	$x^{-2\nu-1}p^{-1}$ $\{(a-p)^{2\nu}\exp(bp)$ $+(a+p)^{2\nu}\exp(-bp)\}$ $\qquad\qquad\qquad x<a$ $\qquad 0 \qquad x>a$ $-3/4<\mathrm{Re}\,\nu<3/4$	$(\tfrac{1}{2}\pi/a)^{\frac{1}{2}}\sec(2\pi\nu)\Big[D_{2\nu-\frac{1}{2}}(u)+D_{2\nu-\frac{1}{2}}(-u)\Big]$ $\cdot\Big[D_{-2\nu-\frac{1}{2}}(v)-D_{-2\nu-\frac{1}{2}}(-v)\Big]$ $\begin{matrix}u\\v\end{matrix} = (2ab\pm 2as)^{\frac{1}{2}}$, $y<b$

$r=(a^2+x^2)^{\frac{1}{2}}$, $p=(a^2-x^2)^{\frac{1}{2}}$ $\qquad\qquad$ $s=(b^2+y^2)^{\frac{1}{2}}$, $\quad q=(b^2-y^2)^{\frac{1}{2}}$

2.4 Logarithmic Functions

	$f(x)$		$g_s(y)$
4.1	$\log x$	$x<1$ $x>1$	$y^{-1}\{Ci(y)-\gamma-\log y\}$
	0		
4.2	$x^{-1}\log x$		$-\tfrac{1}{2}\pi(\gamma+\log y)$
4.3	$x^{-\frac{1}{2}}\log x$		$(\tfrac{1}{2}\pi/y)^{\frac{1}{2}}\{\tfrac{1}{2}\pi-\gamma-\log(4y)\}$
4.4	$x^{\nu-1}\log x$ $-1<Re\nu<1$		$\tfrac{1}{2}\pi y^{-1}\{\Gamma(1-\nu)\}^{-1}sec(\tfrac{1}{2}\pi\nu)\{\Psi(\nu)-\log y+\tfrac{1}{2}\pi cot(\tfrac{1}{2}\pi\nu)\}$
4.5	$\log(a-x)$ 0	$x<a$ $x>a$	$y^{-1}\Big(\log a-\sin(ay)Si(ay)-\cos(ay)\{Ci(ay)$ $\hspace{4cm}-\gamma-\log y\}\Big)$
4.6	$\log(a-x)$ 0 $a>b$	$x<b$ $x>b$	$y^{-1}\Big(\log a-\cos(by)\log(a-b)+\cos(ay)\{Ci(ay-by)$ $\hspace{1cm}-Ci(ay)\}+\sin(ay)\{Si(ay-by)-Si(ay)\}\Big)$
4.7	$\log(a+x)$ 0	$x<b$ $x>b$	$y^{-1}\Big(\log a-\cos(by)\log(a+b)+\cos(ay)\{Ci(ay+by)$ $\hspace{1cm}-Ci(ay)\}+\sin(ay)\{si(ay+by)-si(ay)\}\Big)$
4.8	$\log(a^2-x^2)$ 0	$x<a$ $x>a$	$y^{-1}\Big(\cos(ay)\{\gamma-\log(2a/y)+Ci(2ay)-2Ci(ay)\}$ $\hspace{0.8cm}+\sin(ay)\{Si(2ay)-2Si(ay)\}+2\log a\Big)$
4.9	$\log(a^2-x^2)$ 0 $a>b$	$x<b$ $x>b$	$y^{-1}\Big(2\log a-\cos(by)\log(a^2-b^2)-\cos(ay)\{2Ci(ay)$ $\hspace{1cm}-Ci(ay-by)-Ci(ay+by)\}-\sin(ay)\{2si(ay)$ $\hspace{1cm}-si(ay-by)-si(ay+by)\}\Big)$
4.10	$\log(2ax-x^2)$ 0	$x<2a$ $x>2a$	$2y^{-1}\sin(ay)\Big(\cos(ay)\{\gamma-\log(2a/y)+Ci(2ay)$ $\hspace{0.8cm}-2Ci(ay)\}+\sin(ay)\{Si(2ay)-2Si(ay)\}+2\log a\Big)$
4.11	$x^{-1}\log(2a-x)$ 0	$x<2a$ $x>2a$	$-\tfrac{1}{2}\pi\log a+si(\tfrac{1}{2}ay)Ci(\tfrac{1}{2}ay)$

	$f(x)$	$g_s(y)$
4.12	$(2ax-x^2)^{n-\frac{1}{2}}$ $\cdot \log(2ax-x^2)$ $\quad x<2a$ $\qquad\qquad 0 \qquad x>2a$ $n=1,2,\ldots$ for $n=0$, $\sum_{m=0}^{n-1}(\ )=0$	$(2n)!\pi(n!)^{-1}\sin(ay)(2y/a)^{-n}\Big(\tfrac{1}{2}\pi Y_n(ay)$ $+\tfrac{1}{2}n!\sum_{m=0}^{n-1}(\tfrac{1}{2}ay)^{m-n}\{m!(n-m)!\}^{-1}J_m(ay)$ $+J_n(ay)\{\log(\tfrac{1}{2}a/y)-\gamma+2\sum_{m=0}^{n-1}(2m+1)^{-1}\}\Big)$
4.13	$x(b^2+x^2)^{-1}\log(ax)$	$\tfrac{1}{2}\pi e^{-by}\log(ab)-\tfrac{1}{4}\pi\{e^{by}\mathrm{Ei}(-by)+e^{-by}\overline{\mathrm{Ei}}(by)\}$
4.14	$x(x^2-a^2)^{-1}\log(bx)$	$\tfrac{1}{2}\pi\Big(\cos(ay)\{\log(ab)-\mathrm{Ci}(ay)\}$ $\qquad\qquad\qquad -\sin(ay)\mathrm{si}(ay)\}\Big)$ (Cauchy principal value)
4.15	$x(x^2-a^2)^{-1}\log(x/a)$	$-\tfrac{1}{2}\pi\{\cos(ay)\mathrm{Ci}(ay)+\sin(ay)\mathrm{si}(ay)\}$
4.16	$e^{-ax}\log x$	$(a^2+y^2)^{-1}\{a\arctan(y/a)-\gamma y-\tfrac{1}{2}y\log(a^2+y^2)\}$
4.17	$x^{\nu-1}e^{-ax}\log x$ $\mathrm{Re}\,\nu>-1$	$\Gamma(\nu)(a^2+y^2)^{-\frac{1}{2}\nu}\Big(\cos\{\nu\arctan(\tfrac{y}{a})\}\arctan(\tfrac{y}{a})$ $+\sin\{\nu\arctan(y/a)\}\{\Psi(\nu)-\tfrac{1}{2}\log(a^2+b^2)\}\Big)$
4.18	$x^{-1}(\log x)^2$	$\tfrac{1}{2}\pi\gamma^2+(1/24)\pi^3+\pi\gamma\log y+\tfrac{1}{2}\pi(\log y)^2$
4.19	$\log(1+a/x)$	$y^{-1}\{\gamma+\log(ay)-\cos(ay)\mathrm{Ci}(ay)-\sin(ay)\mathrm{si}(ay)\}$
4.20	$\log\lvert(a+x)/(b-x)\rvert$	$y^{-1}\Big(\log(a/b)+\cos(by)\mathrm{Ci}(by)-\cos(ay)\mathrm{Ci}(ay)$ $+\sin(by)\mathrm{Si}(by)-\sin(ay)\mathrm{Si}(ay)$ $+\tfrac{1}{2}\pi\{\sin(by)+\sin(ay)\}\Big)$
4.21	$\log\lvert(a+x)/(a-x)\rvert$	$\pi y^{-1}\sin(ay)$
4.22	$x^{-2}\log\lvert(a+x)/(a-x)\rvert$	$\pi a^{-1}\{1-\cos(ay)-ay\,\mathrm{si}(ay)\}$
4.23	$\log\{(a^2+x^2)/(b^2+x^2)\}$	$y^{-1}\Big(2\log(a/b)+e^{by}\mathrm{Ei}(-by)-e^{ay}\mathrm{Ei}(-ay)$ $+e^{-by}\overline{\mathrm{Ei}}(by)-e^{-ay}\overline{\mathrm{Ei}}(ay)\Big)$

	$f(x)$	$g_s(y)$
4.24	$\log\left\lvert (x^2+a^2)/(x^2-b^2) \right\rvert$	$y^{-1}\Big(2\,\log(a/b)-e^{-ay}\mathrm{Ei}\,(ay)-e^{ay}\mathrm{Ei}\,(-ay)$ $+2\cos(by)\,\mathrm{Ci}\,(by)+2\sin(by)\,\mathrm{Si}\,(by)\Big)$
4.25	$\log(1+a^2/x^2)$	$y^{-1}\Big(2\gamma+2\,\log(ay)-e^{ay}\mathrm{Ei}\,(-ay)-e^{-ay}\overline{\mathrm{Ei}}\,(ay)\Big)$
4.26	$\log\lvert 1-a^2/x^2\rvert$	$2y^{-1}\Big(\gamma+\log(ay)-\cos(ay)\,\mathrm{Ci}\,(ay)$ $-\sin(ay)\,\mathrm{Si}\,(ay)\Big)$
4.27	$\log\left(\dfrac{a^2+(b+x)^2}{a^2+(b-x)^2}\right)$	$2\pi y^{-1}e^{-ay}\sin(by)$
4.28	$x^{-1}\log(1+a^2x^2)$	$-\pi\mathrm{Ei}\,(-y/a)$
4.29	$x^{-1}\log\lvert 1-a^2x^2\rvert$	$-\pi\mathrm{Ci}\,(y/a)$
4.30	$x^{-1}\log\lvert 1-a^2/x^2\rvert$	$\pi\{\gamma+\log(ay)-\mathrm{Ci}\,(ay)\}$
4.31	$x^{-1}\log\lvert (a^2+x^2)/(b^2-x^2)\rvert$	$\pi\{\mathrm{Ci}\,(by)-\mathrm{Ei}\,(-ay)-\log(b/a)\}$
4.32	$x^{-1}\log\left(\dfrac{c^2+a^2x^2}{c^2+b^2x^2}\right)$	$\pi\Big(\mathrm{Ei}\,(-cy/b)-\mathrm{Ei}\,(-cy/a)\Big)$
4.33	$\begin{array}{ll}(x^2-a^2)^{-\frac12}\log(x^2-a^2), & x>a \\ 0 & x<a\end{array}$	$-\tfrac12\pi\Big(\{\gamma+\log(2y/a)\}J_0(ay)+\tfrac12\pi Y_0(ay)\Big)$
4.34	$(x^2+2ax)^{-\frac12}\log(x^2+2ax)$	$-\tfrac12\pi\Big(\{\gamma+\log(2y/a)\}\{J_0(ay)\cos(ay)$ $+Y_0(ay)\sin(ay)\}+\tfrac12\pi\{Y_0(ay)\cos(ay)$ $-J_0(ay)\sin(ay)\}\Big)$
4.35	$\begin{array}{ll}(2ax-x^2)^{-\frac12}\log(2ax-x^2) & \\ & x<2a \\ 0 & x>2a\end{array}$	$\tfrac12\pi\sin(ay)\Big(\pi Y_0(ay)-2\{\gamma+\log(2y/a)\}J_0(ay)\Big)$
4.36	$\begin{array}{ll}x^{-1}\log\{x+(x^2-a^2)^{\frac12}\} & x>a \\ 0 & x<a\end{array}$	$\tfrac12\pi\Big\{\log a+\displaystyle\int_{ay}^{\infty}t^{-1}J_0(t)\,dt\Big\}$

	$f(x)$	$g_s(y)$
4.37	$\begin{aligned} & 0 \qquad\qquad x<a \\ & \log\left[\tfrac{1}{2}x^{-\frac{1}{2}}\{(x+a)^{\frac{1}{2}}+(x-a)^{\frac{1}{2}}\}\right] \\ & \qquad\qquad\qquad x>a \end{aligned}$	$\tfrac{1}{2}y^{-1}\Big[\mathrm{Ci}(ay)-\log 2-\tfrac{1}{2}\pi Y_0(ay)\Big]$
4.38	$\begin{aligned} & 0 \qquad\qquad x<1 \\ & (1+x)^{-\frac{1}{2}}\log\{x+(x^2-1)^{\frac{1}{2}}\} \\ & \qquad\qquad\qquad x>1 \end{aligned}$	$(\pi/y)^{\frac{1}{2}}\Big[\mathrm{Ci}(2y)\sin(y-\tfrac{1}{4}\pi)-\mathrm{si}(2y)\cos(y-\tfrac{1}{4}\pi)\Big]$
4.39	$x^{-1}\log\{a+(a^2+x^2)^{\frac{1}{2}}\}$	$\tfrac{1}{2}\pi\left[\log a + \displaystyle\int_{ay}^{\infty} t^{-1}\{I_0(t)-\mathbf{L}_0(t)\}dt\right]$
4.40	$\log\{\tfrac{1}{2}+\tfrac{1}{2}(1+a/x)^{\frac{1}{2}}\}$	$\begin{aligned} &\tfrac{1}{4}\pi y^{-1}\Big[\sin(\tfrac{1}{2}ay)J_0(\tfrac{1}{2}ay)-\cos(\tfrac{1}{2}ay)Y_0(\tfrac{1}{2}ay) \\ &+2\pi^{-1}\{\gamma+\log(\tfrac{1}{4}ay)\}\Big] \end{aligned}$
4.41	$\log\left[\tfrac{1}{2}x^{-1}\{x+(a^2+x^2)^{\frac{1}{2}}\}\right]$	$y^{-1}\Big[K_0(ay)+\gamma+\log(\tfrac{1}{2}ay)\Big]$
4.42	$(a^2+x^2)^{-\frac{1}{2}}\log\{x+(a^2+x^2)^{\frac{1}{2}}\}$	$\tfrac{1}{2}\pi\Big[K_0(ay)+\log a\{I_0(ay)-\mathbf{L}_0(ay)\}\Big]$
4.43	$(a^2+x^2)^{-\frac{1}{2}}\log\{(a^2+x^2)^{\frac{1}{2}}-x\}$	$\tfrac{1}{2}\pi\Big[\log a\{I_0(ay)-\mathbf{L}_0(ay)\}-K_0(ay)\Big]$
4.44	$\begin{aligned} & (a^2+x^2)^{-\frac{1}{2}} \\ & \cdot\log\left[(1+x^2/a^2)^{\frac{1}{2}}+x/a\right] \end{aligned}$	$\tfrac{1}{2}\pi K_0(ay)$
4.45	$\begin{aligned} & (a^2+x^2)^{-\frac{1}{2}}\exp\{-b(a^2+x^2)^{\frac{1}{2}}\} \\ & \cdot\log\{x/a+(1+x^2/a^2)^{\frac{1}{2}}\} \end{aligned}$	$\arctan(y/a)K_0\{a(b^2+y^2)^{\frac{1}{2}}\}$
4.46	$\log(1+e^{-ax})$	$\begin{aligned} &y^{-1}\{\log 2 -\tfrac{1}{4}\Psi(1+\tfrac{1}{2}iy/a)-\tfrac{1}{4}\Psi(1-\tfrac{1}{2}iy/a) \\ &+\tfrac{1}{4}\Psi(\tfrac{1}{2}+\tfrac{1}{2}iy/a)+\tfrac{1}{4}\Psi(\tfrac{1}{2}-\tfrac{1}{2}iy/a)\} \end{aligned}$
4.47	$\log(1-e^{-ax})$	$-y^{-1}\{\gamma+\tfrac{1}{2}\Psi(1+\tfrac{1}{2}iy/a)+\tfrac{1}{2}\Psi(1-\tfrac{1}{2}iy/a)\}$

2.5   Trigonometric Functions

	$f(x)$	$g_s(y)$	
5.1	$x^{-1}\sin(ax)$	$\tfrac{1}{2}\log\lvert(y+a)/(y-a)\rvert$	
5.2	$x^{-2}\sin(ax)$	$\tfrac{1}{2}\pi y$	$y<a$
		$\tfrac{1}{2}\pi a$	$y>a$
5.3	$x^{\nu-1}\sin(ax),\,-2<\mathrm{Re}\,\nu<1$	$\tfrac{1}{4}\pi\{\Gamma(1-\nu)\}^{-1}\mathrm{cosec}(\tfrac{1}{2}\pi\nu)\{\lvert y-a\rvert^{-\nu}-(y+a)^{-\nu}\}$	
5.4	$(1-x^2)^{-1}\sin(\pi x)$	$\sin y$	$y<\pi$
		0	$y>\pi$
5.5	$(b^2+x^2)^{-1}\sin(ax)$	$\tfrac{1}{2}\pi b^{-1}e^{ab}\sinh(by)$	$y<a$
		$\tfrac{1}{2}\pi b^{-1}e^{-by}\sinh(ab)$	$y>a$
5.6	$e^{-bx}\sin(ax)$	$\tfrac{1}{2}b\left[\{b^2+(a-y)^2\}^{-1}-\{b^2+(y+a)^2\}^{-1}\right]$	
5.7	$x^{-1}e^{-bx}\sin(ax)$	$\tfrac{1}{4}\log\left[\{b^2+(y+a)^2\}/\{b^2+(y-a)^2\}\right]$	
5.8	$\exp(-bx^2)\sin(ax)$	$\tfrac{1}{2}(\pi/b)^{\tfrac{1}{2}}\exp\left[-\tfrac{1}{4}(a^2+y^2)/b\right]\sinh(\tfrac{1}{2}ay/b)$	
5.9	$x^{-1}\sin^2(ax)$	$\tfrac{1}{4}\pi$	$x<2a$
		$\tfrac{1}{8}\pi$	$y=2a$
		0	$y>2a$
5.10	$x^{-1}\sin(ax)\sin(bx)$	0	$y<\lvert a-b\rvert$
		$\tfrac{1}{4}\pi$	$\lvert a-b\rvert<y<a+b$
		0	$y>a+b$
5.11	$x^{-2}\sin^2(ax)$	$\tfrac{1}{4}\{(y+2a)\log(y+2a)+(y-2a)\log\lvert y-2a\rvert-\tfrac{1}{2}y\,\log y\}$	
5.12	$x^{-3}\sin^3(ax)$	$\tfrac{1}{4}\pi y(2a-\tfrac{1}{2}y)$	$y<2a$
		$\tfrac{1}{4}\pi a^2$	$y>2a$

	$f(x)$	$g_s(y)$
5.13	$x^{-\nu}\sin(ax)\sin(bx)$ $a>b,\ 0<\mathrm{Re}\,\nu<4$	$\tfrac{1}{4}\Gamma(1-\nu)\cos(\tfrac{1}{2}\pi\nu)\Big[(y+a-b)^{\nu-1}-(y+a+b)^{\nu-1}$ $-\mathrm{sgn}(a-b-y)\,|y-a+b|^{\nu-1}$ $+\mathrm{sgn}(a+b-y|y-a-b|^{\nu-1}\Big]$
5.14	$x^{-4}\sin^3(ax)$	$1/(24)\,\pi y(9a^2-y^2)$ $\qquad\qquad y<a$ $1/(48)\,\pi\{24a^3-(3a-y)^3\}$ $\qquad a\le y\le 3a$ $\tfrac{1}{2}\pi a^3$ $\qquad\qquad\qquad\quad y\ge 3a$
5.15	$\left(x^{-1}\sin(ax)\right)^{2m}$ $m=1,2,\ldots$	$(-1)^m 2m2^{-2m}\Big[(m!)^{-2}y\,\log(y/a)$ $+\sum_{n=1}^{m}(-1)^n\{F_1(n)+F_2(n)\}\Big]$ $F_1=\{(m-n)!\,(m+n)!\}^{-1}(y-2an)^{2m-1}\log|2n-\tfrac{y}{a}|$ $F_2=\{(m-n)!\,(m+n)!\}^{-1}(y+2an)^{2m-1}\log(2n+\tfrac{y}{a})$
5.16	$\{x^{-1}\sin(ax)\}^{2m+1}$ $m=0,1,2,\ldots$	$(-1)^m 2^{-2m-1}(2m+1)\Big(\sum_{n=0}^{m}(-1)^n\{G_1(n)-G_2(n)\}\Big)$ $G_1=\{(m+1+n)!\,(m-n)!\}^{-1}\{(2n+1)a+y\}^{2m}\log(2n+1+\tfrac{y}{a})$ $G_2=\{(m+1+n)!\,(m-n)!\}^{-1}\{(2n+1)a-y\}^{2m}\log|2n+1-\tfrac{y}{a}|$
5.17	$e^{-ax}(\sin x)^{2n}$ $n=0,1,2,\ldots$	$-(-4)^{-n-1}(2n+1)^{-1}(A_n^{-1}+B_n^{-1})$ $\begin{aligned}A_n\\B_n\end{aligned}=\begin{pmatrix}n+\tfrac{1}{2}y\pm\tfrac{1}{2}ia\\ 2n+1\end{pmatrix}$
5.18	$e^{-ax}(\sin x)^{2n-1}$ $n=1,2,\ldots$	$-in^{-1}(-4)^{-n-1}(A_n^{-1}-B_n^{-1})$ $\begin{aligned}A_n\\B_n\end{aligned}=\begin{pmatrix}n-\tfrac{1}{2}+\tfrac{1}{2}y\pm\tfrac{1}{2}ia\\ 2n\end{pmatrix}$
5.19	$(a^2+x^2)^{-1}$ $\cdot\mathrm{cosec}(bx)$	$\tfrac{1}{2}\pi a^{-1}\sinh(by)\,\mathrm{cosech}(ab)\ ,\qquad y<a$ Cauchy principal value
5.20	$\{\sin(\pi x)\}^{\nu-1}$ $\qquad\qquad x<1$ $\qquad 0\qquad x>1$ $\mathrm{Re}\,\nu>0$	$2^{1-\nu}\sin(\tfrac{1}{2}y)\,\Gamma(\nu)\{\Gamma(\tfrac{1}{2}+\tfrac{1}{2}\nu+\tfrac{1}{2}i\tfrac{y}{\pi})\,\Gamma(\tfrac{1}{2}+\tfrac{1}{2}\nu-\tfrac{1}{2}i\tfrac{y}{\pi})\}^{-1}$

	$f(x)$	$g_s(y)$
5.21	$x^{-1}\cos(ax)$	$\begin{array}{ll} 0 & y<a \\ \tfrac{1}{4}\pi & y=a \\ \tfrac{1}{2}\pi & y>a \end{array}$
5.22	$x^{\nu-1}\cos(ax)$ $-1\ \mathrm{Re}\,\nu<1$	$\tfrac{1}{2}\Gamma(\nu)\sin(\tfrac{1}{2}\pi\nu)\left[(y+a)^{-\nu}+\mathrm{sgn}(y-a)\,|y-a|^{-\nu}\right]$
5.23	$x(b^2+x^2)^{-1}\cos(ax)$	$\begin{array}{ll} -\tfrac{1}{2}\pi e^{-ab}\sinh(by) & y<a \\ \tfrac{1}{2}\pi e^{-by}\cosh(ab) & y>a \end{array}$
5.24	$x^{-1}(1-2a\,\cos x+a^2)^{-1}$ $a<1$	$\begin{array}{c} \tfrac{1}{2}\pi(1+a)^{-1}(1-a)^{-2}(1+a-2a^{1-[y]}) \\ y\neq 0,1,2,\ldots \\ \tfrac{1}{2}\pi(1+a)^{-1}(1-a)^{-2}(1+a-a^y-a^{1+y}) \\ y=0,1,2,\ldots \end{array}$
5.25	$x(a^2+x^2)^{-1}\sec(bx)$	$-\tfrac{1}{2}\pi\,\mathrm{cosech}(ab)\sinh(ay) \qquad y<b$ (Cauchy principal value
5.26	$x^{-1}(a^2-x^2)^{-1}\sec(bx)$	$0\ ,\ y<b\ ,\qquad$ (Cauchy principal value)
5.27	$x^{-1}(a^2+x^2)^{-1}\sec(bx)$	$\tfrac{1}{4}\pi a^{-2}\,\mathrm{sech}(ab)\sinh(ay) \qquad y<b$ (Cauchy principal value)
5.28	$(\cosh a-\cos x)^{-1}\quad x<\pi$ $0 \qquad\qquad x>\pi$	$y\,\mathrm{cosech}(a)\cos(\pi y)\{\pi y^{-1}\mathrm{cosec}(\pi y)$ $-\sum_{n=0}^{\infty}(-1)^n\varepsilon_n(y^2-n^2)^{-1}e^{-na}\}$
5.29	$\sin(ax^2)$	$(\tfrac{1}{2}\pi/a)^{\tfrac{1}{2}}\{\cos u\,C(u)+\sin u\,S(u)\}\quad u=\tfrac{1}{4}y^2/a$
5.30	$\cos(ax^2)$	$(\tfrac{1}{2}\pi/a)^{\tfrac{1}{2}}\{\sin u\,C(u)-\cos u\,S(u)\}\quad u=\tfrac{1}{4}y^2/a$
5.31	$x^{-1}\sin(ax^2)$	$\tfrac{1}{2}\pi\{C(u)-S(u)\} \qquad\qquad u=\tfrac{1}{4}y^2/a$
5.32	$x^{-1}\cos(ax^2)$	$\tfrac{1}{2}\pi\{C(u)+S(u)\} \qquad\qquad u=\tfrac{1}{4}y^2/a$

	$f(x)$	$g_s(y)$
5.33	$x^{-\frac{1}{2}}\sin(ax^2)$	$-\frac{1}{2}\pi(\frac{1}{2}\frac{y}{a})^{\frac{1}{2}}\sin(u-\frac{3}{8}\pi)J_{\frac{1}{4}}(u)$ $\qquad\qquad u=\frac{1}{8}y^2/a$
5.34	$x^{-\frac{1}{2}}\cos(ax^2)$	$\frac{1}{2}\pi(\frac{1}{2}\frac{y}{a})^{\frac{1}{2}}\cos(u-\frac{3}{8}\pi)J_{\frac{1}{4}}(u)$ $\qquad\qquad u=\frac{1}{8}y^2/a$
5.35	$x^{\frac{1}{2}}\sin(ax^2)$	$\frac{1}{4}\pi(\frac{1}{2}y/a)^{3/2}\{\cos(u)J_{-\frac{1}{4}}(v)-\sin(u)J_{3/4}(v)\}$ $\qquad\qquad u=v-\frac{1}{8}\pi,v=\frac{1}{8}y^2/a$
5.36	$x^{\frac{1}{2}}\cos(ax^2)$	$\frac{1}{4}\pi(\frac{1}{2}y/a)^{3/2}\{\cos(u)J_{3/4}(v)+\sin(u)J_{-\frac{1}{4}}(v)\}$, $\qquad\qquad u,v,\text{ above}$
5.37	$x^{\nu-1}\sin(ax^2)$ $-3<\mathrm{Re}\nu<2$	$-\frac{1}{4}i(\frac{1}{2}\pi)^{\frac{1}{2}}(2a)^{-\frac{1}{2}\nu}\sec(\frac{1}{2}\pi\nu)\left(e^w\{D_{\nu-1}(v)-D_{\nu-1}(-v)\}\right.$ $\left.-e^{-w}\{D_{\nu-1}(u)-D_{\nu-1}(-u)\}\right)$ $w=\frac{1}{4}i(z^2-\pi\nu)$, $\dfrac{u}{v}=z(\pm i)^{\frac{1}{2}}$, $z=y(2a)^{-\frac{1}{2}}$
5.38	$x^{\nu-1}\cos(ax^2)$ $-1<\mathrm{Re}\nu<2$	$-\frac{1}{4}(\frac{1}{2}\pi)^{\frac{1}{2}}(2a)^{-\frac{1}{2}\nu}\sec(\frac{1}{2}\pi\nu)\left(e^w\{D_{\nu-1}(v)-D_{\nu-1}(-v)\}\right.$ $\left.+e^{-w}\{D_{\nu-1}(u)-D_{\nu-1}(-u)\}\right)$, u,v,w,z as above
5.39	$\sin(a^3x^3)$	$\dfrac{3^{-\frac{1}{2}}}{6a}\pi(\frac{y}{a})^{\frac{1}{2}}\{J_{1/3}(z)+J_{-1/3}(z)-3^{\frac{1}{2}}\pi^{-1}K_{1/3}(z)\}$, $\qquad\qquad z=2(\frac{1}{3}\frac{y}{a})^{3/2}$
5.40	$\cos(a^3x^3)$	$\dfrac{\pi}{6a}(\frac{y}{a})^{\frac{1}{2}}\left(I_{-\nu}(z)+I_\nu(z)-J_{-\nu}(z)+J_\nu(z)\right.$ $+2\{\mathbf{J}_\nu(z)-\mathbf{J}_{-\nu}(z)\}$ $\left.+i\mathbf{J}_\nu(iz)-i\mathbf{J}_{-\nu}(iz)\}\right)$, $\nu=1/3$, z as before
5.41	$x^\nu\sin(ax^c)$ $c<1$ $-c-2<\mathrm{Re}\nu<0$ $x^\nu\sin(ax^c)$ $c>1$ $-c-2<\mathrm{Re}\nu<c-1$	$y^{-\nu-c-1}\displaystyle\sum_{n=0}^{\infty}(-1)^n\cos\left[\frac{1}{2}\pi\{\nu+(2n+1)c\}\right]\Gamma\{1+\nu+(2n+1)c\}$ $\cdot y^{-2nc}a^{2n+1}/(2n+1)!$, $|a|<\infty$, $y>0$ $c^{-1}a^{-(\nu+2/c}\displaystyle\sum_{n=0}^{\infty}(-1)^n\sin\{\frac{1}{2}\pi(2+\nu+2n)/c\}\Gamma\{(2+\nu+2n)/c\}$ $\cdot a^{-2n/c}y^{2n+1}/(2n+1)!$, $|y|<\infty$, $a>0$
5.42	$x^\nu\cos(ax^c)$ $c<1$ $-2<\mathrm{Re}\nu<0$	$y^{-\nu-1}\displaystyle\sum_{n=0}^{\infty}(-1)^n\cos\{\frac{1}{2}\pi(\nu+2nc)\}\Gamma(1+\nu+2nc)y^{-2nc}$ $a^{2n}/(2n)!$, $|a|<\infty$, $y>0$

	$f(x)$	$g_s(y)$
5.43	$x^\nu \cos(ax^c)$, $c>1$ $-2<\mathrm{Re}\,\nu<c-1$	$c^{-1}a^{-(\nu+2)/c}\sum\limits_{n=0}^{\infty}(-1)^n\cos\left[\tfrac{1}{2}\pi\{(2+\nu+2n)/c\}\right]a^{-2\frac{n}{c}}$ $\cdot\,\Gamma\{(2+\nu+2n)/c\}y^{2n+1}/(2n+1)!$, $a>0$, $|y|<\infty$
5.44	$x^{-1}\sin(a/x)$	$\tfrac{1}{2}\pi Y_0\{2(ay)^{\frac{1}{2}}\}+K_0\{2(ay)^{\frac{1}{2}}\}$
5.45	$x^{-1}\cos(a/x)$	$\tfrac{1}{2}\pi J_0\{2(ay)^{\frac{1}{2}}\}$
5.46	$x^{-2}\sin(a/x)$	$\tfrac{1}{2}\pi(y/a)^{\frac{1}{2}}J_1\{2(ay)^{\frac{1}{2}}\}$
5.47	$x^{-\frac{1}{2}}\sin(a/x)$	$\tfrac{1}{2}(\tfrac{1}{2}\pi/y)^{\frac{1}{2}}(\sin z-\cos z+e^{-z})$, $z=2(ay)^{\frac{1}{2}}$
5.48	$x^{-1}\cos(a/x)$	$\tfrac{1}{2}(\tfrac{1}{2}\pi/y)^{\frac{1}{2}}(\sin z+\cos z+e^{-z})$, $z=2(ay)^{\frac{1}{2}}$
5.49	$x^{-3/2}\sin(a/x)$	$\tfrac{1}{2}(\tfrac{1}{2}\pi/a)^{\frac{1}{2}}\sin z-\cos z+e^{-z})$, $z=2(ay)^{\frac{1}{2}}$
5.50	$x^{-3/2}\cos(a/x)$	$\tfrac{1}{2}(\tfrac{1}{2}\pi/a)^{\frac{1}{2}}(\sin z+\cos z-e^{-z})$, $z=2(ay)^{\frac{1}{2}}$
5.51	$x^{\nu-1}\sin(a/x)$ $-2<\mathrm{Re}\,\nu<2$	$\tfrac{1}{2}\pi(a/y)^{\frac{1}{2}}\left[\sin(\tfrac{1}{2}\pi\nu)J_\nu(z)+\cos(\tfrac{1}{2}\pi\nu)Y_\nu(z)\right.$ $\left.+2\pi^{-1}\cos(\tfrac{1}{2}\pi\nu)K_\nu(z)\right]$, $\quad z=2(ay)^{\frac{1}{2}}$
5.52	$x^{\nu-1}\cos(a/x)$ $-2<\mathrm{Re}\,\nu<1$	$\tfrac{1}{2}\pi(a/y)^{\frac{1}{2}}\left[\cos(\tfrac{1}{2}\pi\nu)J_\nu(z)-\sin(\tfrac{1}{2}\pi\nu)Y_\nu(z)\right.$ $\left.+2\pi^{-1}\sin(\tfrac{1}{2}\pi\nu)K_\nu(z)\right]$, $\quad z=2(ay)^{\frac{1}{2}}$
5.53	$x^{-1}\sin(a/x)\log(bx)$	$\left[K_0(z)+\tfrac{1}{2}\pi Y_0(z)\right]\log\{b(a/y)^{\frac{1}{2}}\}$, $z=2(ay)^{\frac{1}{2}}$
5.54	$x^{-1}\cos(a/x)\log(bx)$	$\tfrac{1}{2}\pi\left[K_0(z)+\tfrac{1}{2}\pi J_0(z)\right]\log\{b(a/y)^{\frac{1}{2}}\}$, $z=2(ay)^{\frac{1}{2}}$
5.55	$x^{-1}\sin(ax^{\frac{1}{2}})$	$\pi\left[C(\tfrac{1}{4}a^2/y)-S(\tfrac{1}{4}a^2/y)\right]$
5.56	$x^{-1}\cos(ax^{\frac{1}{2}})$	$\pi\left[\tfrac{1}{2}-\{C(\tfrac{1}{4}a^2/y)\}^2-S\{(\tfrac{1}{4}a^2/y)\}^2\right]$
5.57	$x^{-\frac{1}{2}}\sin(ax^{\frac{1}{2}})$	$(2\pi/y)^{\frac{1}{2}}\{\cos z\,C(z)+\sin z\,S(z)\}$, $z=\tfrac{1}{4}a^2/y$

	$f(x)$	$g_s(y)$
5.58	$x^{-\frac{1}{2}}\cos(ax^{\frac{1}{2}})$	$(\pi/y)^{\frac{1}{2}}\cos(\frac{1}{4}\pi+\frac{1}{4}a^2/y)$
5.59	$x^{-3/4}\sin(ax^{\frac{1}{2}})$	$\pi(\frac{1}{2}a/y)^{\frac{1}{2}}\cos(\pi/8+\frac{1}{8}a^2/y)J_{\frac{1}{4}}(\frac{1}{8}a^2/y)$
5.60	$x^{-3/4}\cos(ax^{\frac{1}{2}})$	$\pi(\frac{1}{2}a/y)^{\frac{1}{2}}\sin(\pi/8-\frac{1}{8}a^2/y_{-\frac{1}{4}}(\frac{1}{8}a^2/y)$
5.61	$x^{\nu-1}\sin(ax^{\frac{1}{2}})$ $0<\mathrm{Re}\,\nu<5/2$	$-\frac{1}{2}i(\frac{1}{2}\pi)^{\frac{1}{2}}(2y)^{-\nu}\sec(\pi\nu)$ $\cdot\Big[\exp\{\frac{1}{4}i(z^2-2\pi\nu)\}\{D_{2\nu-1}(v)-D_{2\nu-1}(-v)\}$ $-\exp\{-\frac{1}{4}i(z^2-2\pi\nu)\}\{D_{2\nu-1}(u)-D_{2\nu-1}(-u)\}\Big]$ $z=a(2y)^{\frac{1}{2}},\quad {u\atop v}=z(\pm i)^{\frac{1}{2}}$
5.62	$x^{\nu-1}\cos(ax^{\frac{1}{2}})$ $0<\mathrm{Re}\,\nu<2$	$\frac{1}{2}i(\frac{1}{2}\pi)^{\frac{1}{2}}(2y)^{-\nu}\mathrm{cosec}(\pi\nu)$ $\cdot\Big[\exp\{\frac{1}{4}i(z^2-2\pi\nu)\}\{D_{2\nu-1}(v)+D_{2\nu-1}(-v)\}$ $-\exp\{-\frac{1}{4}i(z^2-2\pi\nu)\}\{D_{2\nu-1}(u)+D_{2\nu-1}(-u)\}$ $z=a(2y)^{\frac{1}{2}},\quad {u\atop v}=z(\pm i)^{\frac{1}{2}}$
5.63	$\exp(-bx^{\frac{1}{2}})\sin(ax^{\frac{1}{2}})$	$\frac{1}{2}ay^{-1}(\frac{1}{2}\pi/y)^{\frac{1}{2}}\exp(-\frac{1}{4}a^2/y)$
5.64	$e^{-bx}\sin(ax^{\frac{1}{2}})$	$-\frac{1}{2}\pi^{\frac{1}{2}}a(b^2+y^2)^{-3/4}\exp\{-\frac{1}{4}a^2b/(b^2+y^2)\}$ $\cdot\sin\{\frac{1}{4}a^2y(b^2+y^2)^{-1}-\frac{3}{2}\arctan(y/b)\}$
5.65	$x^{-1}\exp(-ax^{\frac{1}{2}})\sin(ax^{\frac{1}{2}})$	$-\frac{1}{2}i\pi\mathrm{Erfc}\{a(2y)^{-\frac{1}{2}}\}\mathrm{Erf}\{ia(2y)^{-\frac{1}{2}}\}$
5.66	$x^{-1}\exp(-ax^{\frac{1}{2}})\cos(ax^{\frac{1}{2}})$	$\frac{1}{2}\pi\mathrm{Erfc}\{a(2y)^{-\frac{1}{2}}\}$
5.67	$x^{-\frac{1}{2}}\exp(-ax^{\frac{1}{2}})$ $\cdot\{\cos(ax^{\frac{1}{2}})+\sin(ax^{\frac{1}{2}})\}$	$(\frac{1}{2}\pi/y)^{\frac{1}{2}}\exp(-\frac{1}{4}a^2/y)$
5.68	$x^{\nu-1}\exp(-ax^{\frac{1}{2}})$ $\cos(ax^{\frac{1}{2}}-\frac{1}{2}\pi\nu)\ ,\ \mathrm{Re}\,\nu>-1$	$(\frac{1}{2}\pi)^{\frac{1}{2}}(2y)^{-\nu}\exp(-\frac{1}{4}a^2/y)$ $\cdot D_{2\nu-1}(ay^{-\frac{1}{2}})$
5.69	$x^{-2/3}\sin(ax^{1/3})$	$\frac{1}{2}\pi3^{-\frac{1}{2}}(\frac{a}{y})^{\frac{1}{2}}\{J_{\frac{1}{3}}(z)+J_{-\frac{1}{3}}(z)-\pi^{-1}3^{\frac{1}{2}}K_{\frac{1}{3}}(z)\},\ z=2(\frac{a}{3})^{\frac{3}{2}}y^{-\frac{1}{2}}$

	$f(x)$	$g_s(y)$
5.70	$x^{-2/3}\cos(ax^{1/3})$	$\frac{1}{2}\pi\left(\frac{a}{y}\right)^{\frac{1}{2}}\Big\{I_{-\nu}(z)+I_{\nu}(z)+J_{-\nu}(z)-J_{\nu}(z)$ $-2\{J_{\nu}(z)-J_{-\nu}(z)-iJ_{\nu}(iz)+iJ_{-\nu}(iz)\}\Big\}$ $\nu=1/3$, $2(\frac{1}{3}a)^{3/2}y^{-\frac{1}{2}}$
5.71	$r^{-1}\sin(br)$	$\int_c^b (t^2+y^2-b^2)^{-\frac{1}{2}}\cos(at)dt \qquad y<b$ $\int_0^b (t^2+y^2-b^2)^{-\frac{1}{2}}\cos(at)dt \qquad y>b$ $c=(b^2-y^2)^{\frac{1}{2}}$
5.72	$r^{-1}\cos(br)$	$-\int_0^{\frac{1}{2}\pi}\cos(ab\cos t)\sinh(ay\sin t)dt \quad y<b$ $\frac{1}{2}\pi I_0(aS)-\int_0^{\frac{1}{2}\pi}\cos(ab\cos t)\sinh(ay\sin t)dt,$ $\qquad\qquad y>b$
5.73	$\exp\{-a^2b(b^2+x^2)^{-1}\}$ $\cdot\sin\{a^2x(b^2+x^2)^{-1}\}$	$\frac{1}{4}\pi ay^{-\frac{1}{2}}e^{-by}J_1(2ay^{\frac{1}{2}})$
5.74	$x(b^2+x^2)^{-1}r^{-1}\sin(cr)$	$\frac{1}{2}\pi(a^2-b^2)^{-\frac{1}{2}}e^{-by}\sin\{c(a^2-b^2)^{\frac{1}{2}}\} \qquad y<c$
5.75	$xr^{-2}\cos\{c(b^2+x^2)^{\frac{1}{2}}\}$	$\frac{1}{2}\pi e^{-ay}\cos\{c(b^2-a^2)^{\frac{1}{2}}\} \qquad y>c$
5.76	$x^{-\frac{1}{2}}r^{-1}\sin(br)$	$\frac{1}{2}\pi(\frac{1}{2}\pi y)^{\frac{1}{2}}J_{-\frac{1}{4}}(u)J_{\frac{1}{4}}(v)$, $\frac{u}{v}=\frac{1}{2}ab\pm\frac{1}{2}as$, $y<b$
5.77	$x^{-\frac{1}{2}}r^{-1}\cos(br)$	$-\frac{1}{2}\pi(\frac{1}{2}\pi y)^{\frac{1}{2}}Y_{-\frac{1}{4}}(u)J_{\frac{1}{4}}(v)$, $\frac{u}{v}=\frac{1}{2}ab\pm\frac{1}{2}as$, $y<b$
5.78	$r^{-1}(r-a)^{\frac{1}{2}}\sin(br)$	$(\frac{1}{2}\pi)^{\frac{1}{2}}ys^{-1}(b+s)^{-\frac{1}{2}}\cos(as+\frac{1}{4}\pi) \qquad y<b$ $(\frac{1}{2}\pi/y)^{\frac{1}{2}}s^{-1}e^{-as}\sin\{\frac{1}{2}\arcsin(\frac{b}{y})\} \qquad y>b$
5.79	$r^{-1}(r-a)^{\frac{1}{2}}\cos(br)$	$-(\frac{1}{2}\pi)^{\frac{1}{2}}ys^{-1}(b+s)^{-\frac{1}{2}}\sin(as+\frac{1}{4}\pi) \qquad y<b$ $(\frac{1}{2}\pi/y)^{\frac{1}{2}}s^{-1}e^{-as}\cos\{\frac{1}{2}\arcsin(\frac{b}{y})\} \qquad y>b$

$$r=(a^2+x^2)^{\frac{1}{2}} \qquad\qquad\qquad s=(b^2-y^2)^{\frac{1}{2}}$$

	$f(x)$	$g_s(y)$
5.80	$r^{-1}\{(r+x)^{\nu}-(r-x)^{\nu}\}$ $\cdot\sin(br)$, $-1<\mathrm{Re}\,\nu<1$	$-\tfrac{1}{2}\pi a^{\nu}\left(\{(b+y)/(b-y)\}^{\frac{1}{2}\nu}-\{(b-y)/(b+y)\}^{\frac{1}{2}\nu}\right)$ $\cdot\{\sin(\tfrac{1}{2}\pi\nu)J_{\nu}(as)+\cos(\tfrac{1}{2}\pi\nu)Y_{\nu}(as)\}$ $y<b$ $a^{\nu}\cos(\tfrac{1}{2}\pi\nu)\left(\{(y+b)/(y-b)\}^{\frac{1}{2}\nu}\right.$ $\left.-\{(y-b)/(y+b)\}^{\frac{1}{2}\nu}\right)K_{\nu}(aS)$ $y>b$
5.81	$r^{-1}\{(r+x)^{\nu}-(r-x)^{\nu}\}$ $\cdot\cos(br)$, $-1<\mathrm{Re}\,\nu<1$	$-\tfrac{1}{2}\pi a^{\nu}\left(\{(b+y)/(b-y)\}^{\frac{1}{2}\nu}-\{(b-y)/(b+y)\}^{\frac{1}{2}\nu}\right)$ $\cdot\{\cos(\tfrac{1}{2}\pi\nu)J_{\nu}(as)-\sin(\tfrac{1}{2}\pi\nu)Y_{\nu}(as)\}$ $y<b$ $a^{\nu}\sin(\tfrac{1}{2}\pi\nu)\left(\{(y+b)/(y-b)\}^{\frac{1}{2}\nu}\right.$ $\left.+\{(y-b)/(y+b)\}^{\frac{1}{2}\nu}\right)K_{\nu}(aS)$ $y>b$
5.82	$r^{-1}\log\{(x+r)/a\}\sin(br)$	$-\tfrac{1}{4}\pi\log\{(b+y)/(b-y)\}Y_{o}(as)$ $y<b$ $\tfrac{1}{2}\log\{(y+b)/(y-b)\}K_{o}(aS)$ $y>b$
5.83	$r^{-1}\log\{(x+r)/a\}\cos(br)$	$-\tfrac{1}{4}\pi\log\{(b+y)/(b-y)\}J_{o}\{a(b^{2}-y^{2})^{\frac{1}{2}}\}$,$y<b$ $\tfrac{1}{2}\pi K_{o}\{a(y^{2}-b^{2})^{\frac{1}{2}}\}$,$y>b$
5.84	$v^{-1}\cos(bv)$	$-\sin(ay)K_{o}(as)$,$y<b$ $\tfrac{1}{2}\pi\{\cos(ay)J_{o}(aS)+\sin(ay)Y_{o}(aS)\}$,$y>b$
5.85	$v^{-3/2}\cos(bv)$	$\tfrac{1}{2}\pi(\tfrac{1}{2}\pi b)^{\frac{1}{2}}J_{-\frac{1}{4}}(\tfrac{1}{2}ay-\tfrac{1}{2}aS)\{\cos(ay)J_{\frac{1}{4}}(\tfrac{1}{2}ay+\tfrac{1}{2}aS)$ $+\sin(ay)Y_{\frac{1}{4}}(\tfrac{1}{2}ay+\tfrac{1}{2}aS)\}$ $y>b$
5.86	$v^{-3/2}\sin(bv)$	$\tfrac{1}{2}\pi(\tfrac{1}{2}\pi b)^{\frac{1}{2}}J_{\frac{1}{4}}(\tfrac{1}{2}ay-\tfrac{1}{2}aS)\{\cos(ay)J_{-\frac{1}{4}}(\tfrac{1}{2}ay+\tfrac{1}{2}aS)$ $+\sin(ay)Y_{\frac{1}{4}}(\tfrac{1}{2}ay+\tfrac{1}{2}aS)\}$ $y>b$
5.87	$x^{-\frac{1}{2}}W^{-1}\cos(bw)$ $x<a$ 0 $x>a$	$\tfrac{1}{2}\pi(\tfrac{1}{2}\pi y)^{\frac{1}{2}}J_{\frac{1}{4}}(\tfrac{1}{2}ap+ab)J_{\frac{1}{4}}(\tfrac{1}{2}ap-\tfrac{1}{2}ab)$
5.88	$xw^{-1}\cos(bw)$ $x<a$ 0 $x>a$	$\tfrac{1}{2}\pi ayr^{-1}J_{o}(ap)$
5.89	$v^{-1}\cos(bv)$ $x<2a$ 0 $x>2a$	$\pi\sin(ay)J_{o}(ap)$

$$r=(a^{2}+x^{2})^{\frac{1}{2}},\ v=(2ax-x^{2})^{\frac{1}{2}},\ V=(2ax+x^{2})^{\frac{1}{2}},\ p=(y^{2}+b^{2})^{\frac{1}{2}},\ s=(b^{2}-y^{2})^{\frac{1}{2}}$$
$$w=(a^{2}-x^{2})^{\frac{1}{2}},\ W=(x^{2}-a^{2})^{\frac{1}{2}}\qquad\qquad S=(y^{2}-b^{2})^{\frac{1}{2}}$$

	$f(x)$		$g_s(y)$
5.90	$v^{-3/2}\cos(bv)$	$x<2a$	$\pi(\tfrac{1}{2}\pi b)^{\tfrac{1}{2}}\sin(ay)J_{-\tfrac{1}{4}}(\tfrac{1}{2}ap+\tfrac{1}{2}ay)J_{-\tfrac{1}{4}}(\tfrac{1}{2}ap-\tfrac{1}{2}ay)$
	0	$x>2a$	
5.91	$v^{-3/2}\sin(bv)$	$x<2a$	$\pi(\tfrac{1}{2}\pi b)^{\tfrac{1}{2}}\sin(ay)J_{\tfrac{1}{4}}(\tfrac{1}{2}ap+\tfrac{1}{2}ay)J_{\tfrac{1}{4}}(ap-\tfrac{1}{2}ay)$
	0	$x>2a$	
5.92	$\sin(bv)$	$x<2a$	$\pi ab\,\sin(ay)p^{-1}J_1(ap)$
	0	$x>2a$	
5.93	0	$x<a$	$\tfrac{1}{2}(\pi y)^{\tfrac{1}{2}}I_{\tfrac{1}{4}}(\tfrac{1}{2}ab-\tfrac{1}{2}as)K_{\tfrac{1}{4}}(\tfrac{1}{2}ab+\tfrac{1}{2}as)\qquad y<b$
	$x^{-\tfrac{1}{2}}w^{-1}\cos(bW)$	$x>a$	
5.94	0	$x<a$	$\tfrac{1}{2}\pi(\tfrac{1}{2}\pi b)^{\tfrac{1}{2}}J_{\tfrac{1}{4}}(\tfrac{1}{2}ay-\tfrac{1}{2}aS)J_{-\tfrac{1}{4}}(\tfrac{1}{2}ay+\tfrac{1}{2}aS)\qquad y>b$
	$W^{-3/2}\sin(bW)$	$x>a$	
5.95	0	$x<a$	$\tfrac{1}{2}\pi(\tfrac{1}{2}\pi b)^{\tfrac{1}{2}}J_{\tfrac{1}{4}}(\tfrac{1}{2}ay+\tfrac{1}{2}aS)J_{-\tfrac{1}{4}}(\tfrac{1}{2}ay-\tfrac{1}{2}aS)\qquad y>b$
	$W^{-3/2}\cos(bW)$	$x>a$	
5.96	$x^{-\tfrac{1}{2}}w^{-1}\sin(bw)$	$x<a$	$\tfrac{1}{2}\pi(\tfrac{1}{2}\pi y)^{\tfrac{1}{2}}J_{\tfrac{1}{4}}(\tfrac{1}{2}ap-\tfrac{1}{2}ab)Y_{\tfrac{1}{4}}(\tfrac{1}{2}ap+\tfrac{1}{2}ab)$
	$-x^{-\tfrac{1}{2}}w^{-1}\exp(-bW)$	$x>a$	
5.97	0	$x<a$	$\tfrac{1}{2}\pi c^{-1}(c^2+a^2)^{-\tfrac{1}{2}}\exp\{-b(c^2+a^2)^{\tfrac{1}{2}}\}\sinh(cy),$
	$(x^2+c^2)^{-1}w^{-1}\cos(bW)$	$x>a$	$y<b$
5.98	$x^{-3}\log|\cos(bx|$		$\tfrac{1}{2}\pi y\left(\tfrac{1}{2}y\,\log 2-b+\tfrac{1}{2}\sum_{n=1}^{m}(-1)^n n^{-1}(y-4bn)\right)$
			$m\leq(\tfrac{1}{2}y/b)<m+1;m=1,2,\dots;\text{For }m=0,\sum(\ )=0$
5.99	$x^{-1}(a^2+x^2)^{-1}$		$\tfrac{1}{2}\pi a^{-2}\left\{\sinh(ay)\log(1-e^{-2ab})+(1-e^{-ay})\right.$
	$\cdot\log|c\sin(bx)|$		$\left.\cdot\log(\tfrac{c}{2})+\sum_{n=1}^{m}n^{-1}\{\cosh(ay-2abn)-\cosh(2abn)\}\right\}$
			$m\leq(\tfrac{1}{2}y/b)<m+1;m=1,2,\dots;\text{For }m=0,\sum(\ )=0$

$r=(a^2+x^2)^{\tfrac{1}{2}}$, $v=(2ax-x^2)^{\tfrac{1}{2}}$, $V=(2ax+x^2)^{\tfrac{1}{2}}$ $\qquad p=(y^2+b^2)^{\tfrac{1}{2}}$, $s=(b^2-y^2)^{\tfrac{1}{2}}$

$w=(a^2-x^2)^{\tfrac{1}{2}}$, $W=(x^2-a^2)^{\tfrac{1}{2}}$ $\qquad\qquad S=(y^2-b^2)^{\tfrac{1}{2}}$

	$f(x)$	$g_s(y)$
5.100	$x^{-1}(a^2+x^2)^{-1}$ $\log\lvert c\,\cos(bx)\rvert$	$\tfrac{1}{2}\pi a^{-2}\Big(\sinh(ay)\log(1+e^{-2ab})$ $+(1-e^{-ay})\log(\tfrac{c}{2})$ $+\sum\limits_{n=1}^{m}(-1)^n n^{-1}\{\cosh(ay-2abn)-\cosh(2abn)\}\Big)$ $m\le(\tfrac{1}{2}y/b)<m+1;m=1,2,\ldots;\text{For }m=0,\sum(\ )=0$
5.101	$x^{-1}\log\Big(a\cosh c\pm a\cosh(bx)\Big)$ $c\ge0$	$\tfrac{1}{2}\pi\Big[\tfrac{1}{2}\{c+\log(\tfrac{1}{2}a)\}-\sum\limits_{n=1}^{m}(\mp1)^n n^{-1}e^{-nc}\Big]$ $m\le y/b<m+1;m=1,2,\ldots;\text{For }m=0,\sum(\ )=0$
5.102	$x^{-1}(a^2+x^2)^{-1}$ $\log\Big(c\cosh u\pm c\cos(bx)\Big)$ $u\ge0$	$\pi a^{-2}\Big(\tfrac{1}{2}(1-e^{-ay})\{u+\log(\tfrac{c}{2})\}+\log(1\pm e^{-ab-u})$ $\cdot\sinh(ay)+\sum\limits_{n=1}^{m}(\mp1)^n n^{-1}e^{-nu}\{\cosh(ay-abn)$ $-\cosh(abn)\}\Big);m\le\tfrac{y}{b} m+1.m=1,2,\ldots;$ $\text{For }m=0,\sum=0$
5.103	$x(a^2+x^2)^{-1}$ $\cdot\log\Big(c\cosh u\pm c\cos(bx)\Big)$ $u\ge0$	$\pi\Big(\tfrac{1}{2}\{u+\log(\tfrac{c}{2})\}-\sinh(ay)\log(1\pm e^{-ab-u})$ $-\sum\limits_{n=1}^{n}(\mp1)^n n^{-1}e^{-nu}\cosh(ay-abn)\Big)$ $m\le\tfrac{y}{b}<m+1;m=1,2,\ldots;\text{For }m=0,\ \sum(\ )=0$
5.104	$\exp(a\sin x)\qquad x<\pi$ $0\qquad\qquad x>\pi$	$-\tfrac{1}{2}i\pi\sec(\tfrac{1}{2}\pi y)\Big[\cos(\tfrac{1}{2}\pi Y)\{J_y(ia)-J_{-y}(ia)\}$ $+i\,\sin(\tfrac{1}{2}\pi y)\{J_{-y}(ia)+J_y(ia)\}\Big]$
5.105	$\exp(-a\sin x)\qquad x<\pi$ $0\qquad\qquad x>\pi$	$\tfrac{1}{2}i\pi\sec(\tfrac{1}{2}\pi y)\Big[\cos(\tfrac{1}{2}\pi y)\{J_y(ia)-J_{-y}(ia)\}$ $-i\,\sin(\tfrac{1}{2}\pi y)\{J_y(ia)+J_{-y}(ia)\}\Big]$
5.106	$\sin(a\sin x)\qquad x<\pi$ $0\qquad\qquad x>\pi$	$\sin(\pi y)s_{0,y}(a)=\tfrac{1}{2}\pi\{J_y(a)-J_{-y}(a)\}$ $=-\tfrac{1}{2}\pi\tan(\tfrac{1}{2}\pi y)\{E_y(a)+E_{-y}(a)\}$
5.107	$\cos(a\sin x)\qquad x<\pi$ $0\qquad\qquad x>\pi$	$-y\{1-\cos(\pi y)\}s_{-1,y}(a)=\tfrac{1}{2}\pi\tan(\tfrac{1}{2}\pi y)$ $\cdot\{J_y(a)+J_{-y}(a)\}=\tfrac{1}{2}\pi\{E_y(a)-E_{-y}(a)\}$

	$f(x)$	$g_s(y)$
5.108	$(\sin x)^{-\frac{1}{2}}\exp(-a \sin x),$ $\quad x<\pi$ $\qquad\qquad\quad 0 \qquad x>\pi$	$\pi(\frac{1}{2}\pi a)^{\frac{1}{2}}\sin(\frac{1}{2}\pi y)\{I_{-\frac{1}{4}-\frac{1}{2}y}(\frac{1}{2}a)I_{-\frac{1}{4}+\frac{1}{2}y}(\frac{1}{2}a)$ $-I_{\frac{1}{4}-\frac{1}{2}y}(\frac{1}{2}a)I_{\frac{1}{4}+\frac{1}{2}y}(\frac{1}{2}a)\}$
5.109	$(\sin x)^{-\frac{1}{2}}\exp(a \sin x)$ $\quad x<\pi$ $\qquad\qquad\quad 0 \qquad x>\pi$	$\pi(\frac{1}{2}\pi a)^{\frac{1}{2}}\sin(\frac{1}{2}\pi y)\{I_{-\frac{1}{4}-\frac{1}{2}y}(\frac{1}{2}a)I_{-\frac{1}{4}+\frac{1}{2}y}(\frac{1}{2}a)$ $+I_{\frac{1}{4}-\frac{1}{2}y}(\frac{1}{2}a)I_{\frac{1}{4}+\frac{1}{2}y}(\frac{1}{2}a)\}$
5.110	$(\sin x)^{-\frac{1}{2}}\sin(a \sin x),$ $\quad x<\pi$ $\qquad\qquad\quad 0 \qquad x>\pi$	$\pi(\frac{1}{2}\pi a)^{\frac{1}{2}}\sin(\frac{1}{2}\pi y)J_{\frac{1}{4}-\frac{1}{2}y}(\frac{1}{2}a)J_{\frac{1}{4}+\frac{1}{2}y}(\frac{1}{2}a)$
5.111	$(\sin x)^{-\frac{1}{2}}\cos(a \sin x),$ $\quad x<\pi$ $\qquad\qquad\quad 0 \qquad x>\pi$	$\pi(\frac{1}{2}\pi a)^{\frac{1}{2}}\sin(\frac{1}{2}\pi y)J_{-\frac{1}{4}-\frac{1}{2}y}(\frac{1}{2}a)J_{-\frac{1}{4}+\frac{1}{2}y}(\frac{1}{2}a)$
5.112	$\exp(ia \cos x)$ $\quad x<\pi$ $\qquad\qquad\quad 0 \qquad x>\pi$	Nielsen,N.:Die Zylinderfunktionen und ihre Anwendungen,Leipzig,1904
5.113	$\log\{\sin(\pi x)\}$ $\quad x<1$ $\qquad\qquad 0 \qquad x>1$	$-2y^{-1}\sin^2(\frac{1}{2}y)\{\gamma+\log 2+\frac{1}{2}\Psi(1+\frac{1}{2}y/\pi)$ $+\frac{1}{2}\Psi(1-\frac{1}{2}y/\pi)\}$
5.114	$\{\sin(\pi x)\}^{\nu-1}$ $\cdot\log\{\operatorname{cosec}(\pi x)\}\quad x<1$ $\qquad\qquad\quad 0 \qquad x>1$	$2^{1-\nu}\Gamma(\nu)\sin(\frac{1}{2}y)\left[\Gamma(\frac{1}{2}+\frac{1}{2}\nu+\frac{1}{2}\frac{y}{\pi})\,\Gamma(\frac{1}{2}+\frac{1}{2}\nu-\frac{1}{2}\frac{y}{\pi})\right]^{-1}$ $\cdot\left[\log 2+\frac{1}{2}\Psi(\frac{1}{2}+\frac{1}{2}\nu+\frac{1}{2}\frac{y}{\pi})+\frac{1}{2}\Psi(\frac{1}{2}+\frac{1}{2}\nu-\frac{1}{2}\frac{y}{\pi})-\Psi(\nu)\right]$
5.115	$(\sin x)^{-3/2}$ $\sin(2a \sin x)\quad x<\pi$ $\qquad\qquad\quad 0 \qquad x>\pi$	$2(\pi a)^{3/2}\sin(\frac{1}{2}\pi y)\left[J_{-\frac{1}{4}+\frac{1}{2}y}(a)J_{-\frac{1}{4}-\frac{1}{2}y}(a)\right.$ $\left.+J_{\frac{3}{4}+\frac{1}{2}y}(a)J_{\frac{3}{4}-\frac{1}{2}y}(a)\right]$
5.116	$z^{-\frac{1}{2}},\ a/b<\sin x<\pi-a/b;$ $0\ ,\qquad \sin x<a/b$	$b^{-1}\sin(\frac{1}{2}\pi y)P_{-\frac{1}{2}+\frac{1}{2}y}(2a^2b^{-2}-1)$

$$z=(b^2\sin^2 x-a^2)$$

2.6   Inverse Trigonometric Functions

	$f(x)$		$g_s(y)$
6.1	$\arcsin(x/a)$	$x<a$	$\frac{1}{2}\pi y^{-1}\{J_0(ay)-\cos(ay)\}$
	0	$x>a$	
6.2	$\arccos(x/a)$	$x<a$	$\frac{1}{2}\pi y^{-1}\{1-J_0(ay)\}$
	0	$x>a$	
6.3	$x^{-2}\arctan(ax)$		$\frac{1}{2}\pi\{a-y\mathrm{Ei}(-y/a)-e^{-y/a}\}$
6.4	$(1-x^2)^{-\frac{1}{2}}\cos(\nu\arccos x)$		$\frac{1}{4}\pi\operatorname{cosec}(\frac{1}{2}\pi\nu)\{\mathbf{J}_\nu(y)-\mathbf{J}_{-\nu}(y)\}$
		$x<1$	
	0	$x>1$	
6.5	$x^{-\frac{1}{2}}(1-x^2)^{-\frac{1}{2}}$		$\frac{1}{2}\pi(\frac{1}{2}\pi y)^{\frac{1}{2}}J_{\frac{1}{2}\nu+\frac{1}{4}}(\frac{1}{2}y)\,J_{-\frac{1}{2}\nu+\frac{1}{4}}(\frac{1}{2}y)$
	$\cos(\nu\arccos x)$	$x<1$	
	0	$x>1$	
6.6	$(1+x^2)^{-\frac{1}{2}\nu}$		$\frac{1}{2}\pi\{\Gamma(\nu)\}^{-1}y^{\nu-1}e^{-y}$
	$\sin(\nu\arctan x),\ \mathrm{Re}\,\nu>0$		
6.7	$x^\nu(1+x^2)^{\frac{1}{2}\nu}$		$\pi^{-\frac{1}{2}}\sin(\pi\nu)\,\Gamma(1+\nu)\,y^{-\nu-\frac{1}{2}}\sin(\frac{1}{2}y)\,K_{\frac{1}{2}+\nu}(\frac{1}{2}y)$
	$\sin(\nu\operatorname{arccot} x),\ -1<\mathrm{Re}\,\nu<\frac{1}{2}$		
6.8	$x^\nu(1+x^2)^{\frac{1}{2}\nu}$		$\frac{1}{2}\pi^{\frac{1}{2}}\Gamma(1+\nu)\,y^{-\nu-\frac{1}{2}}\Big(\cosh(\frac{1}{2}y)\,I_{-\nu-\frac{1}{2}}(\frac{1}{2}y)$
	$\cos(\nu\operatorname{arccot} x),\ -1<\mathrm{Re}\,\nu<\frac{1}{2}$		$\qquad -\sinh(\frac{1}{2}y)\,I_{\nu+\frac{1}{2}}(\frac{1}{2}y)\Big)$
6.9	$\arctan(a/x)$		$\pi y^{-1}e^{-\frac{1}{2}ay}\sinh(\frac{1}{2}ay)$
6.10	$\arctan\{ax(b^2+x^2)^{-1}\}$		$\pi y^{-1}\exp\{-y(b^2+\frac{1}{4}a^2)^{\frac{1}{2}}\}\sinh(\frac{1}{2}ay)$
6.11	$\arctan\,(a/x)^n\ ,n=1,3,5$		$\frac{1}{2}\pi y^{-1}\Big(1+\sum_{m=1}^{n}(-1)^m\exp\{-ay\,\sin(m-\frac{1}{2})\pi/n\}$
			$\qquad\cdot\cos\{ay\,\cos(m-\frac{1}{2})\pi/n\}$

	$f(x)$	$g_s(y)$
6.12	$\arctan(x^{-\frac{1}{2}})$	$\frac{1}{2}\pi y^{-1}\Big[1-\{1-C(u)-S(u)\}\cos u$ $-\{C(u)-S(u)\}\sin u\Big]$, $\quad u=a^2y$
6.13	$(a^2-x^2)^{-\frac{1}{2}}\arccos(\frac{x}{a})$, $x<a$ $(x^2-a^2)^{-\frac{1}{2}}\log u \quad x>a$ $u=\{x+(x^2-a^2)^{\frac{1}{2}}\}/a$	$\frac{1}{4}\pi^2\{H_0(ay)-Y_0(ay)\}$
6.14	$(a^2-x^2)^{-\frac{1}{2}}\arcsin(\frac{x}{a})$, $x<a$ $(x^2-a^2)^{-\frac{1}{2}}\log u \quad x>a$ $u=\{x-(x^2-a^2)^{\frac{1}{2}}\}/a$	$\frac{1}{4}\pi^2 Y_0(ay)$

2.7 Hyperbolic Functions

	$f(x)$	$g_s(y)$
7.1	$\mathrm{sech}(ax)$	$-\frac{1}{2}\pi a^{-1}\tanh(\frac{1}{2}\pi y/a)$ $-\frac{1}{2}ia^{-1}\{\Psi(\frac{1}{4}+\frac{1}{4}iy/a)-\Psi(\frac{1}{4}-\frac{1}{4}iy/a)\}$
7.2	$\{\mathrm{sech}(ax)\}^2$	$\frac{1}{4}ya^{-2}\Big[\Psi(\frac{1}{4}iy/a)+\Psi(-\frac{1}{4}iy/a)$ $-\psi(\frac{1}{2}+\frac{1}{4}iy/a)-\psi(\frac{1}{2}-\frac{1}{4}iy/a)\Big]$
7.3	$x\,\mathrm{sech}(ax)$	$\frac{1}{4}(\pi/a)^2\sinh(\frac{1}{2}\pi y/a)\,\mathrm{sech}^2(\frac{1}{2}\pi y/a)$
7.4	$x^{-1}\mathrm{sech}(ax)$	$\arctan\{\sinh(\frac{1}{2}\pi y/a)\}$
7.5	$x\,\mathrm{cosech}(ax)$	$\frac{1}{4}ia^{-2}\{\Psi'(\frac{1}{2}+\frac{1}{2}iy/a)-\Psi'(\frac{1}{2}-\frac{1}{2}iy/a)\}$
7.6	$\mathrm{cosech}(ax)$	$\frac{1}{2}\pi a^{-1}\tanh(\frac{1}{2}\pi y/a)$
7.7	$x^{-1}-\mathrm{cosech}\,x$	$\pi(1+e^{\pi y})^{-1}$
7.8	$x^{-1}(x^{-1}-\mathrm{cosech}\,x)$	$i\,\log\Gamma(\frac{1}{2}-\frac{1}{2}iy)-i\,\log\Gamma(\frac{1}{2}+\frac{1}{2}iy)-y\{\log(\frac{1}{2}y)-1\}$
7.9	$x^{-1}\{1-\mathrm{sech}(ax)\}$	$\mathrm{arccot}\{\sinh(\frac{1}{2}\pi y/a)\}$
7.10	$1-\tanh(ax)$	$y^{-1}-\frac{1}{2}\pi a^{-1}\mathrm{cosech}(\frac{1}{2}\pi y/a)$

	$f(x)$	$g_s(y)$
7.11	$1-\coth(ax)$	$y^{-1}-\tfrac{1}{2}\pi a^{-1}\coth(\tfrac{1}{2}\pi y/a)$
7.12	$x\,\mathrm{sech}^2 x$	$-\tfrac{1}{2}\pi\dfrac{d}{dy}\{y\,\mathrm{cosech}(\tfrac{1}{2}\pi y)\}$
7.13	$x^{-1}\tanh(ax)$	$\tfrac{1}{2}\pi-i\,\log\Gamma(\tfrac{1}{2}+\tfrac{1}{4}iy/a)-i\,\log\Gamma(1-\tfrac{1}{4}iy/a)$ $+\,i\,\log\Gamma(\tfrac{1}{2}-\tfrac{1}{4}iy/a)+i\,\log\Gamma(1+\tfrac{1}{4}iy/a)$
7.14	$x^{\nu-1}(\mathrm{cosech}\,x-x^{-1})$ $-2<\mathrm{Re}\,\nu<1$	$\cos(\tfrac{1}{2}\pi\nu)\,\Gamma(\nu-1)y^{1-\nu}$ $-i2^{-\nu}\Gamma(\nu)\{\zeta(\nu,\tfrac{1}{2}-\tfrac{1}{2}iy)-\zeta(\nu,\tfrac{1}{2}+\tfrac{1}{2}iy)\}$
7.15	$(\tfrac{1}{4}+x^2)^{-1}\mathrm{cosech}(\pi x)$	$\pi\sinh(\tfrac{1}{2}\pi y)-\cosh(\tfrac{1}{2}y)\arctan\{\sinh(\tfrac{1}{2}y)\}$
7.16	$(k^2+x^2)^{-1}\mathrm{cosech}(ax)$	$\tfrac{1}{2}\pi(ka)^{-1}\sinh(\pi y/a)$ $\cdot\displaystyle\int_0^\infty\Big(\cosh(\pi t/a)+\cosh(\pi y/a)\Big)^{-1}e^{-kt}dt$
7.17	$\{\cosh(ax)+\cosh b\}^{-1}$	$-\pi\,\mathrm{cosech}\,b\Big(a^{-1}\cos(by/a)\,\mathrm{cosech}(\pi y/a)$ $-\pi^{-1}y\displaystyle\sum_{n=0}^\infty(-1)^n\varepsilon_n(y^2+a^2n^2)^{-1}e^{-nb}\Big)$
7.18	$\{\cosh(ax)+\cos b\}^{-1}$ $-\pi<b<\pi$	$-2y\cos b\displaystyle\sum_{n=1}^\infty(-1)^n(y^2+a^2n^2)^{-1}\sin(nb)$
7.19	$\sinh(\tfrac{1}{2}ax)$ $\{\cos b+\cosh(ax)\}^{-1},$ $\qquad\qquad -\pi<b<\pi$	$\tfrac{1}{2}\pi a^{-1}\mathrm{cosec}(\tfrac{1}{2}b)\sinh(by/a)\,\mathrm{sech}(\pi y/a)$
7.20	$\begin{array}{ll}0 & x<a\\ (\cosh x-\cosh a)^{-\frac{1}{2}}, & x>a\end{array}$	$2^{-\frac{1}{2}}\pi\tanh(\pi y)P_{-\frac{1}{2}+iy}(\cosh a)$
7.21	$\sinh(ax)$ $\cdot\{\cos b+\cosh(cx)\}^{-1}$	$\pi c^{-1}\mathrm{cosec}\,b\Big(\sin\{(\pi+b)\tfrac{a}{c}\}\sinh\{(\pi-b)\tfrac{y}{c}\}$ $-\sin\{(\pi-b)\tfrac{a}{c}\}\sinh\{(\pi+b)\tfrac{y}{c}\}\Big)$ $\cdot\{\cosh(2\pi y/c)-\cos(2\pi a/c)\}^{-1}$

	$f(x)$	$g_s(y)$
7.22	$0 \qquad x<a$ $(\cosh x-\cosh a)^{-\nu} x>a$ $0<\mathrm{Re}\,\nu<1$	$(2\pi)^{-\frac{1}{2}}\Gamma(1-\nu)(\sinh a)^{\frac{1}{2}-\nu}\sinh(\pi y)$ $\cdot\Gamma(\nu+iy)\Gamma(\nu-iy)p^{\frac{1}{2}-\nu}_{-\frac{1}{2}+iy}(\cosh a)$
7.23	$\sinh x\left[(1+\cosh x)^{-3/2}\right.$ $\left.-(a+\cosh x)^{-3/2}, a>-1\right]$	$-2^{-3/2}i\pi\mathrm{sech}(\pi y)\{P_{-\frac{1}{2}+iy}(a)-P_{\frac{1}{2}-iy}(a)\}$
7.24	$\cosh(ax)\,\mathrm{cosech}(bx)$ $a<b$	$\frac{1}{2}\pi b^{-1}\sinh(\pi y/b)\{\cosh(\pi y/b)+\cos(\pi a/b)\}^{-1}$
7.25	$\sinh(ax)\,\mathrm{sech}(bx)$ $a<b$	$\pi b^{-1}\sin(\frac{1}{2}\pi a/b)\sinh(\frac{1}{2}\pi y/b)$ $\cdot\{\cosh(\pi y/b)+\cos(\pi a/b)\}^{-1}$
7.26	$\sinh(ax)\,\mathrm{cosech}(bx)$ $a<b$	$\frac{1}{2}\pi b^{-1}\sinh(\pi y/b)\{\cosh(\pi y/b)+\cos(\pi a/b)\}^{-1}$ $+\frac{1}{2}ib^{-1}\left(\Psi\{\frac{1}{2}+\frac{1}{2}(a+iy)/b\}-\Psi\{\frac{1}{2}+\frac{1}{2}(a-iy)/b\}\right)$
7.27	$\cosh(ax)\,\mathrm{sech}(bx)$ $a<b$	$\frac{1}{4}ib^{-1}\left(\Psi\{\frac{3}{4}+\frac{1}{4}(a+iy)/b\}-\Psi\{\frac{3}{4}+\frac{1}{4}(a-iy)/b\}\right.$ $+\Psi\{\frac{3}{4}-\frac{1}{4}(a-iy)/b\}-\Psi\{\frac{3}{4}-\frac{1}{4}(a+iy)/b\}$ $\left.-2\pi i\,\sinh(ay/b)\{\cosh(ay/b)+\cos(\pi a/b)\}^{-1}\right)$
7.28	$x^{-1}\sinh(ax)\,\mathrm{cosech}(bx)$ $a\leq b$	$\arctan\left(\tanh(\frac{1}{2}\pi y/b)\tan(\frac{1}{2}\pi a/b)\right)$
7.29	$x^{-1}\cosh(ax)\,\mathrm{sech}(bx), a\leq b$	$\arctan\{\sinh(\frac{1}{2}\pi y/b)\sec(\frac{1}{2}\pi a/b)\}$
7.30	$x^{-1}\sinh(ax)\,\mathrm{sech}(bx), a\leq b$	$\arctan\{\tanh(\frac{1}{2}\pi y/b)\tan(\frac{1}{2}\pi a/b)\}$ $-i\left(\log\Gamma\{\frac{3}{4}-\frac{1}{4}(a-iy)/b\}-\log\Gamma\{\frac{3}{4}-\frac{1}{4}(a+iy)/b\}\right.$ $\left.+\log\Gamma\{\frac{3}{4}+\frac{1}{4}(a-iy)/b\}-\log\Gamma\{\frac{3}{4}+\frac{1}{4}(a+iy)/b\}\right)$
7.31	$x^{-1}[1-\cosh(ax)\,\mathrm{sech}(bx)]$ $a\leq b$	$\arctan\{\cos(\frac{1}{2}\pi a/b)\,\mathrm{cosech}(\frac{1}{2}\pi y/b)\}$
7.32	$(1+x^2)^{-1}\mathrm{cosech}(\frac{1}{2}\pi x)$	$e^y\arctan(e^{-y})-e^{-y}\arctan(e^y)$

	$f(x)$	$g_s(y)$
7.33	$(1+x^2)^{-1}\mathrm{cosech}(\pi x)$	$\sinh y \, \log(1+e^{-y})-\tfrac{1}{2}ye^{-y}$
7.34	$\sinh(ax)\{\mathrm{sech}(bx)\}^2$ $a<2b$	$\pi b^{-2}\{y \sin u \cosh v-a\cos u \sinh v\}$ $\cdot\{\cosh(2v)-\cos(2u)\}^{-1}; \;\; u=\tfrac{1}{2}\pi\tfrac{a}{b}, \;\; v=\tfrac{1}{2}\pi\tfrac{y}{b}$
7.35	$x^{-1}\mathrm{cosech}(bx)\sinh^2(ax)$ $a\leq\tfrac{1}{2}b$	$\tfrac{1}{2}\arctan z$ $z=2\sin^2 u \, \sinh v\{\cosh^2 v-2\sin^2 u\}^{-1}$ $u=\tfrac{\pi a}{2b}, \;\; v=\tfrac{\pi y}{2b}$
7.36	$(a^2+x^2)^{-1}\mathrm{cosech}(\pi x)$	$-\tfrac{1}{2}a^{-2}-\tfrac{1}{2}\pi a^{-1}\mathrm{cosec}(\pi a)e^{-ay}$ $+\tfrac{1}{2}a^{-2}\{{}_2F_1(1,-a;1-a;-e^{-y})$ $+{}_2F_1(1,a;1+a;-e^{-y})\}$
7.37	$x(1+x^2)^{-1}\mathrm{sech}(\tfrac{1}{4}\pi x)$	$2^{-\frac{1}{2}}\Big(\pi e^{-y}+\sinh y \, \log\{(2^{\frac{1}{2}}\cosh y+1)$ $/(2^{\frac{1}{2}}\cosh y-1)\}\Big)$ $-2^{\frac{1}{2}}\cosh\{\arctan(2^{-\frac{1}{2}}\mathrm{cosech}\, y)\}.$
7.38	$\{\sinh(ax)\}^{-\nu}$ $0<\mathrm{Re}\,\nu<2$	$\pi 2^{\nu}a^{-1}\cos(\tfrac{1}{2}\pi\nu)\,\Gamma(1-\nu)\sinh(\tfrac{1}{2}\pi y/a)$ $\cdot\{\Gamma(1-\tfrac{1}{2}\nu-\tfrac{1}{2}iy/a)\,\Gamma(1-\tfrac{1}{2}\nu+\tfrac{1}{2}iy/a)\}^{-1}$ $\cdot\{\cosh(\tfrac{1}{2}\pi y/a)-\cos(\pi\nu)\}^{-1}$
7.39	$e^{-ax}\{\sinh(bx)\}^{\nu}$ $\mathrm{Re}\,\nu>-2$, $b\,\mathrm{Re}\,\nu<\mathrm{Re}\,a$	$-ib^{-1}2^{-\nu-2}\Gamma(1+\nu)\left[\dfrac{\Gamma\{-\tfrac{1}{2}\nu+\tfrac{1}{2}(a-iy)/b\}}{\Gamma\{1+\tfrac{1}{2}\nu+\tfrac{1}{2}(a-iy)/b\}}\right.$ $\left.-\dfrac{\Gamma\{-\tfrac{1}{2}\nu+\tfrac{1}{2}(a+iy)/b\}}{\Gamma\{1+\tfrac{1}{2}\nu+\tfrac{1}{2}(a+iy)/b\}}\right]$
7.40	$(e^{bx}-1)^{-1}\sinh(ax)\,;a>b$	$\tfrac{1}{2}\pi b^{-1}\sinh(2\pi y/b)\{\cosh(2\pi y/b)$ $-\cos(2\pi a/b)\}^{-1}-\tfrac{1}{2}y(a^2+y^2)^{-1}$ $+\tfrac{1}{2}ib^{-1}\Big(\Psi\{1+(a+iy)/b\}-\Psi\{1+(a-iy)/b\}\Big)$
7.41	$e^{-ax}\mathrm{cosech}(bx)\,;a>b$	$-\tfrac{1}{2}ib^{-1}\Big(\Psi\{\tfrac{1}{2}+\tfrac{1}{2}(a+iy)/b\}-\Psi\{\tfrac{1}{2}+\tfrac{1}{2}(a-iy)/b\}\Big)$
7.42	$e^{-ax}\mathrm{cosech}(ax)$	$\tfrac{1}{2}\pi a^{-1}\coth(\tfrac{1}{2}\pi y/a)-y^{-1}$

	$f(x)$	$g_s(y)$
7.43	$x^{-1}e^{-bx}\sinh(ax)$; $b>a$	$\tfrac{1}{2}\arctan\{2ay(y^2+b^2-a^2)^{-1}\}$
7.44	$\sin(\pi x^2)\coth(\pi x)$	$\tfrac{1}{2}\sin(\tfrac{1}{4}\pi+\tfrac{1}{4}y^2/\pi)\tanh(\tfrac{1}{2}y)$
7.45	$\cos(\pi x^2)\coth(\pi x)$	$\tfrac{1}{2}\{1-\cos(\tfrac{1}{4}\pi+\tfrac{1}{4}y^2/\pi)\tanh(\tfrac{1}{2}y)$
7.46	$(x^2+k^2)^{-1}\cosh(ax)$ $\cdot\operatorname{cosech}(bx\qquad b\geq a$	$-\tfrac{1}{2}\pi k^{-1}\cos(ak)\operatorname{cosec}(bk)e^{-ky}$ $+\pi b^{-1}\sum\limits_{n=0}^{\infty}(-1)^n\{k^2-(n\tfrac{\pi}{b})^2\}^{-1}\cos(n\pi\tfrac{a}{b})e^{-n\pi y/b}$
7.47	$(x^2+k^2)^{-1}\sinh(ax)$ $\cdot\operatorname{sech}(bx)\qquad b\geq a$	$\tfrac{1}{2}\pi k^{-1}\sin(ak)\sec(bk)e^{-ky}$ $+\pi b^{-1}\sum\limits_{n=0}^{\infty}(-1)^n\left(k^2-\{(n+\tfrac{1}{2})\pi/b\}^2\right)^{-1}$ $\cdot\sin\{(n+\tfrac{1}{2})\pi a/b\}\exp\{-(n+\tfrac{1}{2})\pi y/b\}$
7.48	$x^{-1}e^{-x^2}\sinh(x^2)$	$\tfrac{1}{4}\pi\operatorname{Erfc}(2^{-3/2}y)$
7.49	$e^{-ax}\begin{array}{c}\tanh(bx^{\frac{1}{2}})\\\coth(bx^{\frac{1}{2}})\end{array}$	Mordell,L.J. Messeng.Math.Vol.49,1920
7.50	$e^{-bx}\sinh(ax^{\frac{1}{2}})$	$\tfrac{1}{2}\pi^{\frac{1}{2}}a(b^2+y^2)^{-3/4}\exp\{\tfrac{1}{4}a^2b(b^2+y^2)^{-1}\}$ $\cdot\sin\left[\tfrac{1}{4}a^2y(b^2+y^2)^{-1}+\tfrac{3}{2}\arctan(y/b)\right]$
7.51	$\begin{array}{ll}x^{-\frac{1}{2}}v^{-1}\cosh(bv) & x<a\\ 0 & x>a\end{array}$	$\tfrac{1}{2}\pi(\tfrac{1}{2}\pi y)^{\frac{1}{2}}J_{\frac{1}{4}}(c)J_{\frac{1}{4}}(d)$; $\dfrac{c}{d}=\tfrac{1}{2}ab\pm\tfrac{1}{2}aS$
7.52	$\begin{array}{ll}\sinh(bu) & x<2a\\ 0 & x>2a\end{array}$	$\begin{array}{ll}\pi ab\ \sin(ay)s^{-1}I_1(as) & y<b\\ \pi ab\ \sin(ay)S^{-1}J_1(aS) & y>b\end{array}$
7.53	$\begin{array}{ll}u^{-1}\cosh(bu) & x<2a\\ 0 & x>2a\end{array}$	$\begin{array}{ll}\pi\sin(ay)I_0(as) & y<b\\ \pi\sin(ay)J_0(aS) & y>b\end{array}$

$u=(2ax-x^2)^{\frac{1}{2}}$, $v=(a^2-x^2)^{\frac{1}{2}}$ $s=(b^2-y^2)^{\frac{1}{2}}$, $S=(y^2-b^2)^{\frac{1}{2}}$

	$f(x)$	$g_s(y)$
7.54	$u^{-3/2}\sinh(bu)$ $x<2a$ 0 $x>2a$	$\pi(\tfrac12\pi b)^{\frac12}\sin(ay)J_{\frac14}(\tfrac12 ay+\tfrac12 aS)J_{\frac14}(\tfrac12 ay-\tfrac12 aS)$
7.55	$u^{-3/2}\cosh(bu)$ $x<2a$ 0 $x>2a$	$\pi(\tfrac12\pi b)^{\frac12}\sin(ay)J_{-\frac14}(\tfrac12 ay+\tfrac12 aS)J_{-\frac14}(\tfrac12 ay-\tfrac12 aS)$
7.56	$\exp(-a\sinh x)$	$yS_{-1,iy}(a)=\tfrac12\pi\,\mathrm{cosech}(\pi y)$ $\cdot\{\mathbf{J}_{iy}(a)+\mathbf{J}_{-iy}(a)-J_{iy}(a)-J_{-iy}(a)\}$
7.57	$\exp(-a\cosh x)$	$\mathrm{cosech}(\pi y)\left(\displaystyle\int_{o}^{\pi}\exp(a\cos t)\cosh)yt)dt\right.$ $\left.-\tfrac12\pi I_{iy}(a)-\tfrac12\pi I_{-iy}(a)\right]$
7.58	$\sinh(\tfrac12 x)\exp(-a\cosh x)$	$\tfrac12\pi^{\frac12}a^{-1}W_{-\frac12,\frac12 iy}(2a)$
7.59	$\sinh x\,\exp(-a\cosh x)$	$ya^{-1}K_{iy}(a)$
7.60	$(\sinh x)^{-\frac12}\exp(-a\,\mathrm{cosech}\,x)$	$i2^{-\frac12}\left[\Gamma(\tfrac12+iy)D_{-\frac12-iy}(u)D_{-\frac12-iy}(v)\right.$ $\left.-\Gamma(\tfrac12-iy)D_{-\frac12+iy}(u)D_{-\frac12+iy}(v)\right]$; $\genfrac{}{}{0pt}{}{u}{v}=(\pm 2ia)^{\frac12}$
7.61	$(\sinh x)^{-\frac12}\exp(-2a\sinh x)$	$\tfrac14\pi i(\pi a)^{\frac12}\left[J_{\frac14-\frac12 iy}(a)Y_{-\frac14-\frac12 iy}(a)\right.$ $-J_{-\frac14-\frac12 iy}(a)Y_{\frac14-\frac12 iy}(a)-J_{\frac14+\frac12 iy}(a)Y_{-\frac14+\frac12 iy}(a)$ $\left.+J_{-\frac14+\frac12 iy}(a)Y_{\frac14+\frac12 iy}(a)\right]$
7.62	$\log\{\tanh(ax)\}$	$-y^{-1}\{\gamma+\log 4+\tfrac12\Psi(\tfrac12+\tfrac14 iy/a)+\tfrac12\Psi(\tfrac12-\tfrac14 iy/a)\}$
7.63	$\mathrm{cosech}(ax)\log\{\sinh(ax)\}$	$-\tfrac12\pi a^{-1}\tanh(\tfrac12\pi y/a)\left[\gamma+\log 2\right.$ $\left.+\tfrac12\Psi(\tfrac12+\tfrac12 iy/a)+\tfrac12\Psi(\tfrac12-\tfrac12 iy/a)\right]$
7.64	0 $x<a$ $(\cosh x-\cosh a)^{-\frac12}$ $\cdot\log(\cosh x-\cosh a)$ $x>a$	$2^{-\frac12}\pi\tanh(\pi y)\left[p_{-\frac12+iy}(z)\{-\gamma-\log 4\right.$ $+\log(\sinh a)-\tfrac12\Psi(\tfrac12+iy)-\tfrac12\Psi(\tfrac12-iy)$ $\left.+\tfrac12 q_{-\frac12+iy}(z)+\tfrac12 q_{-\frac12-iy}(z)\right]$; $z=\cosh a$

$u=(2ax-x^2)^{\frac12}, v=(a^2-x^2)^{\frac12}$ $s=(b^2-y^2)^{\frac12}$, $S-(y^2-b^2)^{\frac12}$

	$f(x)$	$g_s(y)$
7.65	$(1+c^2\sinh^2 x)^{-\frac{1}{2}\nu}$ $\cdot\sin\{\nu\arctan(c\sinh x)\}$ $\mathrm{Re}\,\nu>0$	$(\tfrac{1}{2}\pi/c)^{\frac{1}{2}}\{\Gamma(\nu)\}^{-1}\sin h(\tfrac{1}{2}\pi y)$ $\cdot\Gamma(\nu-iy)\Gamma(\nu+iy)(1-c^2)^{\frac{1}{4}-\frac{1}{2}\nu}P_{-\frac{1}{2}+iy}^{\frac{1}{2}-\nu}(1/c)$
7.66	$\sin(a\sinh x)$	$\sin h(\tfrac{1}{2}\pi y)K_{iy}(a)$
7.67	$\cos(a\sinh x)$	$\tfrac{1}{4}\pi\operatorname{cosech}(\tfrac{1}{2}\pi y)\left[-I_{-iy}(a)-I_{iy}(a)\right.$ $\left.+y\sec(\tfrac{1}{2}\pi y)\{\mathbf{J}_{iy}(ia)+\mathbf{J}_{-iy}(a)\}\right]$
7.68	$(\sinh x)^{-\frac{1}{2}}\sin(2a\sinh x)$	$-\tfrac{1}{2}i(\pi a)^{\frac{1}{2}}\left(I_{\frac{1}{4}-\frac{1}{2}iy}(a)K_{\frac{1}{4}+\frac{1}{2}iy}(a)\right.$ $\left.-I_{\frac{1}{4}+\frac{1}{2}iy}(a)K_{\frac{1}{4}-\frac{1}{2}iy}(a)\right]$
7.69	$(\sinh x)^{-\frac{1}{2}}\cos(2a\sinh x)$	$-\tfrac{1}{2}i(\pi a)^{\frac{1}{2}}\left(I_{-\frac{1}{4}-\frac{1}{2}iy}(a)K_{\frac{1}{4}-\frac{1}{2}iy}(a)\right.$ $\left.-I_{-\frac{1}{4}+\frac{1}{2}iy}(a)K_{\frac{1}{4}+\frac{1}{2}iy}(a)\right]$
7.70	$(\sin x)^{-\frac{1}{2}}$ $\sinh(2a\sin x)\qquad x<\pi$ $\qquad 0 \qquad\qquad x>\pi$	$(a\pi)^{\frac{1}{2}}\sin(\tfrac{1}{2}\pi y)I_{\frac{1}{4}-\frac{1}{2}iy}(a)I_{\frac{1}{4}+\frac{1}{2}iy}(a)$
7.71	$(\sin x)^{-\frac{1}{2}}$ $\cosh(2a\sin x)\qquad x<\pi$ $\qquad 0 \qquad\qquad x>\pi$	$(a\pi)^{\frac{1}{2}}\sin(\tfrac{1}{2}\pi y)I_{-\frac{1}{4}+\frac{1}{2}y}(a)I_{-\frac{1}{4}-\frac{1}{2}y}(a)$
7.72	$(\sin x)^{-3/2}$ $\sinh(2a\sin x($ $\qquad x<\pi$ $\qquad 0 \qquad\qquad x>\pi$	$2(\pi a)^{3/2}\sin(\tfrac{1}{2}\pi y)$ $\cdot\left(I_{-\frac{1}{4}-\frac{1}{2}y}(a)I_{-\frac{1}{4}+\frac{1}{2}y}(a)-I_{\frac{3}{4}-\frac{1}{2}y}(a)J_{\frac{3}{4}+\frac{1}{2}y}(a)\right]$
7.73	$\exp(-a\cosh x)\sin(b\sinh x)$	$\sinh\{y\arctan(b/a)\}K_{iy}\{(a^2+b^2)^{\frac{1}{2}}\}$
7.74	$\sin(a\cosh x)\sin(b\sinh x)$	$\tfrac{1}{4}\pi i\,\sin\left[\tfrac{1}{2}y\log\{(a+b)/(a-b)\}\right]\operatorname{cosech}(\tfrac{1}{2}\pi y)$ $\cdot\left[J_{iy}\{(a^2-b^2)^{\frac{1}{2}}\}-J_{-iy}\{(a^2-b^2)^{\frac{1}{2}}\}\right]$ $\qquad\qquad\qquad\qquad\qquad a>b$ $\sin\{\tfrac{1}{2}y\log(\tfrac{b+a}{b-a})\}\cosh(\tfrac{1}{2}\pi y)K_{iy}\{(b^2-a^2)^{\frac{1}{2}}\}\;a<b$

	$f(x)$	$g_s(y)$
7.75	$\cos(a\cosh x)\sin(b\sinh x)$	$-\tfrac14\pi\sin\left[\tfrac12 y\,\log\{(a+b)/(a-b)\}\right]\mathrm{sech}(\tfrac12\pi y)$
		$\cdot\left[J_{iy}\{(a^2-b^2)^{\tfrac12}\}+J_{iy}\{(a^2-b^2)^{\tfrac12}\}\right] a>b$
		$\cos\{\tfrac12 y\,\log(\tfrac{b+a}{b-a})\}\sinh(\tfrac12\pi y)K_{iy}(b^2-a^2)^{\tfrac12}\ a<b$
7.76	$\mathrm{arccot}\{\sinh(ax)\}$	$\tfrac12\pi y^{-1}\{1-\mathrm{sech}(\tfrac12\pi y/a)\}$

2.8   Orthogonal Polynomials

	$f(x)$	$g_s(y)$
8.1	$P_n(x/a)\qquad x<a$ $0\qquad x>a$	$\tfrac12\pi^{\tfrac12}(1-n)(1+\tfrac12 n)\{\Gamma(\tfrac32-\tfrac12 n)\,\Gamma(1+\tfrac12 n)\}^{-1}$ $\cdot(y/a)^{-\tfrac12}s_{\tfrac12,\,n+\tfrac12}(ay)$
8.2	$P_{2n+1}(x/a)\qquad x<a$ $0\qquad x>a$	$(-1)^n(\tfrac12\pi a/y)^{\tfrac12}J_{2n+3/2}(ay)$
8.3	$x^{\nu-1}P_n(x)\qquad x<1$ $0\qquad x>1$ $\mathrm{Re}(\nu+n)>-1$	$\pi^{\tfrac12}2^{-\nu-1}\Gamma(1+\nu)\left[\Gamma(1+\tfrac12\nu-\tfrac12 n)\,\Gamma(\tfrac32+\tfrac32\nu+\tfrac32 n)\right]^{-1}$ $\cdot y\,{}_2F_3(\tfrac12+\tfrac12\nu,1+\tfrac12\nu;\tfrac32,1+\tfrac12\nu-\tfrac12 n,\tfrac32+\tfrac12\nu+\tfrac12 n;-\tfrac14 y^2)$
8.4	$P_n(a-bx^2),a-1<bx^{\tfrac12}<a+1$ 0 otherwise $a>1$	$\pi(2b)^{-\tfrac12}J_{n+\tfrac12}(u)J_{n+\tfrac12}(v)$ $\genfrac{}{}{0pt}{}{u}{v}=\tfrac12 yb^{-\tfrac12}\{(a+1)^{\tfrac12}\pm(a-1)^{\tfrac12}\}$
8.5	$(a^2+x^2)^{-n-1}$ $P_{2n+1}\{x(a^2+x^2)^{-\tfrac12}\}$	$(-1)^n\{(2n+1)!\}^{-1}y^{2n+1}K_o(ay)$
8.6	$(\mathrm{sech}\,x)^{2n+2}P_{2n+1}(\tanh x)$	$(-1)^n\pi^{-1}2^{2n}\{(2n+1)!\}^{-2}\sinh(\tfrac12\pi y)$ $\cdot\left[\Gamma(n+1+\tfrac12 iy)\,\Gamma(n+1-\tfrac12 iy)\right]$
8.7	$z^{-n-1}P_{2n+1}(az^{-\tfrac12}\sinh x)$ $z=b^2+a^2\sinh x$	$(-1)^n\pi^{-1}2^{2n}a^{-2n-2}\{(2n+1)!\}^{-2}\sinh(\tfrac12\pi y)$ $\cdot\left[\Gamma(n+1+\tfrac12 iy)\,\Gamma(n+\tfrac12-\tfrac12 iy)\right]^2$ $\cdot{}_2F_1(n+1+\tfrac12 iy,n+1-\tfrac12 iy;2n+2;1-b^2/a^2)$

	$f(x)$	$g_s(y)$
8.8	$(a^2-x^2)^{-\frac{1}{2}}T_{2n+1}\left(\frac{x}{a}\right),$ $x<a$ 0 $x>a$	$(-1)^n\frac{1}{2}\pi J_{2n+1}(ay)$
8.9	$x^{-\frac{1}{2}}(a^2-x^2)^{-\frac{1}{2}}T_n(x/a)$ $x<a$ 0 $x>a$	$\frac{1}{2}\pi(\frac{1}{2}\pi y)^{\frac{1}{2}}J_{\frac{1}{2}n+\frac{1}{4}}(\frac{1}{2}ay)J_{-\frac{1}{2}n+\frac{1}{4}}(\frac{1}{2}ay)$
8.10	$x(a^2+x^2)^{-\frac{1}{2}n}T_n\{a(a^2+x^2)^{-\frac{1}{2}}\}$	$\frac{1}{2}\pi(n-1)!y^{n-1}e^{-ay}\{a-(n-1)y^{-1}\}$
8.11	$\sin\{b(a^2-x^2)^{\frac{1}{2}}\}$ $\cdot U_{2n+1}(x/a)$ $x<a$ 0 $x>a$	$(-1)^n\frac{1}{2}\pi ab(b^2+y^2)^{-\frac{1}{2}}$ $\cdot U_{2n+1}\{y(b^2+y^2)^{-\frac{1}{2}}\}J_{2n+2}\{a(b^2+y^2)^{\frac{1}{2}}\}$
8.12	$(a^2-x^2)^{-\frac{1}{2}}\cos\{b(a^2-x^2)^{\frac{1}{2}}\}$ $\cdot T_{2n+1}(x/a)$ $x>a$ 0 $x<a$	$(-1)^n\frac{1}{2}\pi T_{2n+1}\{y(b^2+y^2)^{-\frac{1}{2}}\}$ $\cdot J_{2n+1}\{a(b^2+y^2)^{\frac{1}{2}}\}$
8.13	$x(a^2-x^2)^{-\frac{1}{2}}\sin\{b(a^2-x^2)^{\frac{1}{2}}\}$ $\cdot U_{2n+1}\{(1-x^2/a^2)^{\frac{1}{2}}\}$ $x>a$ 0 $x<a$	$(-1)^n\frac{1}{2}\pi ay(b^2+y^2)^{-\frac{1}{2}}U_{2n+1}\{b(b^2+y^2)^{-\frac{1}{2}}\}$ $\cdot J_{2n+2}\{a(b^2+y^2)^{\frac{1}{2}}\}$
8.14	$x(a^2-x^2)^{-\frac{1}{2}}\cos\{(a^2-x^2)^{\frac{1}{2}}\}$ $\cdot U_{2n}\{(1-x^2/a^2)^{\frac{1}{2}}\}$ $x<a$ 0 $x>a$	$(-1)^n\frac{1}{2}\pi ay(b^2+y^2)^{-\frac{1}{2}}U_{2n}\{b(b^2+y^2)^{-\frac{1}{2}}\}$ $\cdot J_{2n+1}\{a(b^2+y^2)^{\frac{1}{2}}\}$
8.15	$(2ax-x^2)^{-\frac{1}{2}}\cos\{b(2ax-x^2)^{\frac{1}{2}}\}$ $\cdot T_{2n}\{\frac{1}{a}(2ax-x^2)^{\frac{1}{2}}\}$ $x<2a$ 0 $x>2a$	$(-1)^n\pi\sin(ay)T_{2n}\{b(b^2+y^2)^{-\frac{1}{2}}\}$ $\cdot J_{2n}\{a(b^2+y^2)^{\frac{1}{2}}\}$
8.16	$(2ax-x^2)^{-\frac{1}{2}}\sin\{b(2ax-x^2)^{\frac{1}{2}}\}$ $\cdot T_{2n+1}\{\frac{1}{a}(2ax-x^2)^{\frac{1}{2}}\}$ $x<2a$ 0 $x>2a$	$(-1)^n\sin(ay)T_{2n+1}\{b(b^2+y^2)^{-\frac{1}{2}}\}$ $\cdot J_{2n+1}\{a(b^2+y^2)^{\frac{1}{2}}\}$
8.17	$(a^2-x^2)^{\nu-\frac{1}{2}}C_{2n+1}\left(\frac{x}{a}\right),$ $x<a$ 0 $x>a$	$(-1)^n\pi a^\nu\Gamma(2n+2\nu+1)\{(2n+1)!\Gamma(\nu)\}^{-1}$ $\cdot(2y)^{-\nu}J_{\nu+2n+1}(ay)$

	$f(x)$	$g_s(y)$
8.18	$(a^2-x^2)^\nu P_{2n+1}^{(\nu,\nu)}(\frac{x}{a})$ $x<a$ 0 $x>a$ $\mathrm{Re}\,\nu>-1$	$(-1)^n \pi^{\frac{1}{2}} 2^{\nu-\frac{1}{2}} \{(2n+1)!\}^{-1}\Gamma(2n+\nu+2)$ $\cdot (y/a)^{-\nu-\frac{1}{2}} J_{\nu+2n+3/2}(ay)$
8.19	$\{(1-x)^\nu(1+x)^\mu$ $-(1+x)^\nu(1-x)^\mu\} P_{2n}^{(\nu,\mu)}(x)$ $x<1$ 0 $x>1$ $\mathrm{Re}(\nu,\mu)>-1$	$(-1)^{n+1} 2^{2n+\nu+\mu}\{(2n)!\}^{-1} B(2n+\mu+1,2n+\nu+1)$ $\cdot y^{2n} i \left(e^{iy}{}_1F_1(2n+\mu+1;4n+\nu+\mu+2;-2iy) \right.$ $\left. -e^{-iy}{}_1F_1(2n+\mu+1;4n+\nu+\mu+2;2iy) \right)$
8.20	$\{(1-x)^\nu(1+x)^\mu$ $+(1+x)^\nu(1-x)^\mu\} P_{2n+1}^{(\nu,\mu)}(x)$ $x<1$ 0 $x>1$ $\mathrm{Re}(\nu,\mu)>-1$	$(-1)^{n+1} 2^{\nu+\mu+2n+1}\{(2n+1)!\}^{-1} y^{2n+1} B(\nu+2n+2$ $,\mu+2n+2) \left(e^{iy}{}_1F_1(\nu+2n+2;\nu+\mu+4n+4;-2iy) \right.$ $\left. +e^{-iy}{}_1F_1(\nu+2n+2;\nu+\mu+4n+4;2iy) \right)$
8.21	$\exp(-\tfrac{1}{2}x^2) He_{2n+1}(2^{\frac{1}{2}}x)$	$(-1)^n (\tfrac{1}{2}\pi)^{\frac{1}{2}} \exp(-\tfrac{1}{2}y^2) He_{2n+1}(2^{\frac{1}{2}}y)$
8.22	$\exp(-\tfrac{1}{2}x^2) He_n(x)$ $\cdot He_{n+2m+1}(x)$	$(-1)^n (\tfrac{1}{2}\pi)^{\frac{1}{2}} n! y^{2m} \exp(-\tfrac{1}{2}y^2) L_n^{2m+1}(y^2)$
8.23	$x^{2n+1}\exp(-\tfrac{1}{2}x^2) L_n^{n+\frac{1}{2}}(x^2)$	$(\tfrac{1}{2}\pi)^{\frac{1}{2}} y^{2n+1}\exp(-\tfrac{1}{2}y^2) L_n^{n+\frac{1}{2}}(\tfrac{1}{2}y^2)$
8.24	$x^{2n}\exp(-\tfrac{1}{2}x^2) L_n^{2m+1}(x^2)$	$(-1)^n (\tfrac{1}{2}\pi)^{\frac{1}{2}} (n!)^{-1} \exp(-\tfrac{1}{2}y^2)$ $\cdot He_n(y) He_{n+2m+1}(y)$
8.25	$e^{-\tfrac{1}{2}x^2} L_n(\tfrac{1}{2}x^2) He_{2n+1}(\tfrac{1}{2}x)$	$(\tfrac{1}{2}\pi)^{\frac{1}{2}} e^{-\tfrac{1}{2}y^2} L_n(\tfrac{1}{2}y^2) He_{2n+1}(\tfrac{1}{2}y)$
8.26	$xe^{-\tfrac{1}{2}x^2}\{L_n^{\frac{1}{4}}(\tfrac{1}{2}x^2)\}^2$	$(\tfrac{1}{2}\pi)^{\frac{1}{2}} ye^{-\tfrac{1}{2}y^2}\{L_n^{\frac{1}{4}}(\tfrac{1}{2}y^2)\}^2$
8.27	$x\exp(-\tfrac{1}{2}x^2) L_n^\nu(\tfrac{1}{2}x^2)$ $\cdot L_n^{\frac{1}{2}-\nu}(\tfrac{1}{2}x^2)$	$(\tfrac{1}{2}\pi)^{\frac{1}{2}} y\exp(-\tfrac{1}{2}y^2) L_n^\nu(\tfrac{1}{2}y^2) L_n^{\frac{1}{2}-\nu}(\tfrac{1}{2}y^2)$

2.9 Gamma- and Related Functions

	$f(x)$	$g_s(y)$
9.1	$\{\Gamma(a+x)\,(b-x)\}^{-1}$ $\{\Gamma(a-x)\Gamma(b+x)\}^{-1}$	$\{\Gamma(a+b-1)\}^{-1}\{2\cos(\tfrac12 y)\}^{a+b-1}\sin\{\tfrac12 y(b-a)\}$ $\qquad\qquad\qquad\qquad\qquad\qquad y<\pi$ $\qquad\qquad 0 \qquad\qquad\qquad\qquad y>\pi$
9.2	$\Psi(a+ix)-\Psi(a-ix)$	$i\pi e^{-ay}(1-e^{-y})^{-1}$
9.3	$\Psi(\tfrac14+ax)\{-\Psi(\tfrac34+ax)\}$	$(4a)^{-1}\left\{\Psi\{\tfrac14+y/(8\pi a)\}-\Psi\{\tfrac34+y/(8\pi a)\}\right\}$
9.4	$x^{-1}\{\Psi(1+x)-\log x\}$	$\pi\log\Gamma(1+\tfrac12 y/\pi)-\tfrac12 y\,\log(\tfrac12 y/\pi)+\tfrac12 y$ $-\pi\log(1+\tfrac12\pi^{-1})-\tfrac12+\tfrac12\log(\tfrac12\pi^{-1})$
9.5	$x^{-1}\{\Psi(\tfrac12+iax)+\Psi(\tfrac12-iax)\}$	$-\pi\left\{\gamma+\log\{4\tanh(\tfrac14 y/a)\}\right\}$
9.6	$x^{-1}\Psi(1+x)$	$\pi\log\Gamma(1+\tfrac12 y/a)-\tfrac12 y\,\log(\tfrac12 y/\pi)+\tfrac12 y-\tfrac12\pi(\gamma+\log y)$

2.10 The Error- and the Fresnel Integral

	$f(x)$	$g_s(y)$
10.1	$x^{\nu-1}\mathrm{Erf}(ax)$ $-2<\mathrm{Re}\,\nu<1$	$\sin(\tfrac12\pi\nu)\Gamma(\nu)y^{-\nu}-\pi^{-\frac12}a^{-\nu-1}(1+\nu)^{-1}\Gamma(1+\tfrac12\nu)$ $\cdot y\,{}_2F_2(\tfrac12+\tfrac12\nu,1+\tfrac12\nu;\tfrac32,\tfrac32+\tfrac12\nu;-\tfrac14 y^2/a^2)$
10.2	$\exp(-a^2x^2)\mathrm{Erf}(iax)$	$\tfrac12 i\pi^{\frac12}a^{-1}\exp(-\tfrac14 y^2/a^2)$
10.3	$x^{-1}\mathrm{Erf}\{(ax)^{\frac12}\}$	$-\tfrac12\pi+2\arctan\left[\{(a^2+y^2)^{\frac12}-a\}^{\frac12}/z\right]$ $z=\{(a^2+y^2)^{\frac12}+a\}^{\frac12}-(2a)^{\frac12}$
10.4	$x^{-\frac12}\mathrm{Erf}\{(ax)^{\frac12}\}$	$(2\pi y)^{-\frac12}\arctan\{(2ay)^{\frac12}/(y-a)\}$ $+\log\left[(a^2+y^2)^{-\frac12}\{y+a+(2ay)^{\frac12}\}\right]$
10.5	$\mathrm{Erf}\{(ax)^{-\frac12}\}$	$y^{-1}\left[1-\exp\{-(2y/a)^{\frac12}\}\cos\{(2y/a)^{\frac12}\}\right]$

	$f(x)$	$g_s(y)$
10.6	$\text{Erfc}(ax)$	$y^{-1}\{1-\exp(-\tfrac{1}{4}y^2/a^2)\}$
10.7	$\exp(-a^2\sinh^2 x)$ $\cdot\text{Erf}(ia\sinh x)$	$\tfrac{1}{2}i\exp(\tfrac{1}{2}a^2)\tanh(\tfrac{1}{2}\pi y)K_{\frac{1}{2}iy}(\tfrac{1}{2}a^2)$
10.8	$x^{\nu-1}\text{Erfc}(ax)$ $\text{Re}\,\nu>-1$	$\pi^{-\frac{1}{2}}a^{-\nu-1}(1+\nu)^{-1}\Gamma(1+\tfrac{1}{2}\nu)y$ $\cdot\,_2F_2(\tfrac{1}{2}+\tfrac{1}{2}\nu;1+\tfrac{1}{2}\nu;3/2;3/2+\tfrac{1}{2}\nu;-\tfrac{1}{4}y^2/a^2)$
10.9	$\exp(a^2x^2)\text{Erfc}(ax)$	$\tfrac{1}{2}\pi^{\frac{1}{2}}a^{-1}\exp(\tfrac{1}{4}y^2/a^2)\text{Erfc}(\tfrac{1}{2}y/a)$
10.10	$\exp(-bx)\text{Erfc}(ab-\tfrac{1}{2}x/a)$ $-\exp(bx)\text{Erfc}(ab+\tfrac{1}{2}x/a)$	$2y(b^2+y^2)^{-1}\exp\{-a^2(b^2+y^2)\}$
10.11	$\text{Erfc}\{(ax)^{\frac{1}{2}}\}$	$y^{-1}-(\tfrac{1}{2}a)^{\frac{1}{2}}(a^2+y^2)^{-\frac{1}{2}}\{(a^2+y^2)^{\frac{1}{2}}-a\}^{-\frac{1}{2}}$
10.12	$x^{-1}\text{Erfc}\{(ax)^{\frac{1}{2}}\}$	$\pi-2\arctan\left[\{(a^2+y^2)^{\frac{1}{2}}-a\}^{\frac{1}{2}}/z\right]$ $z=\{(a^2+y^2)^{\frac{1}{2}}+a\}^{\frac{1}{2}}-(2a)^{\frac{1}{2}}$
10.13	$\exp(ax)\text{Erfc}\{(ax)^{\frac{1}{2}}\}$	$\{1+(2y/a)\}^{-\frac{1}{2}}\{a+y+(2ay)^{\frac{1}{2}}\}^{-1}$
10.14	$x^{-1}\text{Erfc}(ax^{-\frac{1}{2}})$	$i\text{Ei}\{-a(-2iy)^{\frac{1}{2}}\}-i\text{Ei}\{-a(2iy)^{\frac{1}{2}}\}$
10.15	$x^{-1}\exp(ax)\text{Erfc}\{(ax)^{\frac{1}{2}}\}$	$2\arctan\left[y^{\frac{1}{2}}\{y^{\frac{1}{2}}+(2a)^{\frac{1}{2}}\}^{-1}\right]$
10.16	$x^{-1}\exp(a^2/x)\text{Erfc}(ax^{-\frac{1}{2}})$	$\tfrac{1}{2}i\pi\left[\mathbf{H}_0\{2a(iy)^{\frac{1}{2}}\}-\mathbf{H}_0\{2a(-iy)^{\frac{1}{2}}\}\right.$ $\left.-Y_0\{2a(iy)^{\frac{1}{2}}\}+Y_0\{2a(-iy)^{\frac{1}{2}}\}\right]$
10.17	$\text{Erf}\{i(ax)^{\frac{1}{2}}\}\text{Erfc}\{(ax)^{\frac{1}{2}}\}$	$i(\tfrac{1}{2}a)^{\frac{1}{2}}(a^2+y^2)^{-\frac{1}{2}}\{a+(a^2+y^2)^{\frac{1}{2}}\}^{-\frac{1}{2}}$
10.18	$\text{Erfc}(u)\text{Erf}(iv)$ ${u \atop v}=a\{(b^2+x^2)^{\frac{1}{2}}\pm b\}$	$i2^{-\frac{1}{2}}a\,\exp(-a^2b)(a^4+y^2)^{-\frac{1}{2}}$ $\cdot\{(a^4+y^2)^{\frac{1}{2}}+a^2\}^{-\frac{1}{2}}\exp\{-b(a^4+y^2)^{\frac{1}{2}}\}$
10.19	$\tfrac{1}{2}-S(ax)$	$\tfrac{1}{2}y^{-1}\left[1-(\tfrac{1}{2}a)^{\frac{1}{2}}(a^2-y^2)^{-\frac{1}{2}}\{a+(a^2-y^2)^{\frac{1}{2}}\}^{\frac{1}{2}}\right],y<a$ $\tfrac{1}{2}y^{-1}\left[1+(\tfrac{1}{2}a)^{\frac{1}{2}}(y^2-a^2)^{-\frac{1}{2}}\{y-(y^2-a^2)^{\frac{1}{2}}\}^{\frac{1}{2}}\right],y>a$

	$f(x)$	$g_s(y)$
10.20	$\frac{1}{2}-C(ax)$	$\frac{1}{2}y^{-1}\left(1-(\frac{1}{2}a)^{\frac{1}{2}}(a^2-y^2)^{-\frac{1}{2}}\{a+(a^2-y^2)^{\frac{1}{2}}\}^{\frac{1}{2}}\right),y<a$ $\frac{1}{2}y^{-1}\left(1-(\frac{1}{2}a)^{\frac{1}{2}}(y^2-a^2)^{-\frac{1}{2}}\{y+(y^2-a^2)^{\frac{1}{2}}\}^{\frac{1}{2}}\right),y>a$
10.21	$x^{-\frac{1}{2}}S(ax)$	$\frac{1}{2}(\frac{1}{2}\pi/y)^{\frac{1}{2}}$ $y<a$ $\qquad\qquad 0$ $y>a$
10.22	$\frac{1}{2}-S(ax^2)$	$\frac{1}{2}y^{-1}\{1-\cos(\frac{1}{4}y^2/a)+\sin(\frac{1}{4}y^2/a)\}$
10.23	$\frac{1}{2}-C(ax^2)$	$\frac{1}{2}y^{-1}\{1-\cos(\frac{1}{4}y^2/a)-\sin(\frac{1}{4}y^2/a)\}$
10.24	$C(ax)-S(ax)$	$\frac{1}{2}a^{\frac{1}{2}}y^{-1}(y-a)^{-\frac{1}{2}}$ $y>a$ $\qquad\qquad 0$ $y<a$
10.25	$C(ax^2)-S(ax^2)$	$y^{-1}\sin(\frac{1}{4}y^2/a)$
10.26	$\sin(ax^2)C(ax^2)$ $-\cos(ax^2)S(ax^2)$	$\frac{1}{2}(\frac{1}{2}\pi/a)^{\frac{1}{2}}\cos(\frac{1}{4}y^2/a)$
10.27	$\cos(ax^2)C(ax^2)$ $+\sin(ax^2)S(ax^2)$	$\frac{1}{2}(\frac{1}{2}\pi/a)^{\frac{1}{2}}\sin(\frac{1}{4}y^2/a)$
10.28	$x^{-1}\{C^2(u)+S^2(u)\}$ $u=ax^2$	$-\frac{1}{2}si(\frac{1}{4}y^2/a)$
10.29	$\{\frac{1}{2}-C(ax^2)\}\cos(ax^2)$ $+\{\frac{1}{2}-S(ax^2)\}\sin(ax^2)\}$	$\frac{1}{2}(\frac{1}{2}\pi/a)^{\frac{1}{2}}\left(\{C(z)+S(z)-1\}\sin z\right.$ $\left.+\{C(z)-S(z)\}\cos z\right)$, $z=\frac{1}{4}y^2/a$
10.30	$S(a/x)$	$\frac{1}{4}y^{-1}(2-e^{-z}-\cos z-\sin z)$, $z=2(ay)^{\frac{1}{2}}$
10.31	$C(a/x)$	$\frac{1}{4}y^{-1}(2-e^{-z}-\cos z+\sin z)$, $z=2(ay)^{\frac{1}{2}}$
10.32	$x^{-1}\{\sin(a/x)C(a/x)$ $-\cos(a/x)S(a/x)\}$	$\frac{1}{4}\pi\{I_o(z)-J_o(z)+\mathbf{H}_o(z)-\mathbf{L}_o(z)\},z=2(ay)^{\frac{1}{2}}$

	$f(x)$	$g_s(y)$
10.33	$x^{-1}\{\cos(a/x)C(a/x)$ $+\sin(a/x)S(a/x)\}$	$\tfrac{1}{4}\pi\{J_0(z)-I_0(z)+L_0(z)+H_0(z)\},\quad z=2(ay)^{\frac{1}{2}}$
10.34	$\tfrac{1}{2}-S(ax^{\frac{1}{2}})$	$\tfrac{1}{2}y^{-1}+\tfrac{1}{4}a^{\frac{1}{2}}y^{-3/2}\sin(\tfrac{1}{8}a^2/y-\tfrac{7}{8}\pi)J_{\frac{1}{4}}(\tfrac{1}{8}a^2/y)$
10.35	$\tfrac{1}{2}-C(ax^{\frac{1}{2}})$	$\tfrac{1}{2}y^{-1}+\tfrac{1}{4}a^{\frac{1}{2}}y^{-3/2}\sin(\tfrac{1}{8}a^2/y-\tfrac{5}{8}\pi)J_{-\frac{1}{4}}(\tfrac{1}{8}a^2/y)$
10.36	$u^{-\frac{1}{2}}\{\sin u\ C(u)$ $-\cos u\ S(u)\},u=a\ \sinh x$	$\tfrac{1}{8}i\pi^{3/2}\operatorname{sech}(\tfrac{1}{2}\pi y)\Big(I_{-\frac{1}{4}+\frac{1}{2}iy}(\tfrac{1}{2}a)I_{\frac{1}{4}+\frac{1}{2}iy}(\tfrac{1}{2}a)$ $-I_{-\frac{1}{4}-\frac{1}{2}iy}(\tfrac{1}{2}a)I_{\frac{1}{4}-\frac{1}{2}iy}(\tfrac{1}{2}a)\Big)$

2.11 The Exponential- and Related Integrals

	$f(x)$	$g_s(y)$	
11.1	$Ei(-ax)$	$-\tfrac{1}{2}y^{-1}\log(1+y^2/a^2)$	
11.2	$Ei(-bx)\qquad\qquad x<a$ $\qquad 0\qquad\qquad x>a$	$\tfrac{1}{2}y^{-1}\Big[Ei\{-a(b+iy)\}+Ei\{-a(b-iy)\}$ $-\log(1+y^2/b^2)-2\cos(ay)Ei(-ab)\Big]$	
11.3	$e^{-ax}Ei(-bx)$ $a\geq-b$	$(a^2+y^2)^{-1}\Big[a\ \arctan\{y/(a+b)\}+y\ \log b$ $-\tfrac{1}{2}y\ \log\{(a+b)^2+y^2\}\Big]$	
11.4	$e^{ax}Ei(-ax)$	$(a^2+y^2)^{-1}\{y\ \log(a/y)-\tfrac{1}{2}\pi a\}$	
11.5	$e^{-ax}\overline{Ei}(bx)$ $a>b$	$(a^2+y^2)^{-1}\Big[a\ \arctan\{y/(a-b)\}+y\ \log b$ $-\tfrac{1}{2}y\ \log\{(a-b)^2+y^2\}\Big]$	
11.6	$e^{-ax}\overline{Ei}(ax)$	$(a^2+y^2)^{-1}\{\tfrac{1}{2}\pi a-y\ \log(y/a)\}$	
11.7	$e^{-au}\overline{Ei}(au)-e^{au}\overline{Ei}(-au)$ $u=\sinh x$	$\pi\tanh(\tfrac{1}{2}\pi y)S_{0,iy}(a)$	
11.8	$si(ax)$	0 $-\tfrac{1}{2}\pi/y$	$y<a$ $y>a$
11.9	$Ci(ax)$	$-\tfrac{1}{2}y^{-1}\log\lvert1-y^2/a^2\rvert$	

	$f(x)$	$g_s(y)$
11.10	$Si(bx)$ $\quad\quad x<a$ $0 \quad\quad\quad x>a$	$y^{-1}\{\tfrac{1}{2}Si(ab+ay)+\tfrac{1}{2}Si(ab-ay)-\cos(ay)Si(ab)\}$
11.11	$e^{-bx}si(ax)$	$\tfrac{1}{2}(b^2+y^2)^{-1}\Big(y\ \arctan\{(y+a)/b\}$ $-y\ \arctan\{(y-a)/b\}+\tfrac{1}{2}b\ \log\{b^2+(y+a)^2\}$ $-\tfrac{1}{2}b\ \log\{b^2+(y-a)^2\}-\pi y\Big)$
11.12	$e^{-bx}Si(ax)$	$\tfrac{1}{2}(b^2+y^2)^{-1}\Big(y\ \arctan\{(y+a)/b\}$ $-y\ \arctan\{(y-a)/b\}+\tfrac{1}{2}b\ \log\{b^2+(y+a)^2\}$ $-\tfrac{1}{2}b\ \log\{b^2+(y-a)^2\}\Big)$
11.13	$(b^2+x^2)^{-1}si(ax)$	$\tfrac{1}{4}\pi b^{-1}\Big(e^{-by}\{\overline{E}i(by)-Ei(-ab)\}$ $+e^{by}\{Ei(-ab)-Ei(-by)\}\Big)$ $\quad\quad y<a$ $\tfrac{1}{4}\pi b^{-1}e^{-by}\{\overline{E}i(ab)-Ei(-ab)\}$ $\quad y>a$
11.14	$Ci(bx)$ $\quad\quad x<a$ $0 \quad\quad\quad x>a$	$\tfrac{1}{2}y^{-1}\Big(Ci(ay+ab)+Ci(|ay-ab|)$ $-2\cos(ay)Ci(ab)-\log(|1-y^2/b^2|)\Big)$ $\quad y\neq b$ $\tfrac{1}{2}b^{-1}\{\gamma+Ci(2ab)+\log(\tfrac{1}{2}ab)$ $-2\ \cos(ab)Ci(ab)\}$ $\quad\quad\quad y=b$
11.15	$x^{-1}Ci(ax)$	0 $\quad\quad\quad y<a$ $-\tfrac{1}{2}\pi\log(y/a)$ $\quad\quad y>a$
11.16	$(b^2+x^2)^{-1}Ci(ax)$	$\tfrac{1}{2}\pi\sinh(by)Ei(-ab)$ $\quad\quad y<a$ $\tfrac{1}{2}\pi\sinh(by)Ei(-by)+\tfrac{1}{4}\pi e^{-by}$ $\cdot\Big(Ei(-by)+\overline{E}i(by)-Ei(-ab)-\overline{E}i(ab)\Big)$ $\quad y>a$
11.17	$e^{-bx}Ci(ax)$	$\tfrac{1}{2}(a^2+y^2)^{-1}\Big(a\ \arctan\{2ay/(a^2+b^2-y^2)\}$ $-\tfrac{1}{2}y\ \log\{(a^2+b^2-y^2)^2+4a^2y^2\}+2y\ \log b\Big)$
11.18	$x^{-1}\{\sin(ax)si(ax)$ $+\cos(ax)Ci(ax)\}$	$-\tfrac{1}{2}\pi\log(1+y/a)$

	$f(x)$	$g_s(y)$
11.19	$\text{si}(ax^2)$	$-\pi y^{-1}\Big[\{C(\tfrac14 y^2/a)\}^2+\{S(\tfrac14 y^2/a)\}^2\Big]$
11.20	$\text{si}(ax)\,\text{Ci}(ax)$	$\qquad\qquad\qquad 0\qquad\qquad\qquad y<2a$ $\tfrac12\pi y^{-1}\log(y/a-1)\qquad\qquad y>2a$
11.21	$\text{si}(a/x)$	$-\tfrac12\pi y^{-1}J_0\{2(ay)^{\frac12}\}$
11.22	$x^{-1}\{\cos(a/x)\,\text{Ci}(a/x)$ $+\sin(a/x)\,\text{si}(a/x)\}$	$-\pi K_0\{2(ay)^{\frac12}\}$
11.23	$x^{-1}\{\sin(a/x)\,\text{Ci}(a/x)$ $-\cos(a/x)\,\text{si}(a/x)\}$	$\tfrac12\pi a^{-1}\Big[1-2(ay)^{\frac12}K_1\{2(ay)^{\frac12}\}\Big]$

2.12   Legendre Functions

	$f(x)$	$g_s(y)$
12.1	$p_\nu(1+ax)$ $-1<\text{Re}\,\nu<0$	$-(2ay/\pi)^{-\frac12}\Big[\sin(\tfrac12\pi\nu-y/a)J_{\nu+\frac12}(y/a)$ $+\cos(\tfrac12\pi\nu-y/a)Y_{\nu+\frac12}(y/a)\Big]$
12.2	$p_\nu(1+ax^2)$ $-1<\text{Re}\,\nu<0$	$(2a)^{-\frac12}K_{\nu+\frac12}(z)\{I_{\nu+\frac12}(z)+I_{-\nu-\frac12}(z)\}z=y(2a)^{-\frac12}$
12.3	$\qquad 0\qquad x<a$ $p_\nu(2x^2/a^2-1),\quad x>a$ $-1<\text{Re}\,\nu<0$	$-\tfrac14\pi a\ \text{sec}(\pi\nu)\Big[\{J_{\nu+\frac12}(\tfrac12 ay)\}^2-\{J_{-\nu-\frac12}(\tfrac12 ay)\}^2\Big]$
12.4	$\qquad 0\qquad x<a+b$ $p_\nu(\dfrac{x^2-a^2-b^2}{2ab}),\,x>a+b$ $-1<\text{Re}\,\nu<0$	$-\tfrac12\pi(ab)^{\frac12}\{J_{\nu+\frac12}(ay)Y_{-\nu-\frac12}(ay)$ $+Y_{\nu+\frac12}(ay)J_{-\nu-\frac12}(ay)\}$
12.5	$\{(a+x)(b+x)\}^{\frac12\nu}$ $\cdot p_\nu\{2(1+x/a)(1+x/b)-1\}$ $-1<\text{Re}\,\nu<0$	$-\tfrac14\pi(ab)^{\frac12\nu+\frac12}\Big[\{J_{\nu+\frac12}(u)J_{\nu+\frac12}(v)$ $-Y_{\nu+\frac12}(u)Y_{\nu+\frac12}(v)\}\cos\{\tfrac12(a+b)y-\pi\nu\}$ $+\{J_{\nu+\frac12}(u)Y_{\nu+\frac12}(v)+J_{\nu+\frac12}(v)Y_{\nu+\frac12}(u)\}$ $\cdot\sin\{\tfrac12(a+b)y-\pi\nu\}\Big]; \quad u=\tfrac12 ay;\ v=\tfrac12 by$

	$f(x)$	$g_s(y)$
12.6	$x^{-1}q_\nu(1+2a^2/x^2)$ $\mathrm{Re}\,\nu>-1$	$\tfrac{1}{2}\pi\Gamma(1+\nu)(ay)^{-1}W_{-\frac{1}{2}-\nu,0}(ay)M_{\frac{1}{2}+\nu,0}(ay)$
12.7	$z^{-n-\frac{1}{2}}Q_{2n}(xz^{-\frac{1}{2}})$ $n=0,1,2,\ldots;z=(a^2+x^2)$	$(-1)^n\tfrac{1}{2}\pi\{(2n)!\}^{-1}y^{2n}K_0(ay)$
12.8	$Q_{2n}(x/a)\qquad x<a$ $q_{2n}(x/a)\qquad x>a$ $n=0,1,2,\ldots$	$(-1)^n(\tfrac{1}{2}\pi)^{3/2}(a/y)^{\frac{1}{2}}J_{2n+\frac{1}{2}}(ay)$
12.9	$p_\nu(1+2a^2\sinh^2 x)$ $a<1\ ,-1<\mathrm{Re}\,\nu<0$	$\tfrac{1}{4}i\ \mathrm{sech}(\tfrac{1}{2}\pi y)\Big(P_\nu^{-\frac{1}{2}iy}(u)\{Q_\nu^{\frac{1}{2}iy}(u)+Q_{-\nu-1}^{\frac{1}{2}iy}(u)\}$ $-P_\nu^{\frac{1}{2}iy}(u)\{Q_\nu^{-\frac{1}{2}iy}(u)+Q_{-\nu-1}^{-\frac{1}{2}iy}(u)\}\Big)$ $u=(1-a^2)^{\frac{1}{2}}$
12.10	$p_\nu(1+2a^2\sinh^2 x)$ $a>1\ ,-1<\mathrm{Re}\,\nu<0$	$-\{4a\ \sin(\pi\nu)\}^{-1}\tanh(\tfrac{1}{2}\pi y)$ $\cdot\Big(P_{-\frac{1}{2}+\frac{1}{2}iy}^{-\nu-\frac{1}{2}}(u)\{Q_{-\frac{1}{2}+\frac{1}{2}iy}^{\nu+\frac{1}{2}}(u)+Q_{-\frac{1}{2}-\frac{1}{2}iy}^{\nu+\frac{1}{2}}(u)\}$ $+P_{-\frac{1}{2}+\frac{1}{2}iy}^{\nu+\frac{1}{2}}(u)\{Q_{-\frac{1}{2}+\frac{1}{2}iy}^{-\nu-\frac{1}{2}}(u)+Q_{-\frac{1}{2}-\frac{1}{2}iy}^{-\nu-\frac{1}{2}}(u)\}\Big)$ $u=(1-a^{-2})^{\frac{1}{2}}$
12.11	$p_\nu(2a^2\sinh^2 x-1),$ $\quad\quad 0\quad \sinh x<\tfrac{1}{a}$ $\quad\quad\quad \sinh x>\tfrac{1}{a}$ $-1<\mathrm{Re}\,\nu<0$	$(2\pi i a)^{-1}\Big(q_{-\frac{1}{2}+\frac{1}{2}iy}^{-\nu-\frac{1}{2}}(u)q_{-\frac{1}{2}+\frac{1}{2}iy}^{\nu+\frac{1}{2}}(u)$ $-q_{-\frac{1}{2}-\frac{1}{2}iy}^{-\nu-\frac{1}{2}}(u)q_{-\frac{1}{2}-\frac{1}{2}iy}^{-\nu+\frac{1}{2}}(u)\Big)$ $u=(1+a^{-2})^{\frac{1}{2}}$
12.12	$(\mathrm{sech}\ x)^{2n+1}Q_{2n}(\tanh x)$ $n=0,1,2,\ldots$	$(-1)^n 2^{2n-2}\{(2n)!\}^{-2}\sinh(\tfrac{1}{2}\pi y)$ $\cdot\{\Gamma(n+\tfrac{1}{2}+\tfrac{1}{2}iy)\ \Gamma(n+\tfrac{1}{2}-\tfrac{1}{2}iy)\}^2$
12.13	$\quad\quad 0\quad\quad x<A$ $P_\nu(b-ax^2)\quad A<x<B$ $\tfrac{2}{\pi}\sin(\pi\nu)q_\nu(ax^2-b),x>B$ $aA^2=b-1;aB^2=b+1;\quad b>1$ $\mathrm{Re}\,\nu>-1$	$\pi(2a)^{-\frac{1}{2}}J_{\nu+\frac{1}{2}}(u)J_{\nu+\frac{1}{2}}(v)$ ${u\atop v}=\tfrac{1}{2}a^{-\frac{1}{2}}y\{(b+1)^{\frac{1}{2}}\pm(b-1)^{\frac{1}{2}}\}$

	$f(x)$	$g_s(y)$
12.14	$(1-x^2)^{-\frac{1}{2}\mu}P_\nu^\mu(x)$ $x<1$ 0 $x>1$ $\mathrm{Re}\,\mu<1$	$\pi^{\frac{1}{2}}2^{\mu-2}\{\Gamma(3/2-\frac{1}{2}\mu-\frac{1}{2}\nu)\,\Gamma(2-\frac{1}{2}\mu+\frac{1}{2}\nu)\}^{-1}$ $\cdot(1-\mu-\nu)(2+\nu-\mu)y^{\mu-\frac{1}{2}}s_{\frac{1}{2}-\mu,\frac{1}{2}+\nu}(y)$
12.15	$u^{2n+1}Q_{2n}(ua\sinh x)$ $n=0,1,2,\ldots;$ $u=(b^2+a^2\sinh^2 x)^{-\frac{1}{2}}$	$(-1)^n 2^{2n-2}a^{-2n-1}\{(2n)!\}^{-2}\sinh(\frac{1}{2}\pi y)$ $\cdot\{\Gamma(n+\frac{1}{2}+\frac{1}{2}iy)\,\Gamma(n+\frac{1}{2}-\frac{1}{2}iy)\}^2$ $\cdot {}_2F_1(n+\frac{1}{2}+\frac{1}{2}iy,n+\frac{1}{2}-\frac{1}{2}iy;2n+1;1-b^2/a^2)$
12.16	$x^{\lambda-1}(1-x^2)^{-\frac{1}{2}\mu}P_\nu^\mu(x)\,,x<1$ 0 $x>1$ $\mathrm{Re}\,\lambda>-1$, $\mathrm{Re}\,\mu<1$	$\pi^{\frac{1}{2}}2^{\mu-\lambda-1}\Gamma(1+\lambda)$ $\cdot\{\Gamma(1+\frac{1}{2}\lambda-\frac{1}{2}\mu-\frac{1}{2}\nu)\,\Gamma(\frac{3}{2}+\frac{1}{2}\lambda+\frac{1}{2}\nu-\frac{1}{2}\mu)\}^{-1}y\,{}_2F_3(\frac{1}{2}+\frac{1}{2}\lambda$ $,1+\frac{1}{2}\lambda;\frac{3}{2},1+\frac{1}{2}\lambda-\frac{1}{2}\mu-\frac{1}{2}\nu,\frac{3}{2}+\frac{1}{2}\lambda-\frac{1}{2}\mu+\frac{1}{2}\nu;-\frac{1}{4}y^2)$
12.17	0 $x<1$ $(x^2-1)^{-\frac{1}{2}\mu}P_\nu^\mu(x)$ $x>1$ $-\frac{1}{2}<\mathrm{Re}\,\mu<1,\mathrm{Re}\,\mu>\mathrm{Re}\,\nu>-1-\mathrm{Re}\,\mu$	$(\frac{1}{2}\pi/y)^{\frac{1}{2}}y^\mu\Big(\sin(\frac{1}{2}\pi\mu-\frac{1}{2}\pi\nu)J_{\nu+\frac{1}{2}}(y)$ $-\cos(\frac{1}{2}\pi\mu-\frac{1}{2}\pi\nu)Y_{\nu+\frac{1}{2}}(y)\Big)$
12.18	$(x^2-1)^{\frac{1}{2}\mu}P_\nu^\mu(x)$ $\mathrm{Re}\,\mu<3/2$, $\mathrm{Re}(\mu+\nu)<1$	$(\pi/y)^{\frac{1}{2}}(\frac{1}{2}y)^{-\mu}\{\Gamma(\frac{1}{2}-\frac{1}{2}\mu-\frac{1}{2}\nu)\,\Gamma(1-\frac{1}{2}\mu+\frac{1}{2}\nu)\}^{-1}$ $\cdot S_{\mu+\frac{1}{2},\nu+\frac{1}{2}}(y)$
12.19	$(x+x^2)^{-\frac{1}{2}\mu}P_\nu^\mu(x)$ $\mathrm{Re}\,\mu<2,-1-\mathrm{Re}\,\mu<\mathrm{Re}\,\nu<\mathrm{Re}\,\mu$	$\frac{1}{2}(\pi/y)^{\frac{1}{2}}y^\mu\Big(\sin\{\frac{1}{2}y+\frac{1}{2}\pi(\mu-\nu)\}J_{\nu+\frac{1}{2}}(\frac{1}{2}y)$ $-\cos\{\frac{1}{2}y+\frac{1}{2}\pi(\mu-\nu)\}Y_{\nu+\frac{1}{2}}(\frac{1}{2}y)\Big)$
12.20	$z^{-\nu}(P_\nu^\mu(zx)+P_\nu^\mu(-zx)\}$ $=-2\pi^{-1}\cot(\frac{1}{2}\pi\nu+\frac{1}{2}\pi\mu)$ $\cdot z^{-\nu}\{Q_\nu^\mu(zx)-Q_\nu^\mu(-zx)\}$ $z=(a^2+x^2)^{-\frac{1}{2}},\mathrm{Re}(\nu\pm\mu)<0$	$\pi^{\frac{1}{2}}2^\mu y^{-\nu-1}\{\Gamma(\frac{1}{2}-\frac{1}{2}\nu-\frac{1}{2}\mu)\}^{-1}\Big(2^\nu\Gamma(1+\frac{1}{2}\nu+\frac{1}{2}\mu)$ $\cdot\{I_\mu(ay)+I_{-\mu}(ay)-2\pi^{-1}\sin(\pi\nu)K_\mu(ay)\}$ $+e^{-\frac{1}{2}i\pi\nu}\{\Gamma(1+\frac{1}{2}\nu-\frac{1}{2}\mu)\}^{-1}s_{\nu+1,\mu}(iay)\Big)$
12.21	$z^{-\nu}\{P_\nu^\mu(zx)-P_\nu^\mu(-zx)\}$ $=2\pi^{-1}\tan(\frac{1}{2}\pi\nu+\frac{1}{2}\pi\mu)$ $\cdot z^{-\nu}\{Q_\nu^\mu(zx)-Q_\nu^\mu(-zx)\}$ $z=(a^2+x^2)^{-\frac{1}{2}},\mathrm{Re}(\nu\pm\mu)<0$	$-2\cos(\frac{1}{2}\pi\nu-\frac{1}{2}\pi\mu)\{\Gamma(-\mu-\nu)\}^{-1}y^{-\nu-1}K_\mu(ay)$

	$f(x)$	$g_s(y)$
12.22	$\{1+(a^2+b^2)/x\}^{\frac{1}{2}\mu}$ $\cdot (a^2+x^2)^{\frac{1}{2}\nu}(b^2+x)^{-\frac{1}{2}\nu-\frac{1}{2}}$ $\cdot P_\nu^\mu\left[ab\{(a^2+x)(b^2+x)\}^{-\frac{1}{2}}\right]$ $\mathrm{Re}\,\mu<2$	$i\left[r^{-1}\exp\{\frac{1}{2}iy(a^2+b^2)\}D_{\mu+\nu}(ar)D_{\mu-\nu-1}(br)\right.$ $\left.-s^{-1}\exp\{-\frac{1}{2}iy(a^2+b^2)\}D_{\mu+\nu}(as)D_{\mu-\nu-1}(bs)\right]$ $r=(2iy)^{\frac{1}{2}}\ ,\ \ s=(-2iy)^{\frac{1}{2}}$
12.23	$\tanh(\pi x)P_{-\frac{1}{2}+ix}(\cosh a)$	$\begin{aligned}(2\cosh y-2\cosh a)^{-\frac{1}{2}} &\quad y>a\\ 0 &\quad y<a\end{aligned}$
12.24	$\mathrm{sech}(\pi x)$ $\cdot\{P_{\frac{1}{2}+ix}(a)-P_{\frac{1}{2}-ix}(a)\}$ $a>-1$	$i2^{\frac{1}{2}}\sinh y\{(1+\cosh y)^{-3/2}-(a+\cosh y)^{-3/2}\}$
12.25	$\Gamma(\frac{1}{2}-\mu+ix)\Gamma(\frac{1}{2}-\mu-ix)$ $\cdot\sinh(\pi x)P_{-\frac{1}{2}+ix}^\mu(\cosh a)$ $\mathrm{Re}\,\mu>-\frac{1}{2}$	$\begin{aligned}\pi(\frac{1}{2}\pi)^{\frac{1}{2}}\{\Gamma(\frac{1}{2}+\mu)\}^{-1}(\sinh a)^{-\mu}&\\ (\cosh y-\cosh a)^{\mu-\frac{1}{2}}&\quad y>a\\ 0&\quad y<a\end{aligned}$
12.26	$\Gamma(\frac{1}{2}-\mu+ix)\Gamma(\frac{1}{2}-\mu-ix)$ $\cdot\sinh(\frac{1}{2}\pi x)P_{-\frac{1}{2}+ix}^\mu(a),a>1$	$(\frac{1}{2}\pi)^{\frac{1}{2}}(a^2-1)^{-\frac{1}{2}\mu}\Gamma(\frac{1}{2}-\mu)(a^2+\sinh^2 y)^{\frac{1}{2}\mu-\frac{1}{4}}$ $\cdot\sin\{(\frac{1}{2}-\mu)\arctan(a^{-1}\sin y)\}$
12.27	$Q_{-\frac{1}{2}+ix}^\mu(\cosh a)$ $-Q_{-\frac{1}{2}-ix}^\mu(\cosh a),\mathrm{Re}\,\mu<\frac{1}{2}$	$\begin{aligned}-i\pi e^{i\mu\pi}(\frac{1}{2}\pi)^{\frac{1}{2}}\{\Gamma(\frac{1}{2}-\mu)\}^{-1}&\\ \cdot(\sinh a)^\mu(\cosh y-\cosh a)^{-\mu-\frac{1}{2}}&\quad y>a\\ 0&\quad y<a\end{aligned}$
12.28	$\tanh(\pi x)\,\mathrm{sech}(\pi x)$ $P_{-\frac{1}{2}+ix}(a)P_{-\frac{1}{2}+ix}(-a)$	$\pi^{-1}\{\cosh^2(\frac{1}{2}y)-a^2\}^{-\frac{1}{2}}$ $\cdot K\left(\sinh(\frac{1}{2}y)\{\cosh^2(\frac{1}{2}y)-a^2\}^{-\frac{1}{2}}\right)\quad a<1$
12.29	$\tanh(\pi x)P_{-\frac{1}{2}+ix}(a)$ $\{Q_{-\frac{1}{2}+ix}(a)+Q_{-\frac{1}{2}-ix}(a)\}$ $a>1$	$\begin{aligned}0\ ,&\quad \sinh(\frac{1}{2}y)<(a^2-1)^{\frac{1}{2}}\\ -uK\left(\{1-(a^2-1)u^2\}^{\frac{1}{2}}\right),&\quad \sinh(\frac{1}{2}y)>(a^2-1)^{\frac{1}{2}}\\ & u=\mathrm{cosech}(\frac{1}{2}y)\end{aligned}$

2.13 Bessel Functions of Arguments x, x^2 and $1/x$

	$f(x)$	$g_s(y)$	
13.1	$J_0(ax)$	0	$y<a$
		$(y^2-a^2)^{-\frac{1}{2}}$	$y>a$
13.2	$J_{2n+1}(ax)$	$(-1)^n(a^2-y^2)^{-\frac{1}{2}}T_{2n+1}(y/a)$,	$y<a$
	$n=0,1,\ldots$	0	$y>a$
13.3	$x^{-1}J_0(ax)$	$\arcsin(y/a)$	$y<a$
		$\frac{1}{2}\pi$	$y>a$
13.4	$x^{-\frac{1}{2}}J_{2n+\frac{1}{2}}(ax)$	$(-1)^n(\frac{1}{2}\pi a)^{-\frac{1}{2}}Q_{2n}(y/a)$	$y<a$
	$n=0,1,\ldots$	$(-1)^n(\frac{1}{2}\pi a)^{-\frac{1}{2}}q_{2n}(y/a)$	$y>a$
13.5	$x^{-\frac{1}{2}}J_{2n+3/2}(ax)$	$(-1)^n(\frac{1}{2}\pi/a)^{\frac{1}{2}}P_{2n+1}(y/a)$	$y<a$
	$n=0,1,\ldots$	0	$y>a$
13.6	$x^{-\frac{1}{2}}\log x\, J_0(ax)$	$(\pi a)^{-\frac{1}{2}}\left[\mathbf{K}\{(\frac{1}{2}-\frac{1}{2}\frac{y}{a})^{\frac{1}{2}}\}-\mathbf{K}\{(\frac{1}{2}+\frac{1}{2}\frac{y}{a})^{\frac{1}{2}}\}\right]$	
		$\cdot\{\gamma+\log 4+\frac{1}{2}\log(a^2-y^2)\}$	$y<a$
		$(2\pi)^{-\frac{1}{2}}(a+y)^{-\frac{1}{2}}\mathbf{K}\{(\frac{1}{2}+\frac{1}{2}\frac{y}{a})^{-\frac{1}{2}}\}$	
		$\cdot\{\pi-2\gamma-4\log 2-\log(y^2-a^2)\}$	$y>a$
13.7	$J_\nu(ax)$	$(a^2-y^2)^{-\frac{1}{2}}\sin\{\nu\arcsin(y/a)\}$	$y<a$
	$\mathrm{Re}\,\nu>-2$	$a^\nu\cos(\frac{1}{2}\pi\nu)(y^2-a^2)^{-\frac{1}{2}}\{y+(y^2-a^2)^{\frac{1}{2}}\}^{-\nu}$	$y>a$
13.8	$x^{-1}J_\nu(ax)$	$\nu^{-1}\sin\{\nu\arcsin(y/a)\}$	$y<a$
	$\mathrm{Re}\,\nu>-1$	$\nu^{-1}a^\nu\sin(\frac{1}{2}\pi\nu)\{y+(y^2-a^2)^{\frac{1}{2}}\}^{-\nu}$	$y>a$
13.9	$x^\nu J_\nu(ax)$	0	$y<a$
	$-1<\mathrm{Re}\,\nu<\frac{1}{2}$	$\pi^{\frac{1}{2}}\{\Gamma(\frac{1}{2}-\nu)\}^{-1}(2a)^\nu(y^2-a^2)^{-\nu-\frac{1}{2}}$	$y>a$

	$f(x)$	$g_s(y)$	
13.10	$x^{1-\nu} J_\nu(ax)$ $\mathrm{Re}\,\nu > \tfrac{1}{2}$	$\pi^{\frac{1}{2}} 2^{1-\nu} a^\nu \{\Gamma(\nu-\tfrac{1}{2})\}^{-1} y (a^2-y^2)^{\nu-3/2}$ 0	$y<a$ $y>a$
13.11	$x^{1+\nu} J_\nu(ax)$ $-3/2 < \mathrm{Re}\,\nu < -\tfrac{1}{2}$	$-\pi^{-\frac{1}{2}} 2^{\nu+1} \Gamma(\tfrac{3}{2}+\nu) a^\nu y (a^2-y^2)^{-\nu-3/2}$ $-2\pi^{-\frac{1}{2}} (2a)^\nu \Gamma(\tfrac{3}{2}+\nu) \sin(\pi\nu) y (y^2-a^2)^{-\frac{3}{2}-\nu}$	$y<a$ $,y>a$
13.12	$x^{-2} J_\nu(ax)$ $\mathrm{Re}\,\nu > 0$	$\tfrac{1}{2} a \nu^{-1} \Big((\nu-1)^{-1} \sin\{(\nu-1)\arcsin(y/a)\}$ $+(\nu+1)^{-1} \sin\{(\nu+1)\arcsin(y/a)\} \Big)$ $\tfrac{1}{2} a \nu^{-1} \cos(\tfrac{1}{2}\pi\nu) \Big((\nu+1)^{-1} a^{\nu+1} \{y+(y^2-a^2)^{\frac{1}{2}}\}^{-\nu-1}$ $- (\nu-1)^{-1} a^{\nu-1} \{y+(y^2-a^2)^{\frac{1}{2}}\}^{1-\nu} \Big)$	$y<a$ $y>a$
13.13	$x^{-\nu} J_{\nu+2n-1}(ax)$ $n=0,1,\dots, \mathrm{Re}\,\nu > -\tfrac{1}{2}$	$\tfrac{1}{2} (-1)^n (\tfrac{2}{a})^\nu (2n+1)! \, \Gamma(\nu) \{\Gamma(1+2n+2\nu)\}^{-1}$ $(a^2-y^2)^{\nu-\frac{1}{2}} C_{2n+1}^\nu (y/a)$ 0	 $y<a$ $y>a$
13.14	$x^\mu J_\nu(ax)$ $-2-\mathrm{Re}\,\nu < \mathrm{Re}\,\mu < \tfrac{1}{2}$	$(2\pi a)^{-\frac{1}{2}} \sin(\tfrac{1}{2}\pi\mu - \tfrac{1}{2}\pi\nu) \Gamma(1+\mu+\nu) \Gamma(1+\mu-\nu)$ $\cdot (a^2-y^2)^{-\frac{1}{2}\mu-\frac{1}{4}} \{P_{\nu-\frac{1}{2}}^{-\frac{1}{2}-\mu}(y/a) - P_{\nu-\frac{1}{2}}^{-\frac{1}{2}-\mu}(-y/a)\},$ $\Gamma(1+\nu+\mu) \cos(\tfrac{1}{2}\pi\nu + \tfrac{1}{2}\pi\mu)(y^2-a^2)^{-\frac{1}{2}\mu-\frac{1}{2}}$ $\cdot P_\mu^{-\nu}\{y(y^2-a^2)^{-\frac{1}{2}}\}$	 $y<a$ $y>a$
13.15	$x^{-1}\{(x+b)^{-\nu} J_{\nu+2n}(x+b)$ $+(x-b)^{-\nu} J_{\nu+2n}(x-b)\}$ $n=0,1,\dots \quad \mathrm{Re}\,\nu > -\tfrac{1}{2}$	$\pi b^{-1} J_{\nu+2n}(b)$	$y>1$
13.16	$(b^2+x^2)^{-1} J_0(ax)$	$b^{-1} \sinh(by) K_0(ab)$	$y<a$
13.17	$x(b^2+x^2)^{-1} J_0(ax)$	$\tfrac{1}{2}\pi e^{-by} I_0(ab)$	$y>a$
13.18	$(b^2+x^2)^{-1}$ $\cdot \{J_\nu(ax)+J_{-\nu}(ax)\}$ $-2 < \mathrm{Re}\,\nu < 2$	$2b^{-1} \cos(\tfrac{1}{2}\pi\nu) \sinh(by) K_\nu(ab)$	$y<a$

	$f(x)$	$g_s(y)$
13.19	$x^{\frac{1}{2}}(b^2+x^2)^{-1}J_{2n+\frac{1}{2}}(ax)$ $n=0,1,\ldots$	$(-1)^n\sinh(by)K_{2n+\frac{1}{2}}(ab)$ $\qquad y<a$
13.20	$x^{\nu}(b^2+x^2)^{-1}J_{\nu}(ax)$ $-1<\mathrm{Re}\,\nu<5/2$	$b^{\nu-1}\sinh(by)K_{\nu}(ab)$ $\qquad y<a$
13.21	$x^{1-\nu}(b^2+x^2)^{-1}J_{\nu}(ax)$ $\mathrm{Re}\,\nu>-3/2$	$\tfrac{1}{2}\pi b^{-\nu}e^{-by}I_{\nu}(ab)$ $\qquad y>a$
13.22	$x^{\nu+2n}(b^2+x^2)^{-1}J_{\nu}(ax)$ $-1<\mathrm{Re}\,(\nu+n)<\tfrac{1}{2}-n,n=0,\pm1,..$	$(-1)^n b^{\nu+2n-1}\sinh(by)K_{\nu}(ab)$ $\qquad y<a$
13.23	$x^{2n+1-\nu}(b^2+x^2)^{-1}J_{\nu}(ax)$ $\mathrm{Re}\,\nu>2n-3/a,n=1,0,1,\ldots$	$(-1)^n\tfrac{1}{2}\pi b^{2n-\nu}e^{-by}I_{\nu}(ab)$ $\qquad y>a$
13.24	$x^{-1}e^{-bx}J_0(ax)$	$\arcsin(2y/z)$ $z=\{b^2+(a+y)^2\}^{\frac{1}{2}}+\{b^2+(a-y)^2\}^{\frac{1}{2}}$
13.25	$x^{-\frac{1}{2}}\log x\,J_{\nu}(ax)$ $\mathrm{Re}\,\nu>-3/2$	$(\tfrac{1}{2}\pi/a)^{\frac{1}{2}}\sec(\pi\nu)\Big[\{P_{\nu-\frac{1}{2}}(y/a)-P_{\nu-\frac{1}{2}}(-y/a)\}$ $\cdot\{\tfrac{1}{2}\pi\cos(\tfrac{1}{4}\pi+\tfrac{1}{2}\pi\nu)-\Psi(\tfrac{1}{2}+\nu)\sin(\tfrac{1}{4}\pi+\tfrac{1}{2}\pi\nu)$ $+\tfrac{1}{2}\sin(\tfrac{1}{4}\pi+\tfrac{1}{2}\pi\nu)\log(a^2-y^2)\}-\sin(\tfrac{1}{4}\pi+\tfrac{1}{2}\pi\nu)$ $\cdot\{Q_{-\nu-\frac{1}{2}}(y/a)-Q_{-\nu-\frac{1}{2}}(-y/a)\}\Big]$ $\qquad y<a$ $(\tfrac{1}{2}\pi a)^{-\frac{1}{2}}q_{\nu-\frac{1}{2}}(y/a)\{\cos(\tfrac{1}{4}\pi-\tfrac{1}{2}\pi\nu)\Psi(\tfrac{1}{2}+\nu)$ $+\tfrac{1}{2}\pi\sin(\tfrac{1}{4}\pi-\tfrac{1}{2}\pi\nu)-\tfrac{1}{2}\cos(\tfrac{1}{4}\pi-\tfrac{1}{2}\pi\nu)\log(y^2-a^2)\}$ $\qquad y>a$
13.26	$x^{\nu}\cos x\,J_{\nu}(x)$ $-1<\mathrm{Re}\,\nu<\tfrac{1}{2}$	$2^{\nu-1}\pi^{\frac{1}{2}}\{\Gamma(\tfrac{1}{2}-\nu)\}^{-1}(y^2+2y)^{-\nu-\frac{1}{2}}$ $\quad y<2$ $2^{\nu-1}\pi^{\frac{1}{2}}\{\Gamma(\tfrac{1}{2}-\nu)\}^{-1}\{(y^2+2y)^{-\nu-\frac{1}{2}}$ $+(y^2-2y)^{-\nu-\frac{1}{2}}\}$ $\qquad y>2$
13.27	$x^{1-\nu}\cos x\,J_{\nu}(x)$ $\mathrm{Re}\,\nu>\tfrac{1}{2}$	$2^{-\nu}\pi^{\frac{1}{2}}\{\Gamma(\tfrac{1}{2}-\nu)\}^{-1}(y-1)(2y-y^2)^{\nu-3/2}$ $\;\; y<2$ 0 $\qquad y>2$

	$f(x)$	$g_s(y)$	
13.28	$x^{-\nu}\sin x\, J_\nu(x)$ $\mathrm{Re}\,\nu > -\tfrac{1}{2}$	$2^{-\nu-1}\pi^{\frac{1}{2}}\{\Gamma(\tfrac{1}{2}+\nu)\}^{-1}(2y-y^2)^{\nu-\frac{1}{2}}$ 0	$y<2$ $y>2$
13.29	$\{J_0(ax)\}^2$	$(\pi a)^{-1}\mathbf{K}(\tfrac{1}{2}y/a)$ $2(\pi y)^{-1}\mathbf{K}(2a/y)$	$y<2a$ $y>2a$
13.30	$J_0(ax)J_0(bx)$	0 $\pi^{-1}(ab)^{-\frac{1}{2}}\mathbf{K}(z)$ $(\pi z)^{-1}(ab)^{-\frac{1}{2}}\mathbf{K}(1/z)$ $z=\tfrac{1}{2}(ab)^{-\frac{1}{2}}\{y^2-(a-b)^2\}^{\frac{1}{2}}$	$y<\lvert a-b\rvert$ $\lvert a-b\rvert<y<a+b$ $y>a+b$
13.31	$x^{-\frac{1}{2}}\{J_0(ax)\}^2$	$(\tfrac{1}{2}\pi/y)^{\frac{1}{2}}\left[P_{-\frac{1}{4}}\{(1-4a^2/y^2)^{\frac{1}{2}}\}\right]^2$	$y>2a$
13.32	$J_{n+\frac{1}{2}}(ax)J_{n+\frac{1}{2}}(bx)$ $n=0,1,2,\ldots$	$\tfrac{1}{2}(ab)^{-\frac{1}{2}}P_n\{(a^2+b^2-y^2)/(2ab)\}$ 0	$y<a+b$ $y>a+b$
13.33	$x^{\frac{1}{2}}\{J_{\frac{1}{4}}(ax)\}^2$	$(\tfrac{1}{2}\pi y)^{-\frac{1}{2}}(4a^2-y^2)^{-\frac{1}{2}}$ 0	$y<2a$ $y>2a$
13.34	$x^{\frac{1}{2}}J_{\frac{1}{4}-\nu}(ax)J_{\frac{1}{4}+\nu}(ax)$	$(\tfrac{1}{2}\pi y)^{-\frac{1}{2}}(4a^2-y^2)^{-\frac{1}{2}}\cos\{2\nu\,\mathrm{arccos}\,(\tfrac{1}{2}y/a)\}$ 0	$y<2a$ $y>2a$
13.35	$x^{-\frac{1}{2}}\{J_\nu(ax)\}^2$ $\mathrm{Re}\,\nu > -\tfrac{1}{4}$	$(\tfrac{1}{2}\pi/y)^{\frac{1}{2}}\Gamma(\tfrac{3}{4}+\nu)\{\Gamma(\tfrac{3}{4}-\nu)\}^{-1}$ $\cdot\left[P_{-\frac{1}{4}}^{-\nu}\{(1-4a^2/y^2)^{\frac{1}{2}}\}\right]^2$	$y>2a$
13.36	$x^{-\frac{1}{2}}J_\nu(ax)J_{\mp\nu}(ax)$	$(\tfrac{1}{2}\pi/y)^{\frac{1}{2}}P_{-\frac{1}{4}}^{\nu}\{(1-4a^2/y^2)^{\frac{1}{2}}\}$ $\cdot P_{-\frac{1}{4}}^{-\nu}\{(1-4a^2/y^2)^{\frac{1}{2}}\}$	$y>2a$
13.37	$\{J_\nu(ax)\}^2-\{J_{-\nu}(ax)\}^2$ $-1<\mathrm{Re}\,\nu<1$	$-a^{-1}\sin(\pi\nu)P_{\nu-\frac{1}{2}}(\tfrac{1}{2}y^2/a^2-1)$ 0	$y>2a$ $y<2a$

	$f(x)$	$g_s(y)$
13.38	$J_\nu(ax)J_\nu(bx)$ $\mathrm{Re}\,\nu>-\tfrac{1}{2}$ $u=-v=(a^2+b^2-y^2)/(2ab)$	$\begin{aligned}&0 &\quad y<\lvert b-a\rvert\\ &\tfrac{1}{2}(ab)^{-\frac{1}{2}}P_{\nu-\frac{1}{2}}(u) &\quad \lvert b-a\rvert<y<b+a\\ &\pi^{-1}(ab)^{-\frac{1}{2}}\cos(\pi\nu)q_{\nu-\frac{1}{2}}(v) &\quad y>b+a\end{aligned}$
13.39	$x^{\nu-\mu}J_\nu(ax)J_\mu(bx)$ $-1<\mathrm{Re}\,\nu<1+\mathrm{Re}\,\mu,\ a>b$	0
13.40	$x^{\nu-\mu-2}J_\nu(ax)J_\mu(bx)$ $0<\mathrm{Re}\,\nu<3+\mathrm{Re}\,\mu\ ,\ a>b$	$2^{\nu-\mu-1}b^\mu a^{-\nu}\{\Gamma(1+\mu)\}^{-1}y\Gamma(\nu),\quad y<a-b$
13.41	$J_\nu(ax)J_\mu(bx)$ $\mathrm{Re}\,(\nu+\mu)>-2$	Watson,G.N. Journal Lond.Math.Soc. Vol.9,p.21,(1936)
13.42	$x^\lambda J_\nu(ax)J_\mu(bx)$ $\mathrm{Re}\,(\lambda+\nu+\mu)>-2$ $\mathrm{Re}\,\lambda<1$	Bailey,W.N. Proc.Lond.Math.Soc. Vol.40,p.37,(1936)
13.43	$x^{-1}\Big(\{J_{n+\frac{1}{2}}(x-b)\}^2$ $-\{J_{n+\frac{1}{2}}(x+b)\}^2\Big)$	$\pi\{J_{n+\frac{1}{2}}(b)\}^2 \qquad\qquad y>2$ $n=0,1,2,\ldots$
13.44	$x^{-1}\Big((x-b)^{-2\nu}$ $\cdot\{J_{\nu+n}(x-b)\}^2+(x+b)^{-2\nu}$ $\cdot\{J_{\nu+n}(x+b)\}^2\Big),\ \mathrm{Re}\,\nu>-1$	$\pi b^{-2\nu}\{J_{\nu+n}(b)\}^2 \qquad\qquad y\geq2$ $n=0,1,2,\ldots$
13.45	$Y_0(ax)$	$2\pi^{-1}(a^2-y^2)^{-\frac{1}{2}}\arcsin(y/a) \qquad y<a$ $2\pi^{-1}(y^2-a^2)^{-\frac{1}{2}}\log\Big(a^{-1}\{y-(y^2-a^2)^{\frac{1}{2}}\}\Big)\ y>a$
13.46	$Y_1(ax)$	$\begin{aligned}&0 &\quad y<a\\ &-a^{-1}y(y^2-a^2)^{-\frac{1}{2}} &\quad y>a\end{aligned}$
13.47	$Y_\nu(ax)$ $-2<\mathrm{Re}\,\nu<2$	$\cot(\tfrac{1}{2}\pi\nu)(a^2-y^2)^{-\frac{1}{2}}\sin\{\nu\arcsin(y/a)\}\ y<a$ $\tfrac{1}{2}\mathrm{cosec}(\tfrac{1}{2}\pi\nu)(y^2-a^2)^{-\frac{1}{2}}\Big(a^{-\nu}\cos(\pi\nu)$ $\cdot\{y-(y^2-a^2)^{\frac{1}{2}}\}^\nu-a^\nu\{y-(y^2-a^2)^{\frac{1}{2}}\}^{-\nu}\Big)\ y>a$

	$f(x)$	$g_s(y)$
13.48	$x^{-1}Y_0(ax)$	$0 \qquad\qquad\qquad\qquad y<a$ $\log\left[a^{-1}\{y-(y^2-a^2)^{\frac12}\}\right] \qquad y>a$
13.49	$x^{-1}\{\text{Ci}(ax)-\log 2$ $-\tfrac12\pi Y_0(ax)\}$	$\pi\log\left[\tfrac12 y^{-\frac12}\{(y+a)^{\frac12}+(y-a)^{\frac12}\}\right] \quad y>a$ $0 \qquad\qquad\qquad\qquad\qquad\qquad\qquad y<a$
13.50	$x^{-1}Y_\nu(ax)$ $-1<\text{Re}\,\nu<1$	$-\nu^{-1}\tan(\tfrac12\pi\nu)\sin\{\nu\arcsin(y/a)\} \quad y<a$ $\tfrac12\nu^{-1}\sec(\tfrac12\pi\nu)\left[a^{-\nu}\cos(\pi\nu)\{y-(y^2-a^2)^{\frac12}\}^\nu\right.$ $\left. -a^\nu\{y-(y^2-a^2)^{\frac12}\}^{-\nu}\right] \qquad\qquad\qquad y>a$
13.51	$x^{1+\nu}Y_\nu(ax)$ $-3/2<\text{Re}\,\nu<3/2$	$0 \qquad\qquad\qquad\qquad\qquad\qquad\qquad y<a$ $\pi^{\frac12}2^{1+\nu}a^\nu\{\Gamma(-\tfrac12-\nu)\}^{-1}y(y^2-a^2)^{-\nu-3/2} \quad y>a$
13.52	$x^{1-\nu}Y_\nu(ax)$ $\tfrac12<\text{Re}\,\nu<3/2$	$2(2a)^{-\nu}\pi^{-\frac12}\sin(\pi\nu)\,\Gamma(\tfrac32-\nu)y(a^2-y^2)^{\nu-3/2},$ $\qquad\qquad\qquad\qquad\qquad\qquad\qquad\qquad y<a$ $-2\pi^{-\frac12}\Gamma(\tfrac32-\nu)(2a)^{-\nu}y(y^2-a^2)^{\nu-3/2} \quad y>a$
13.53	$x^\mu Y_\nu(ax)$ $\text{Re}\,\mu<\tfrac12,\ \text{Re}(\mu\pm\nu)>-2$	$-(2\pi a)^{-\frac12}\cos(\tfrac12\pi\mu-\tfrac12\pi\nu)\,\Gamma(1+\mu+\nu)\,\Gamma(1+\mu-\nu)$ $\cdot(a^2-y^2)^{-\frac12\mu-\frac14}\left[P_{\nu-\frac12}^{-\mu-\frac12}(\tfrac{y}{a})-P_{\nu-\frac12}^{-\mu-\frac12}(-\tfrac{y}{a})\right] \quad y<a$ $-(\tfrac12\pi a)^{-\frac12}(y^2-a^2)^{-\frac12\mu-\frac14}\left(\sin(\tfrac12\pi\mu+\tfrac12\pi\nu)\right.$ $\cdot e^{-i\pi(\frac12+\mu)}q_{\nu-\frac12}^{\frac12+\mu}(\tfrac{y}{a})+\cos(\tfrac12\pi\mu-\tfrac12\pi\nu)$ $\left. \cdot\Gamma(1+\mu+\nu)\,\Gamma(1+\mu-\nu)\,p_{\nu-\frac12}^{-\frac12-\mu}(y/a)\right] \qquad y>a$
13.54	$\cos(ax-\tfrac12\nu\pi)J_\nu(ax)$ $+\sin(ax-\tfrac12\nu\pi)Y_\nu(ax)$ $-1<\text{Re}\,\nu<1$	$\tfrac12(2a)^{-\nu}(y+2a)^{-\frac12}y^{-\frac12}$ $\cdot\left(\{(y+2a)^{\frac12}+y^{\frac12}\}^{2\nu}+\{(y+2a)^{\frac12}-y^{\frac12}\}^{2\nu}\right)$
13.55	$(b^2+x^2)^{-1}$ $\cdot\{J_0(ax)\log x+\tfrac12\pi Y_0(ax)\}$	$b^{-1}\log b\,\sinh(by)K_0(ab) \qquad\qquad y<a$
13.56	$(b^2+x^2)^{-1}$ $\cdot\{Y_{-\nu}(ax)-Y_\nu(ax)\}$ $-2<\text{Re}\,\nu<2$	$2b^{-1}\sin(\tfrac12\pi\nu)\sinh(by)K_\nu(ab) \qquad\qquad y<a$

	$f(x)$	$g_s(y)$
13.57	$x^{\frac{1}{2}}(b^2+x^2)^{-1}$ $\cdot\{J_\nu(ax)\cos(\frac{1}{2}\pi\nu-\frac{1}{4}\pi)$ $-Y_\nu(ax)\sin(\frac{1}{2}\pi\nu-\frac{1}{4}\pi)\}$ $-\frac{5}{2}<\text{Re}\nu<\frac{5}{2}$	$b^{-\frac{1}{2}}\sinh(by)K_\nu(ab)$ $y<a$
13.58	$J_0(ax)Y_0(ax)$	$-2(\pi y)^{-1}\mathbf{K}\{(1-4a^2/y^2)^{\frac{1}{2}}\}$ $y>2a$ 0 $y<2a$
13.59	$J_0(ax)Y_0(bx)$ $+Y_0(ax)J_0(bx)$	0 $y<a+b$ $-4\pi^{-1}v^{-\frac{1}{2}}\mathbf{K}(u/v)$ $y>a+b$ $\frac{u}{v}=y^2-(a\pm b)^2$
13.60	$J_\nu(ax)Y_{-\nu}(bx)$ $+Y_\nu(ax)J_{-\nu}(bx)$	0 $y<a+b$ $-(ab)^{-\frac{1}{2}}P_{\nu-\frac{1}{2}}\{(y^2-a^2-b^2)/(2ab)\}$ $y>a+b$
13.61	$x^{\frac{1}{2}}J_{\frac{1}{4}}(ax)Y_{\frac{1}{4}}(ax)$	0 $y<2a$ $-(\frac{1}{2}\pi y)^{-\frac{1}{2}}(y^2-4a^2)^{-\frac{1}{2}}$ $y>2a$
13.62	$x^{\frac{1}{2}}J_{\frac{1}{4}-\nu}(ax)Y_{\frac{1}{4}+\nu}(ax)$ $\text{Re}\nu<\frac{5}{4}$	$-(\frac{1}{2}\pi y)^{-\frac{1}{2}}(4a^2-y^2)^{-\frac{1}{2}}\sin\{2\nu\arccos(\frac{1}{2}\frac{y}{a})\}$, $y<2a$ $-(\frac{1}{2}\pi y)^{-\frac{1}{2}}(2a)^{-2\nu}(y^2-4a^2)^{-\frac{1}{2}}$ $\cdot\{(y+(y^2-4a^2)^{\frac{1}{2}}\}^{2\nu}$ $y>2a$
13.63	$x^{-1}\Big(J_{n+\frac{1}{2}}(x-b)Y_{n+\frac{1}{2}}(x-b)$ $+J_{n+\frac{1}{2}}(x+b)Y_{n+1}(x+b)\Big)$	$\pi J_{n+\frac{1}{2}}(b)Y_{n+\frac{1}{2}}(b)$ $y\geq 2$
13.64	$x^{-\frac{1}{2}}\Big(\{J_\nu(z)\}^2+\{Y_\nu(z)\}^2$ $z=ax$, $-\frac{5}{4}<\text{Re}\nu<\frac{5}{4}$	$(\frac{1}{2}\pi y)^{-\frac{1}{2}}\{P_{-\frac{1}{4}}^\nu(u)Q_{-\frac{1}{4}}^\nu(u)-P_{-\frac{1}{4}}^{-\nu}(u)Q_{-\frac{1}{4}}^{-\nu}(u)\}$, $y>2a$ $u=(1-4y^2/a^2)^{\frac{1}{2}}$
13.65	$J_0(ax^2)$	$\frac{1}{8}\pi ya^{-1}J_{\frac{1}{4}}(u)\{J_{\frac{1}{4}}(u)-Y_{\frac{1}{4}}(u)\}$ $u=\frac{1}{8}y^2/a$
13.66	$\sin(\frac{1}{4}ax^2)J_0(\frac{1}{4}ax^2)$	$(2a/\pi)^{-\frac{1}{2}}\sin(\frac{1}{4}\pi-\frac{1}{4}y^2/a)J_0(\frac{1}{8}y^2/a)$

	$f(x)$	$g_s(y)$
13.67	$\cos(\tfrac{1}{2}ax^2)J_o(\tfrac{1}{2}ax^2)$	$(2a/\pi)^{-\frac{1}{2}}\cos(\tfrac{1}{4}\pi-\tfrac{1}{4}y^2/a)J_o(\tfrac{1}{4}y^2/a)$
13.68	$x^{\frac{1}{2}}J_{\frac{1}{4}}(ax^2)$	$\tfrac{1}{2}a^{-1}(\tfrac{1}{2}\pi y)^{\frac{1}{2}}J_{\frac{1}{4}}(\tfrac{1}{4}y^2/a)$
13.69	$x^{\frac{1}{2}}J_{-\frac{1}{4}}(ax^2)$	$\tfrac{1}{2}a^{-1}(\tfrac{1}{2}\pi y)^{\frac{1}{2}}\{H_{\frac{1}{4}}(\tfrac{1}{4}y^2/a)+J_{\frac{1}{4}}(\tfrac{1}{4}y^2/a)\}$
13.70	$x^{3/2}J_{\frac{1}{4}}(ax^2)$	$\tfrac{1}{4}a^{-11/4}y(\tfrac{1}{2}\pi y)^{\frac{1}{2}}J_{-3/4}(\tfrac{1}{4}y^2/a)$
13.71	$x^{3/2}J_{3/4}(ax^2)$	$\tfrac{1}{4}a^{-11/4}y(\tfrac{1}{2}\pi y)^{\frac{1}{2}}J_{-\frac{1}{4}}(\tfrac{1}{4}y^2/a)$
13.72	$x^{\frac{1}{2}}\exp(-ax^2)J_{\frac{1}{4}}(bx^2)$	$\tfrac{1}{2}(\tfrac{1}{2}\pi y)^{\frac{1}{2}}(a^2+b^2)^{-\frac{1}{2}}\exp\{-\tfrac{1}{4}ay^2(a^2+b^2)^{-1}\}$ $\cdot J_{\frac{1}{4}}\{\tfrac{1}{4}by^2(a^2+b^2)^{-1}\}$
13.73	$x^{\frac{1}{2}}\sin(ax^2)J_{\frac{1}{4}}(ax^2)$	$-\tfrac{1}{2}(ay)^{-\frac{1}{2}}\sin\{(\tfrac{1}{8}y^2/a)-3\pi/8\}$
13.74	$x^{\frac{1}{2}}\cos(ax^2)J_{\frac{1}{4}}(ax^2)$	$\tfrac{1}{2}(ay)^{-\frac{1}{2}}\cos\{(\tfrac{1}{8}y^2/a)-3\pi/8\}$
13.75	$x^{1/3}\cos(ax^2)J_{1/3}(ax^2)$	$2^{-73/6}a^{-5/6}y^{1/3}\left[\cos(z+\pi/12)J_{1/3}(z)\right.$ $\left.-\sin(z+\pi/12)Y_{1/3}(z)\right]; \; z=y^2/(16a)$
13.76	$x^{1/3}\sin(ax^2)J_{1/3}(ax^2)$	$-2^{-73/6}a^{-5/6}y^{1/3}\left[\sin(z+\pi/12)J_{1/3}(z)\right.$ $\left.+\cos(z+\pi/12)Y_{1/3}(z)\right]; \; z=y^2/(16a)$
13.77	$x^{\frac{1}{2}}\{J_{1/8}(ax^2)\}^2$	$-\tfrac{1}{2}a^{-1}(\tfrac{1}{2}\pi y)^{\frac{1}{2}}J_{1/8}(z)Y_{1/8}(z); \; z=y^2/(16a)$
13.78	$x^{\frac{1}{2}}J_{1/8}(ax^2)J_{-1/8}(ax^2)$	$\tfrac{1}{4}a^{-1}(\tfrac{1}{2}\pi y)^{\frac{1}{2}}J_{1/8}(z)\left[\sin(\pi/8)J_{1/8}(z)\right.$ $\left.-\cos(\pi/8)Y_{1/8}(z)\right] \; ; \; z=y^2/(16a)$
13.79	$x^{\frac{1}{2}}Y_{\frac{1}{4}}(ax^2)$	$-\tfrac{1}{2}a^{-1}(\tfrac{1}{2}\pi y)^{\frac{1}{2}}H_{\frac{1}{4}}(\tfrac{1}{4}y^2/a)$
13.80	$x^{3/2}Y_{3/4}(ax^2)$	$-\tfrac{1}{4}a^{-2}(\tfrac{1}{2}\pi y)^{\frac{1}{2}}H_{-\frac{1}{4}}(\tfrac{1}{4}y^2/a)$
13.81	$x^{\frac{1}{2}}J_{1/8}(ax^2)Y_{1/8}(ax^2)$	$-\tfrac{1}{4}a^{-1}(\tfrac{1}{2}\pi y)^{\frac{1}{2}}\{J_{1/8}(y^2/16a)\}^2$

	$f(x)$	$g_s(y)$
13.82	$x^{-1}e^{-a/x}J_\nu(b/x)$ $\mathrm{Re}\,\nu>-1$	$i\left(J_\nu(ui^{\frac{1}{2}})K_\nu(vi^{\frac{1}{2}})-J_\nu(ui^{-\frac{1}{2}})K_\nu(vi^{-\frac{1}{2}})\right)$ ${}^{\,v}_{\,u}=(2y)^{\frac{1}{2}}\{(a^2+b^2)^{\frac{1}{2}}\pm a\}^{\frac{1}{2}}$
13.83	$x^{-1}e^{-a/x}Y_\nu(b/x)$ $-1<\mathrm{Re}\,\nu<1$	$i\left(Y_\nu(ui^{\frac{1}{2}})K_\nu(vi^{\frac{1}{2}})-Y_\nu(ui^{-\frac{1}{2}})K_\nu(vi^{-\frac{1}{2}})\right)$ ${}^{\,v}_{\,u}=(2y)^{\frac{1}{2}}\{(a^2+b^2)^{\frac{1}{2}}\pm a\}^{\frac{1}{2}}$
13.84	$x^{-1}\sin(a/x)J_\nu(b/x)$ $\mathrm{Re}\,\nu>-2$	$\frac{1}{2}\pi J_\nu(uy^{\frac{1}{2}})\{\sin(\frac{1}{2}\pi\nu)J_\nu(vy^{\frac{1}{2}})$ $+\cos(\frac{1}{2}\pi\nu)Y_\nu(vy^{\frac{1}{2}})\}+\cos(\frac{1}{2}\pi\nu)I_\nu(uy^{\frac{1}{2}})K_\nu(vy^{\frac{1}{2}})$ ${}^{\,u}_{\,v}=(a+b)^{\frac{1}{2}}\pm(a-b)^{\frac{1}{2}}$
13.85	$x^{-1}\cos(a/x)J_\nu(b/x)$ $\mathrm{Re}\,\nu>-1$	$\frac{1}{2}\pi J_\nu(uy^{\frac{1}{2}})\{\cos(\frac{1}{2}\pi\nu)J_\nu(vy^{\frac{1}{2}})$ $-\sin(\frac{1}{2}\pi\nu)Y_\nu(vy^{\frac{1}{2}})\}-\sin(\frac{1}{2}\pi\nu)I_\nu(uy^{\frac{1}{2}})K_\nu(vy^{\frac{1}{2}})$ ${}^{\,u}_{\,v}=(a+b)^{\frac{1}{2}}\pm(a-b)^{\frac{1}{2}}$
13.86	$x^{-\frac{1}{2}}\sin(a/x)J_{2n-\frac{1}{2}}(a/x)$ $n=0,1,2,\ldots$	$\frac{1}{2}(-1)^{n-1}(\pi/y)^{\frac{1}{2}}J_{4n-1}\{2(2ay)^{\frac{1}{2}}\}$
13.87	$x^{-\frac{1}{2}}\cos(a/x)J_{2n+\frac{1}{2}}(a/x)$ $n=0,1,2,\ldots$	$\frac{1}{2}(-1)^{n-1}(\pi/y)^{\frac{1}{2}}J_{4n+1}\{2(2ay)^{\frac{1}{2}}\}$
13.88	$x^{-1}J_\nu(a/x)$	$i\{J_\nu(u)K_\nu(u)-J_\nu(v)K_\nu(v)\}\quad {}^{\,u}_{\,v}=(\pm2iay)^{\frac{1}{2}}$
13.89	$x^{-1}\sin(a/x)Y_\nu(b/x)$ $-2<\mathrm{Re}\,\nu<2$	$\frac{1}{2}\pi Y_\nu(u)\left(\cos(\frac{1}{2}\pi\nu)Y_\nu(v)+\sin(\frac{1}{2}\pi\nu)J_\nu(v)\right)$ $-K_\nu(v)\{\sin(\frac{1}{2}\pi\nu)J_\nu(u)+2\pi^{-1}\cos(\frac{1}{2}\pi\nu)K_\nu(u)\}$ $a>b \qquad\qquad {}^{\,u}_{\,v}=y^{\frac{1}{2}}\{(a+b)^{\frac{1}{2}}\pm(a-b)^{\frac{1}{2}}\}$
13.90	$x^{-1}\cos(a/x)Y_\nu(b/x)$ $-2<\mathrm{Re}\,\nu<2$	$-\frac{1}{2}\pi Y_\nu(u)\left(\sin(\frac{1}{2}\pi\nu)Y_\nu(v)-\cos(\frac{1}{2}\pi\nu)J_\nu(v)\right)$ $-K_\nu(v)\{\cos(\frac{1}{2}\pi\nu)I_\nu(u)+2\pi^{-1}\sin(\frac{1}{2}\pi\nu)K_\nu(u)\}$ ${}^{\,u}_{\,v}=y^{\frac{1}{2}}\{(a+b)^{\frac{1}{2}}\pm(a-b)^{\frac{1}{2}}\};\quad a>b$

	$f(x)$	$g_s(y)$
13.91	$x^{2\mu}J_{2\nu}(a/x)$ $-\frac{5}{4}<\mathrm{Re}_\mu<\mathrm{Re}\nu$	$\pi^{\frac{1}{2}}4^{\mu-2\nu}a^{2\nu}\Gamma(1+\mu-\nu)\{\Gamma(1+2\mu)\Gamma(\frac{1}{2}+\nu-\mu)\}^{-1}$ $y^{2\nu-2\mu-1}{}_0F_3(1+2\nu,\nu-\mu,\frac{1}{2}+\nu-\mu;a^2y^2/16)$ $+2^{-2\mu-3}a^{2+2\mu}\{\Gamma(\nu+\mu+2)\}^{-1}\Gamma(\nu-\mu-1)y$ $\cdot {}_0F_3(\frac{3}{2},2+\mu-\nu,2+\nu+\mu;a^2y^2/16)$

2.14 Bessel Functions of Argument $(ax^2+bx+c)^{\frac{1}{2}}$

	$f(x)$	$g_s(y)$
14.1	$J_0(ax^{\frac{1}{2}})$	$y^{-1}\cos(\frac{1}{4}a^2/y)$
14.2	$x^{-1}J_0(ax^{\frac{1}{2}})$	$-\mathrm{si}(\frac{1}{4}a^2/y)$
14.3	$\log(bx)J_0(ax^{\frac{1}{2}})$	$2y^{-1}\{\cos u\,\log(\frac{1}{4}\frac{a}{y}b^{\frac{1}{2}})-\frac{1}{2}\mathrm{Ci}(u)-\frac{1}{2}\sin u\,\mathrm{si}(u)\}$ $\qquad\qquad u=\frac{1}{4}a^2/y$
14.4	$x^{-\frac{1}{2}}J_1(ax^{\frac{1}{2}})$	$2a^{-1}\sin(\frac{1}{4}a^2/y)$
14.5	$x^{-\frac{1}{2}}e^{-bx}J_1(2ax^{\frac{1}{2}})$	$a^{-1}\exp\{-a^2b(b^2+y^2)^{-1}\}\sin\{a^2y(b^2+y^2)^{-1}\}$
14.6	$J_\nu(ax^{\frac{1}{2}})$ $\mathrm{Re}\nu>-4$	$\frac{1}{4}ay^{-1}(\pi/y)^{\frac{1}{2}}\Big(\cos(u-\frac{1}{4}\pi\nu)J_{\frac{1}{2}\nu-\frac{1}{2}}(u)$ $-\sin(u-\frac{1}{4}\pi\nu)J_{\frac{1}{2}\nu+\frac{1}{2}}(u)\Big);\qquad u=a^2/(8y)$
14.7	$x^{-\frac{1}{2}}J_\nu(ax^{\frac{1}{2}})$ $\mathrm{Re}\nu>-3$	$(\pi/y)^{\frac{1}{2}}\sin(\frac{1}{4}\pi+\frac{1}{4}\pi\nu-\frac{1}{8}a^2/y)J_{\frac{1}{2}\nu}(\frac{1}{8}a^2/y)$
14.8	$x^{\frac{1}{2}}J_\nu(ax^{\frac{1}{2}})$ $-2<\mathrm{Re}\nu<\frac{1}{2}$	$y^{-1}(\frac{1}{2}a/y)^\nu\cos(\frac{1}{2}\pi\nu-\frac{1}{4}a^2/y)$
14.9	$x^{\frac{1}{2}\nu}e^{-ax}J_\nu\{2(bx)^{\frac{1}{2}}\}$ $\mathrm{Re}\nu>-2$	$b^{\frac{1}{2}\nu}(a^2+y^2)^{-\frac{1}{2}\nu-\frac{1}{2}}\exp\{-ab(a^2+y^2)^{-1}\}$ $\cdot\sin\{(\nu+1)\arctan(y/a)-by(a^2+y^2)^{-1}\}$
14.10	$J_\nu(ax^{\frac{1}{2}})J_\nu(bx^{\frac{1}{2}})$ $\mathrm{Re}\nu>-2$	$y^{-1}\cos\{\frac{1}{4}y^{-1}(a^2+y^2)-\frac{1}{2}\pi\nu\}J_\nu(\frac{1}{2}ab/y)$

	$f(x)$	$g_s(y)$	
14.11	$J_\nu(ax)J_{2\nu}(bx^{\frac{1}{2}})$ $\mathrm{Re}\,\nu > -1$	$(a^2-y^2)^{-\frac{1}{2}}\sin\{\tfrac{1}{4}b^2 y(a^2-y^2)^{-1}\}$ $\cdot J_\nu\{\tfrac{1}{4}ab^2(a^2-y^2)^{-1}\}$ $(y^2-a^2)^{-\frac{1}{2}}\cos\{\tfrac{1}{4}b^2 y(y^2-a^2)^{-1}\}$ $\cdot J_\nu\{\tfrac{1}{4}ab^2(y^2-b^2)^{-1}\}$	$y<a$ $y>a$
14.12	$J_0(bp)$	$-s^{-1}\sin(ay)\exp(-as)$ $s^{-1}\cos(ay-aS)$	$y<b$ $y>b$
14.13	$p^{-1}J_{2\nu}(bp)$	$\tfrac{1}{2}\pi J_\nu(\tfrac{1}{2}ay-\tfrac{1}{2}aS)\left[J_\nu(\tfrac{1}{2}ay+\tfrac{1}{2}aS)\cos(ay-\tfrac{1}{2}\pi\nu)\right.$ $\left. -Y_\nu(\tfrac{1}{2}ay+\tfrac{1}{2}aS)\sin(ay-\tfrac{1}{2}\pi\nu)\right]$	 $y>b$
14.14	$p^\nu J_\nu(bp)$ $-2<\mathrm{Re}\,\nu<\tfrac{1}{2}$	$-(2a/\pi)^{\frac{1}{2}}(ab)^\nu\sin(ay)s^{-\nu-\frac{1}{2}}K_{\nu+\frac{1}{2}}(as)$ $-\tfrac{1}{2}(\tfrac{1}{2}\pi a)^{\frac{1}{2}}(ab)^\nu S^{-\nu-\frac{1}{2}}\left[J_{\nu+\frac{1}{2}}(aS)\sin(\pi\nu-ay)\right.$ $\left. +Y_{\nu+\frac{1}{2}}(aS)\cos(\pi\nu-ay)\right]$	$y<b$ $y>b$
14.15	$\left(1+\dfrac{a}{x}\right)^{-\frac{1}{2}\nu}J_\nu\{b(x^2+ax)^{\frac{1}{2}}\}$ $\mathrm{Re}\,\nu>-2$	$s^{-1}e^{-\frac{1}{2}as}\sin\{\nu\arctan(y/s)-\tfrac{1}{2}ay\}$ $b^\nu s^{-1}(y+S)^{-\nu}\cos(\tfrac{1}{2}aS-\tfrac{1}{2}ay+\tfrac{1}{2}\pi\nu)$	$y<b$ $y>b$
14.16	$xr^{-1}J_1(br)$	$b^{-\frac{1}{2}}ys^{-1}\cos(as)$ 0	$y<b$ $y>b$
14.17	$xr^\nu J_\nu(br)$ $\mathrm{Re}\,\nu<-\tfrac{1}{2}$	$(\tfrac{1}{2}\pi a)^{\frac{1}{2}}(ab)^\nu ays^{-\nu-3/2}Y_{\nu+3/2}(as)$ $-(2a/\pi)^{\frac{1}{2}}a(ab)^\nu\sin(\pi\nu)S^{-\nu-3/2}$ $\cdot K_{\nu+3/2}(aS)$	$y<b$ $y>b$
14.18	$xr^{-\nu}J_\nu(br)$ $\mathrm{Re}\,\nu>\tfrac{1}{2}$	$(\tfrac{1}{2}\pi a)^{\frac{1}{2}}(ab)^{-\nu}ays^{\nu-3/2}J_{\nu-3/2}(as)$ 0	$y<b$ $y>b$
14.19	$x^{-1}r^{-\nu}J_\nu(br)$ $\mathrm{Re}\,\nu>-3/2$	$\tfrac{1}{2}\pi a^{-\nu}J_\nu(ab)$	$y>b$

$$r=(a^2+x^2)^{\frac{1}{2}} \qquad p=(x^2+2ax)^{\frac{1}{2}} \qquad s=(b^2-y^2)^{\frac{1}{2}} \qquad S=(y^2-b^2)^{\frac{1}{2}}$$

	$f(x)$	$g_s(y)$
14.20	$x(k^2+x^2)^{-1}r^{-\nu}J_\nu(br)$ $\mathrm{Re}\,\nu>-3/2$	$\frac{1}{2}\pi e^{-ky}(a^2-k^2)^{-\frac{1}{2}\nu}J_\nu\{(a^2-k^2)^{\frac{1}{2}}\}$ $y>b$
14.21	$x^{2n-1}(k^2+x^2)^{-1}r^{-\nu}J\nu(br)$ $n=0,1,2,\ldots$	$(-1)^{n+1}\frac{1}{2}\pi k^{2n-2}e^{-ky}(a^2-k^2)^{-\frac{1}{2}\nu}$ $\cdot J_\nu\{b(a^2-k^2)^{\frac{1}{2}}\}$ $y>b$
14.22	$x^{2n-1}(k^2+x^2)^{-m-1}r^{-\nu}$ $\cdot J_\nu(br)$;$m,n=0,1,2,\ldots$ $\mathrm{Re}\,\nu>2n-2m-7/2$	$(-1)^{n+m+1}2^{-m-1}(m!)^{-1}$ $\cdot\left(\dfrac{d}{k\,dk}\right)^m\left\{k^{-2n-2}e^{-ky}(a^2-k^2)^{-\frac{1}{2}\nu}J_\nu\{b(a^2-k^2)^{\frac{1}{2}}\}\right\}$ $y>b$
14.23	$J_o(bu)$ $x<2a$ 0 $x>2a$	$2\sin(ay)w^{-1}\sin(aw)$
14.24	$u^{-1}J_\nu(bu)$ $x<2a$ 0 $x>2a$ $\mathrm{Re}\,\nu>-3$	$\pi\sin(ay)J_{\frac{1}{2}\nu}(\frac{1}{2}aw+\frac{1}{2}ay)J_{\frac{1}{2}\nu}(\frac{1}{2}aw-\frac{1}{2}ay)$
14.25	$u^\nu J_\nu(bu)$ $x>2a$ 0 $x<2a$ $\mathrm{Re}\,\nu>-2$	$(2\pi a)^{\frac{1}{2}}(ab)^\nu\sin(ay)w^{-\nu-\frac{1}{2}}J_{\nu+\frac{1}{2}}(aw)$
14.26	$v^{-1}J_\nu(bv)$ $x>a$ 0 $x<a$ $\mathrm{Re}\,\nu>-1$	$\frac{1}{2}\pi J_{\frac{1}{2}\nu}(\frac{1}{2}aw+\frac{1}{2}ay)J_{\frac{1}{2}\nu}(\frac{1}{2}aw-\frac{1}{2}ay)$
14.27	$xv^\nu J_\nu(bv)$ $x<a$ 0 $x>a$ $\mathrm{Re}\,\nu>-1$	$(\frac{1}{2}\pi a)^{\frac{1}{2}}a(ab)^\nu yw^{-\nu-3/2}J_{\nu+3/2}(aw)$
14.28	$\log\{(a+x)/(a-x)\}J_o(bv)$ $x<a$ 0 $x>a$	$w^{-1}\Big(\cos(aw)\{\mathrm{Ci}(aw+ay)-\mathrm{Ci}(aw-ay)\}$ $+\sin(aw)\{\mathrm{Si}(aw+ay)-\mathrm{Si}(aw-ay)\}$ $-2\cos(aw)\log\{(y+w)/b\}\Big)$

$u=(2ax-x^2)^{\frac{1}{2}},r=(a^2+x^2)^{\frac{1}{2}}$ $w=(b^2+y^2)^{\frac{1}{2}}$

$v=(a^2-x^2)^{\frac{1}{2}},V=(x^2-a^2)^{\frac{1}{2}}$ $s=(b^2-y^2)^{\frac{1}{2}},\; S=(y^2-b^2)^{\frac{1}{2}}$

	$f(x)$		$g_s(y)$	
14.29	0	$x<a$	0	$y<b$
	$J_0(bV)$	$x>a$	$s^{-1}\cos(aS)$	$y>b$
14.30	0	$x<a$	$s^{-1}e^{-as}\sin\{\nu\arcsin(y/b)\}$	$y<b$
	$\left(\dfrac{x-a}{x+a}\right)^{\frac{1}{2}\nu}J_0(bV)$	$x>a$	$b^{\nu}S^{-1}(y+S)^{-\nu}\cos(\tfrac{1}{2}\pi\nu+aS)$	$y>b$
	$\mathrm{Re}\,\nu>-1$			
14.31	0	$x<a$	$\tfrac{1}{2}\pi J_{\frac{1}{2}\nu}(\tfrac{1}{2}ay-\tfrac{1}{2}aS)J_{-\frac{1}{2}\nu}(\tfrac{1}{2}ay+\tfrac{1}{2}aS)$	$y>b$
	$V^{-1}J_{\nu}(bV)$	$x>a$		
	$\mathrm{Re}\,\nu>-1$			
14.32	0	$x<a$	0	$y<b$
	$V^{\nu}J_{\nu}(bV)$	$x>a$	$(\tfrac{1}{2}\pi a)^{\frac{1}{2}}(ab)^{\nu}s^{-\nu-\frac{1}{2}}J_{-\nu-\frac{1}{2}}(aS),$	$y>b$
	$\mathrm{Re}\,\nu>-2$			
14.33	0	$x<a$	$-a(2a/\pi)^{\frac{1}{2}}(ab)^{\nu}s^{-\nu-3/2}K_{\nu+3/2}(as)$	$y<b$
	$xV^{\nu}J_{\nu}(bV)$	$x>a$	$(\tfrac{1}{2}\pi a)^{\frac{1}{2}}a(ab)^{\nu}s^{-\nu-3/2}Y_{-\nu-3/2}(aS)$	$y>b$
	$-1<\mathrm{Re}\,\nu<5/2$			
14.34	0	$x<a$	$k^{-1}(a^2+k^2)^{\frac{1}{2}\nu}\sinh(ky)$	
	$(x^2+k^2)^{-1}V^{\nu}J_{\nu}(bV)$	$x>a$	$\cdot K_{\nu}\{b(a^2+k^2)^{\frac{1}{2}}\}$	$y<b$
	$-1<\mathrm{Re}\,\nu<5/2$			
14.35	0	$x<a$	$(-1)^{n+1}k^{-1}(a^2+k^2)^{\frac{1}{2}\nu+n-1}\sinh(ky)$	
	$(x^2+k^2)^{-1}V^{\nu+2n-2}J_{\nu}(bV)$		$\cdot K_{\nu}\{b(a^2+k^2)^{\frac{1}{2}}\}$	$y>b$
		$x>a$		
	$n=0,1,2,\ldots$			
	$-n-1<\mathrm{Re}\,\nu<-2n+9/2$			
14.36	$P_{2n+1}(x/a)J_0(bv)$	$x<a$	$(-1)^n(\tfrac{1}{2}\pi a)^{\frac{1}{2}}P_{2n+1}(y/w)$	
	0	$x>a$	$w^{-\frac{1}{2}}J_{2n+3/2}(aw)$	
	$n=0,1,2,\ldots$			

$u=(2ax-x^2)^{\frac{1}{2}}, v=(a^2-x^2)^{\frac{1}{2}}$ $w=(b^2+y^2)^{\frac{1}{2}}, s=(b^2-y^2)^{\frac{1}{2}}, S=(y^2-b^2)^{\frac{1}{2}}$

$V=(x^2-a^2)^{\frac{1}{2}}$

	$f(x)$	$g_s(y)$	
14.37	$r^{-\frac{1}{2}}P_{2n+1}(x/r)$ $\cdot J_{2n+3/2}(br)$ $n=0,1,2,\ldots$	$(-1)^n(\tfrac{1}{2}\pi/b)^{\frac{1}{2}}P_{2n+1}(y/b)J_0(as)$ 0	$y<b$ $y>b$
14.38	$T_{2n+1}(x/r)J_{2n+1}(br)$ $n=0,1,2,\ldots$	$(-1)^n s^{-1}\cos(as)T_{2n+1}(y/b)$ 0	$y<b$ $y>b$
14.39	$r^{-1}U_{2n+1}(x/r)J_{2n+2}(br)$ $n=0,1,2,\ldots$	$(-1)^n(ab)^{-1}\sin(as)U_{2n+1}(y/b)$ 0	$y<b$ $y>b$
14.40	$xr^{-1}U_{2n}(a/r)J_{2n+1}(br)$ $n=0,1,2,\ldots$	$(-1)^n b^{-\frac{1}{4}}ys^{-1}\cos(as)U_{2n}(s/b)$ 0	$y<b$ $y>b$
14.41	$xr^{-1}U_{2n+1}(a/r)J_{2n+2}(br)$ $n=0,1,2,\ldots$	$(-1)^n b^{-\frac{1}{4}}ys^{-1}\sin(as)U_{2n+1}(s/b)$ 0	$y<b$ $y>b$
14.42	$r^{-\nu}C_{2n+1}^{\nu}(x/r)$ $\cdot J_{\nu+2n+1}(br)$ $n=0,1,2,\ldots$	$(-1)^n(\tfrac{1}{2}\pi a)^{\frac{1}{2}}(ab)^{-\nu}s^{\nu-\frac{1}{2}}C_{2n+1}^{\nu}(y/b)$ $J_{\nu-\frac{1}{2}}(as)$ 0	$y<b$ $y>b$
14.43	$(a^2-x^2)^{\frac{1}{2}\nu-\frac{1}{4}}C_{2n+1}^{\nu}(x/a)$ $\cdot J_{\nu-\frac{1}{2}}\{b(a^2-x^2)^{\frac{1}{2}}\}$ $\quad x<a$ $0 \qquad x>a$	$(-1)^n(\tfrac{1}{2}\pi/b)^{\frac{1}{2}}(ab)^{\nu}(b^2+y^2)^{-\frac{1}{4}\nu}$ $C_{2n}^{\nu}\{y(b^2+y^2)^{-\frac{1}{2}}\}J_{\nu+2n+1}\{a(b^2+y^2)^{\frac{1}{2}}\}$	
14.44	$Y_0(ax^{\frac{1}{2}})$	$(\pi y)^{-1}\left(\cos u\, \text{Ci}(u)+\sin u\{\pi+\text{si}(u)\right),u=\dfrac{a^2}{4y}$	
14.45	$J_0(ax^{\frac{1}{2}})Y_0(ax^{\frac{1}{2}})$	$\tfrac{1}{2}y^{-1}\{\sin u J_0(u)+\cos u Y_0(u)\},u=\tfrac{1}{2}a^2/y$	
14.46	$J_{\nu}(ax^{\frac{1}{2}})Y_{-\nu}(bx^{\frac{1}{2}})$ $+J_{-\nu}(bx^{\frac{1}{2}})Y_{\nu}(ax^{\frac{1}{2}}),-2<\text{Re}\nu<2$	$y^{-1}\{\sin(\tfrac{1}{2}\pi\nu+u)J_{\nu}(v)+\cos(\tfrac{1}{2}\pi\nu+u)Y_{\nu}(v)\}$ $u=\tfrac{1}{4}(a^2+b^2)/y, \quad v=\tfrac{1}{2}ab/y$	
14.47	$xr^{\nu}Y_{\nu}(br)$ $\text{Re}\nu>-\tfrac{1}{2}$	$-(\tfrac{1}{2}\pi a)^{\frac{1}{2}}(ab)^{\nu}s^{-\nu-3/2}ayJ_{\nu+3/2}(as)$ $-(2a/\pi)^{\frac{1}{2}}(ab)^{\nu}ayS^{-\nu-3/2}K_{\nu+3/2}(aS)$	$y<b$ $y>b$

$$v=(a^2-x^2)^{\frac{1}{2}},r=(a^2+x^2)^{\frac{1}{2}} \qquad w=(b^2+y^2)^{\frac{1}{2}},\; s=(b^2-y^2)^{\frac{1}{2}},\; S=(y^2-b^2)^{\frac{1}{2}}$$

	$f(x)$	$g_s(y)$
14.48	$xr^{-\nu}Y_\nu(br)$ $\mathrm{Re}\,\nu>\tfrac{1}{2}$	$(\tfrac{1}{2}\pi a)^{\tfrac{1}{2}}(ab)^{-\nu}ay\ s^{\nu-3/2}Y_{\nu-3/2}(as)\quad y<b$ $-(2a/\pi)^{\tfrac{1}{2}}(ab)^{-\nu}ay\ s^{\nu-3/2}K_{\nu-3/2}(aS)\,y>b$
14.49	$0\qquad\qquad x<a$ $Y_0(bV)\qquad x>a$	$-(\pi S)^{-1}\Big(\cos(aS)\{Ci(ay+aS)-Ci(ay-aS)\}$ $\qquad +\sin(aS)\{Si(ay+aS)+Si(ay-aS)\}\Big]y>b$
14.50	$0\qquad\qquad x<a$ $V^{-1}Y_{2\nu}(bV)\quad x>a$ $-\tfrac{1}{2}<\mathrm{Re}\,\nu<\tfrac{1}{2}$	$\tfrac{1}{4}\pi\sec(\pi\nu)\Big(J_\nu(\tfrac{1}{2}ay-\tfrac{1}{2}aS)\,Y_\nu(\tfrac{1}{2}ay-\tfrac{1}{2}aS)$ $-\cos(2\pi\nu)J_\nu(\tfrac{1}{2}ay-aS)\,Y_\nu(\tfrac{1}{2}ay+\tfrac{1}{2}aS)$ $-\sin(2\pi\nu)J_\nu(\tfrac{1}{2}ay-\tfrac{1}{2}aS)\,J_\nu(\tfrac{1}{2}ay+\tfrac{1}{2}aS)\Big\}\ y>b$
14.51	$(k^2+x^2)^{-1}V^{2n+\nu-1}Y_\nu(bV)$ $\qquad\qquad x>a$ $0\qquad\qquad x<a$ $n=0,1,\ldots,\tfrac{1}{2}-n<\mathrm{Re}\,\nu<\tfrac{7}{2}-2n$	$(-1)^{n-1}k^{-1}\sinh(ky)(a^2+k^2)^{n-\tfrac{1}{2}+\tfrac{1}{2}\nu}$ $\cdot K_\nu\{b(a^2+k^2)^{\tfrac{1}{2}}\}\qquad\qquad y<b$
14.52	$x^{\tfrac{1}{2}}J_{\tfrac{1}{4}}(br+ab)J_{\tfrac{1}{4}}(br-ab)$	$(\tfrac{1}{2}\pi y)^{-\tfrac{1}{2}}u^{-1}\cos(au)\qquad\qquad y<2b$ $0\qquad\qquad\qquad\qquad\qquad y>2b$
14.53	$x^{\tfrac{1}{2}}J_{\tfrac{1}{4}}(br-ab)Y_{\tfrac{1}{4}}(br+ab)$	$(\tfrac{1}{2}\pi y)^{-\tfrac{1}{2}}u^{-1}\sin(au)\qquad\qquad y<2b$ $-(\tfrac{1}{2}\pi y)^{\tfrac{1}{2}}U^{-1}\exp(-aU)\qquad y>2b$

2.15 Bessel Functions of Trigonometric and Hyperbolic Arguments

15.1	$J_\nu(a\sin x)\qquad x<\pi$ $0\qquad\qquad x>\pi$ $\mathrm{Re}\,\nu>-2$	$\pi\sin(\tfrac{1}{2}\pi y)J_{\tfrac{1}{2}\nu-\tfrac{1}{2}y}(\tfrac{1}{2}a)J_{\tfrac{1}{2}\nu+\tfrac{1}{2}y}(\tfrac{1}{2}a)$
15.2	$\mathrm{cosec}\,x\,J_{2\nu}(2a\sin x),$ $\qquad\qquad x<\pi$ $0\qquad\qquad x>\pi$ $\mathrm{Re}\,\nu>0$	$\tfrac{1}{2}\pi a\nu^{-1}\sin(\tfrac{1}{2}\pi y)\{J_{\nu-\tfrac{1}{2}+\tfrac{1}{2}y}(a)J_{\nu-\tfrac{1}{2}-\tfrac{1}{2}y}(a)$ $+J_{\nu+\tfrac{1}{2}+\tfrac{1}{2}y}(a)J_{\nu+\tfrac{1}{2}-\tfrac{1}{2}y}(a)\}$
15.3	$(\sin x)^{-m}J_m(2a\sin x),$ $\qquad\qquad x<\pi$ $0\qquad\qquad x>\pi$ $m=0,1,2,\ldots$	$m!a^m\pi\sin(\tfrac{1}{2}\pi y)$ $\cdot\displaystyle\sum_{n=0}^{m}\varepsilon_n\{(m+n)!(m-n)!\}^{-1}J_{n-\tfrac{1}{2}y}(a)J_{n+\tfrac{1}{2}y}(a)$

$r=(a^2+x^2)^{\tfrac{1}{2}},V=(x^2-a^2)^{\tfrac{1}{2}},\quad s=(b^2-y^2)^{\tfrac{1}{2}},S=(y^2-b^2)^{\tfrac{1}{2}},u=(4b^2-y^2)^{\tfrac{1}{2}},U=(y^2-4b^2)^{\tfrac{1}{2}}$

	$f(x)$	$g_s(y)$
15.4	$J_0(2a \sinh x)$	$\pi^{-1}\sinh(\tfrac{1}{2}\pi y)\{K_{\frac{1}{2}iy}(a)\}^2$
15.5	$J_{2\nu}(2a \sinh x)$ $\text{Re}\,\nu>-1$	$\tfrac{1}{2}i\left[I_{\nu+\frac{1}{2}iy}(a)K_{\nu-\frac{1}{2}iy}(a)-I_{\nu-\frac{1}{2}iy}(a)K_{\nu+\frac{1}{2}iy}(a)\right]$
15.6	$(\sinh x)^{-m}J_m(2a \sinh x)$ $m=0,1,2,\ldots$	$\tfrac{1}{2}im!\,a^m\sum_{m=o}^{n}\varepsilon_n\{(m+n)!\,(m-n)!\}^{-1}$ $\cdot\left[I_{n+\frac{1}{2}iy}(a)K_{n-\frac{1}{2}iy}(a)-I_{n-\frac{1}{2}iy}(a)K_{n+\frac{1}{2}iy}(a)\right]$
15.7	$J_0\{a(\sinh x)^{\frac{1}{2}}\}$	$i\,K_{iy}(u)\{J_{iy}(u)-J_{-iy}(u)\};\qquad u=a2^{-\frac{1}{2}}$
15.8	$\begin{array}{ll}0 & \sinh x<\frac{b}{a}\\[4pt] J_0(a^2\sinh^2 x-b^2)^{\frac{1}{2}} \\[4pt] & \sinh x>\frac{b}{a}\end{array}$	$\pi^{-1}\sinh(\tfrac{1}{2}\pi y)K_{\frac{1}{2}iy}(u)K_{\frac{1}{2}iy}(v)$ $\begin{array}{l}u\\v\end{array}=\tfrac{1}{2}\{(b^2+a^2)^{\frac{1}{2}}\pm b\}$
15.9	$Y_0\{2a \sinh(\tfrac{1}{2}x)\}$	$\pi^{-1}\left[K_{iy}(a)\dfrac{d}{dy}\{I_{iy}(a)+I_{-iy}(a)\}\right.$ $\left.-\{I_{iy}(a)+I_{-iy}(a)\}\dfrac{d}{dy}K_{iy}(a)\right]$
15.10	$Y_1(2a \sinh x)$	$-\pi^{-1}\sinh(\tfrac{1}{2}\pi y)K_{\frac{1}{2}+\frac{1}{2}iy}(a)K_{\frac{1}{2}-\frac{1}{2}iy}(a)$
15.11	$I_0(2a \sinh x)-L_0(2a \sinh x)$	$\tfrac{1}{2}\pi\tanh(\tfrac{1}{2}\pi y)\left[\{I_{\frac{1}{2}iy}(a)\}^2+\{Y_{\frac{1}{2}iy}(a)\}^2\right]$

2.16   Bessel Functions of Variable Order

16.1	$\sin(\pi x)J_{\nu-x}(a)J_{\nu+x}(a)$ $\text{Re}\,\nu>-1$	$\tfrac{1}{4}J_{2\nu}\{2a \sin(\tfrac{1}{2}y)\}\qquad\qquad y<2\pi$ $0\qquad\qquad\qquad\qquad\qquad y>2\pi$
16.2	$\text{sech}(\pi x)\cos(\tfrac{1}{2}\pi x)$ $\cdot\{J_{ix}(a)-J_{-ix}(a)\}$	$-i\sin(a \cosh y)\{C(u)+S(u)\}$ $-i\cos(a \cosh y)\cdot\{C(u)-S(u)\};$ $\qquad\qquad\qquad u=2a \sinh^2(\tfrac{1}{2}y)$
16.3	$\text{sech}(\pi x)\sinh(\tfrac{1}{2}\pi x)$ $\cdot\{J_{ix}(a)+J_{-ix}(a)\}$	$\sin(a \cosh y)\{C(u)-S(u)\}-\cos(a \cosh y)$ $\cdot\{C(u)+S(u)\}\;;\; u=2a \sinh^2(\tfrac{1}{2}y)$

	$f(x)$	$g_s(y)$
16.4	$\tanh(\pi x)$ $\left[\{J_{ix}(a)\}^2 + \{Y_{ix}(a)\}^2\right]$	$I_0\{2a\,\sinh(\tfrac{1}{2}y)\} - \mathbf{L}_0\{2a\,\sinh(\tfrac{1}{2}y)\}$

2.17 Modified Bessel Functions of Arguments x, x^2 and $1/x$

	$f(x)$	$g_s(y)$
17.1	$e^{-ax}I_0(ax)$	$(2y)^{-\frac{1}{2}}(y^2+4a^2)^{-\frac{1}{2}}\{(y^2+4a^2)^{\frac{1}{2}}+y\}^{\frac{1}{2}}$
17.2	$e^{-bx}I_0(ax),\qquad b>a$	$(2u)^{-\frac{1}{2}}\{u^{\frac{1}{2}}+a^2+y^2-b^2\}^{\frac{1}{2}}$ $u = 4b^2y^2+(b^2-a^2-y^2)^2$
17.3	$x^\nu e^{-bx}I_\nu(ax)\qquad b>a$ $\mathrm{Re}\,\nu>-1$	$\pi^{-\frac{1}{2}}(2a)^\nu\Gamma(\tfrac{1}{2}+\nu)(z^2+4b^2y^2)^{-\frac{1}{4}-\frac{1}{2}\nu}$ $\cdot\pi^{-\frac{1}{2}}\sin\{(\tfrac{1}{2}+\nu)\arctan(2by/z)\}$ $z = b^2-a^2-y^2$
17.4	$K_0(ax)$	$(a^2-y^2)^{-\frac{1}{2}}\log\{\tfrac{y}{a}+(1+y^2/a^2)^{\frac{1}{2}}\}$
17.5	$xK_0(ax)$	$\tfrac{1}{2}\pi y(a^2+y^2)^{-3/2}$
17.6	$x^{-\frac{1}{2}}K_0(ax)$	$(\tfrac{1}{2}\pi u)^{\frac{1}{2}}\left[\mathbf{K}\{(\tfrac{1}{2}+\tfrac{1}{2}uy)^{\frac{1}{2}}\}-\mathbf{K}\{(\tfrac{1}{2}-\tfrac{1}{2}uy)^{\frac{1}{2}}\}\right]$ $u = (a^2+y^2)^{-\frac{1}{2}}$
17.7	$x^{\frac{1}{2}}K_0(ax)$	$\tfrac{1}{2}u(\tfrac{1}{2}\pi u)^{\frac{1}{2}}\left[2\mathbf{E}\{(\tfrac{1}{2}-\tfrac{1}{2}uy)^{\frac{1}{2}}\}-2\mathbf{E}\{(\tfrac{1}{2}+\tfrac{1}{2}uy)^{\frac{1}{2}}\}\right.$ $\left.+\mathbf{K}\{(\tfrac{1}{2}+\tfrac{1}{2}uy)^{\frac{1}{2}}\}-\mathbf{K}\{(\tfrac{1}{2}-\tfrac{1}{2}uy)^{\frac{1}{2}}\}\right], u=(a^2+y^2)^{-\frac{1}{2}}$
17.8	$x^{-\frac{1}{2}}\log x\,K_0(ax)$	$(\tfrac{1}{2}\pi)^{\frac{1}{2}}(a^2+y^2)^{-\frac{1}{4}}\{\gamma+\log 4-\tfrac{1}{2}\pi+\tfrac{1}{2}\log(a^2+y^2)\}$ $\cdot\{\mathbf{K}(u)-\mathbf{K}(v)\};\,{}^v_u=\left[\tfrac{1}{2}(a^2+y^2)^{-\frac{1}{2}}\{(a^2+y^2)^{\frac{1}{2}}\pm y\}\right]^{\frac{1}{2}}$
17.9	$x^{-1}\{\gamma+\log(\tfrac{1}{2}ax)+K_0(ax)\}$	$\tfrac{1}{2}\pi\log\left[\tfrac{1}{2}y^{-1}\{y+(a^2+y^2)^{\frac{1}{2}}\}\right]$
17.10	$x^{2n}K_0(ax), n=0,1,2,\ldots$	$(-1)^n(2n)!\,(a^2+y^2)^{-n-\frac{1}{2}}Q_{2n}\{y(a^2+y^2)^{-\frac{1}{2}}\}$
17.11	$x^{2n+1}K_0(ax); n=0,1,2,\ldots$	$(-1)^n\tfrac{1}{2}\pi(2n+1)!\,(a^2+y^2)^{-n-1}$ $\cdot P_{2n+1}\{y(a^2+y^2)^{-\frac{1}{2}}\}$

	$f(x)$	$g_s(y)$
17.12	$\sinh(ax)K_o(ax)$	$\pi a(2y)^{-\frac{1}{2}}(4a^2+y^2)^{-\frac{1}{2}}\{(4a^2+y^2)^{\frac{1}{2}}+y\}^{-\frac{1}{2}}$
17.13	$K_\nu(ax)$ $-2<\mathrm{Re}\nu<2$	$\frac{1}{2}\pi a^{-\nu}\mathrm{cosec}(\frac{1}{2}\pi\nu)(a^2+y^2)^{-\frac{1}{2}}$ $\cdot\left(\{(y^2+a^2)^{\frac{1}{2}}+y\}^\nu-\{(y^2+a^2)^{\frac{1}{2}}-y\}^\nu\right]$
17.14	$x^{\nu+1}K_\nu(ax)$ $\mathrm{Re}\nu>-3/2$	$\pi^{\frac{1}{2}}(2a)^\nu\Gamma(\frac{3}{2}+\nu)y(a^2+y^2)^{-\nu-3/2}$
17.15	$x^\mu K_\nu(ax)$ $\mathrm{Re}(\pm\nu)>-2$	$\frac{1}{2}\pi\mathrm{cosec}(\frac{1}{2}\pi\mu-\frac{1}{2}\pi\nu)\Gamma(1+\mu+\nu)(a^2+y^2)^{-\frac{1}{2}-\frac{1}{2}\mu}$ $\cdot\left(P_\mu^{-\nu}\{y(a^2+y^2)^{-\frac{1}{2}}\}-P_\mu^{-\nu}\{-y(a^2+y^2)^{-\frac{1}{2}}\}\right]$
17.16	$x^{-\frac{1}{2}}\log xK_\nu(ax)$ $-3/2<\mathrm{Re}\nu<3/2$ $u=(1+a^2/y^2)^{-\frac{1}{2}}$	$\frac{1}{2}\pi\mathrm{cosec}(\frac{1}{2}\pi\nu+\frac{1}{4}\pi)\Gamma(\frac{1}{2}+\nu)(a^2+y^2)^{-\frac{1}{4}}$ $\cdot\left(\{P_{-\frac{1}{2}}^{-\nu}(u)-P_{-\frac{1}{2}}^{-\nu}(-u)\}\{\frac{1}{2}\log(a^2+y^2)-\Psi(\frac{1}{2}+\nu)\right.$ $-\frac{1}{2}\pi\cot(\frac{1}{2}\pi\nu+\frac{1}{4}\pi)\}\Big]$
17.17	$x^{\nu+\mu}K_\nu(ax)K_\mu(bx)$	$-\frac{1}{4}\pi^2\int_o^\infty t^{\nu+\mu}\{J_\nu(at)J_\mu(bt)$ $-Y_\nu(at)Y_\mu(bt)\}e^{-ty}dt$
17.18	$I_o(ax)K_o(ax)$	$(y^2+4a^2)^{-\frac{1}{2}}K\{y(y^2+4a^2)^{-\frac{1}{2}}\}$
17.19	$x^{\frac{1}{4}}I_{\frac{1}{4}}(ax)K_{\frac{1}{4}}(ax)$	$(\frac{1}{2}\pi/y)^{\frac{1}{2}}(4a^2+y^2)^{-\frac{1}{2}}$
17.20	$x^{\frac{1}{2}}I_{\frac{1}{4}-\frac{1}{2}\nu}(ax)K_{\frac{1}{4}+\frac{1}{2}\nu}(ax)$ $\mathrm{Re}\nu<5/2$	$(\frac{1}{2}\pi/y)^{\frac{1}{2}}(2a)^{-\nu}(4a^2+y^2)^{-\frac{1}{2}}\{y+(4a^2+y^2)^{\frac{1}{2}}\}^\nu$
17.21	$\{I_\nu(ax)+I_{-\nu}(ax)\}K_\nu(ax)$ $-1<\mathrm{Re}\nu<1$	$\frac{1}{2}\pi a^{-1}P_{\nu-\frac{1}{2}}(1+\frac{1}{2}y^2/a^2)$
17.22	$xI_\nu(bx)K_\nu(ax)$ $a>b,\ \mathrm{Re}\nu>-3/2$	$-\frac{1}{2}(ab)^{-3/2}y(u^2-1)^{-\frac{1}{2}}Q_{\nu-\frac{1}{2}}^1(u)$ $u=(a^2+b^2+y^2)/(2ab)$
17.23	$xK_\nu(ax)K_\nu(bx)$ $-3/2<\mathrm{Re}\nu<3/2$	$\frac{1}{4}\pi^2(ab)^{-3/2}(\frac{1}{4}-\nu^2)\sec(\pi\nu)y(u^2-1)^{-\frac{1}{2}}P_{\nu-\frac{1}{2}}^{-1}(u)$ $u=(a^2+b^2+y^2)/(2ab)$

	$f(x)$	$g_s(y)$
17.24	$x^{-\frac{1}{2}}I_\nu(ax)K_\nu(ax)$ $\quad$ $\mathrm{Re}\,\nu>-3/4$	$\frac{1}{2}(\frac{1}{2}\pi y)^{\frac{1}{2}}e^{-\frac{1}{4}i\pi}p_{\nu-\frac{1}{2}}^{-\frac{1}{4}}(u)q_{\nu-\frac{1}{2}}^{\frac{1}{4}}(u)$ $u=(1+\frac{1}{4}y^2/a^2)^{\frac{1}{2}}$
17.25	$x^{\frac{1}{2}}\{K_\nu(ax)\}^2$ $\quad$ $-5/4<\mathrm{Re}\,\mu<5/4$	$\frac{1}{2}\pi a^{-1}(\frac{1}{2}\pi y)^{\frac{1}{2}}\Gamma(\frac{5}{4}+\nu)\,\Gamma(\frac{5}{4}-\nu)\,(4a^2+y^2)^{-\frac{1}{2}}$ $\cdot p_{\nu-\frac{1}{2}}^{-3/4}(u)p_{\nu-\frac{1}{2}}^{\frac{1}{4}}(u)\ ;\ u=(1+\frac{1}{4}y^2/a^2)^{\frac{1}{2}}$
17.26	$x^{-\frac{1}{2}}\{K_\nu(ax)\}^2$ $\quad$ $-3/4<\mathrm{Re}\,\nu<3/4$	$\frac{1}{2}\pi a^{-1}(\frac{1}{2}\pi y)^{\frac{1}{2}}\Gamma(\frac{3}{4}+\nu)\,\Gamma(\frac{3}{4}-\nu)\left(p_{\nu-\frac{1}{2}}^{-\frac{1}{4}}(u)\right)^2$ $u=(1+\frac{1}{4}y^2/a^2)^{\frac{1}{2}}$
17.27	$x^{\frac{1}{2}}K_{\nu-\frac{1}{2}}(ax)K_{\nu+\frac{1}{2}}(ax)$ $\quad$ $-5<\mathrm{Re}\,\nu<5/2$	$\frac{1}{2}\pi a^{-1}(\frac{1}{2}\pi y)^{\frac{1}{2}}(4a^2+y^2)^{-\frac{1}{2}}p_{-\nu}^{-\frac{1}{4}}(u)p_\nu^{-\frac{1}{4}}(u)$ $u=(1+\frac{1}{4}y^2/a^2)^{\frac{1}{2}}$
17.28	$\exp(-ax^2)I_o(ax^2)$	$\frac{1}{2}(\frac{1}{2}\pi/a)^{\frac{1}{2}}\exp\{-y^2/(16a)\}I_o\{y^2/(16a)\}$
17.29	$x^{\frac{1}{2}}\exp(-ax^2)I_{\frac{1}{4}}(ax^2)$	$\frac{1}{2}(ay)^{-\frac{1}{2}}\exp(-ay^2/8)$
17.30	$x^{1/3}\exp(-ax^2)I_{1/3}(ax^2)$	$\frac{1}{2}\pi^{-\frac{1}{2}}a^{-1}(\frac{1}{2}ay^2)^{1/6}\exp(-u)K_{\frac{1}{3}}(u);u=(\frac{y}{4})^2/a$
17.31	$x^{\frac{1}{2}}\exp(-ax^2)I_{\frac{1}{4}}(bx^2)$ $\quad$ $a>b$	$\frac{1}{2}(\frac{1}{2}\pi y)^{\frac{1}{2}}u\exp(-\frac{1}{4}ay^2u^2)I_{\frac{1}{4}}(\frac{1}{4}by^2u^2)$ $u=(a^2-b^2)^{-\frac{1}{2}}$
17.32	$x^{2\nu}\exp(-\frac{1}{4}ax^2)I_\nu(\frac{1}{4}ax^2)$ $\quad$ $-\frac{1}{2}<\mathrm{Re}\,\nu<\frac{1}{2}$	$\Gamma(\frac{1}{2}+\nu)\{\Gamma(1-\nu)\}^{-1}2^{\frac{1}{2}+3/2\nu}a^{-\frac{1}{2}\nu}y^{-\nu-1}$ $\cdot\exp(-\frac{1}{4}y^2/a)M_{3/2\nu,-\frac{1}{2}\nu}(\frac{1}{4}y^2/a)$
17.33	$x^{1-2\nu}\exp(-\frac{1}{4}ax^2)I_\nu(\frac{1}{4}ax^2)$ $\quad$ $\mathrm{Re}\,\nu>0$	$2^{\frac{1}{2}-3/2\nu}a^{\frac{1}{2}\nu-\frac{1}{2}}y^{\nu-1}\exp(-\frac{1}{4}y^2/a)$ $\cdot W_{\frac{1}{2}-3/2\nu,\frac{1}{2}\nu-\frac{1}{2}}(\frac{1}{2}y^2/a)$
17.34	$x^{\frac{1}{2}}K_{\frac{1}{4}}(ax^2)$	$\frac{1}{2}\pi a^{-1}(\frac{1}{2}\pi y)^{\frac{1}{2}}\{I_{\frac{1}{4}}(u)-\mathbf{L}_{\frac{1}{4}}(u)\};\ u=\frac{1}{4}y^2/a$
17.35	$x^{1/3}\exp(-ax^2)K_{1/3}(ax^2)$	$\frac{1}{2}a^{-1}\pi^{3/2}(\frac{1}{2}ay^2)^{1/6}e^{-u}I_{1/3}(u);\ u=(\frac{1}{4}y)^2/a$

	$f(x)$	$g_s(y)$
17.36	$x^{1/3}\exp(ax^2)K_{1/3}(ax^2)$	$\tfrac{1}{2}a^{-1}\pi^{\frac{1}{2}}(\tfrac{1}{2}ay^2)^{\frac{1}{6}}e^{u}K_{\frac{1}{3}}(u);\ u=(\tfrac{1}{4}y)^2/a$
17.37	$x^{3/2}K_{3/4}(ax^2)$	$\tfrac{1}{4}a^{-2}(\tfrac{1}{2}\pi y)^{3/2}\{I_{-\frac{1}{4}}(u)-\mathbf{L}_{-\frac{1}{4}}(u)\};\ u=\tfrac{1}{8}y^2/a$
17.38	$x^{2\nu+1}\exp(-\tfrac{1}{4}ax^2 K_\nu(\tfrac{1}{4}ax^2)$ $\mathrm{Re}\,\nu>-3/4$	$2\pi\Gamma(\tfrac{3}{2}+2\nu)\{\Gamma(2+\nu)\}^{-1}2^{-\frac{1}{2}+3/2\nu}a^{-\frac{1}{2}-\frac{1}{2}\nu}$ $\cdot y^{-1-\nu}\exp(-\tfrac{1}{8}y^2/a)M_{\frac{1}{2}+3/2\nu,\frac{1}{2}+\frac{1}{2}\nu}(\tfrac{1}{8}y^2/a)$
17.39	$x^{2\nu+1}\exp(\tfrac{1}{4}ax^2)K_\nu(\tfrac{1}{4}ax^2)$ $-3/4<\mathrm{Re}\,\nu<1/4$	$4\pi\Gamma(\tfrac{3}{2}+2\nu)\{\Gamma(\tfrac{1}{2}-\nu)\}^{-1}2^{-\frac{1}{2}+3/2\nu}a^{-\frac{1}{2}-\frac{1}{2}\nu}$ $\cdot y^{-1-\nu}\exp(\tfrac{1}{8}y^2/a)W_{-\frac{1}{2}-3/2\nu,\frac{1}{2}+\frac{1}{2}\nu}(\tfrac{1}{8}y^2/a)$
17.40	$x^{\frac{1}{2}}I_{1/8}(ax^2)K_{1/8}(ax^2)$	$\tfrac{1}{4}a^{-1}(\tfrac{1}{2}\pi y)^{\frac{1}{2}}I_{1/8}(u)K_{1/8}(u)\ ;\ u=(\tfrac{1}{4}y)^2/a$
17.41	$K_0(a/x)$	$-\pi(\tfrac{1}{2}a/y)^{\frac{1}{2}}\{K_1(u)Y_0(u)+K_0(u)Y_1(u)\};$ $u=(2ay)^{\frac{1}{2}}$
17.42	$x^{-1}K_0(a/x)$	$\pi J_0(u)K_0(u)\qquad u=(2ay)^{\frac{1}{2}}$
17.43	$x^{-3}K_0(a/x)$	$\pi a^{-1}yJ_1(u)K_1(u)\qquad u=(2ay)^{\frac{1}{2}}$
17.44	$x^{-1}K_\nu(a/x)$	$\pi K_\nu(u)\{\cos(\tfrac{1}{2}\pi\nu)J_\nu(u)-\sin(\tfrac{1}{2}\pi\nu)Y_\nu(u)\}$ $u=(2ay)^{\frac{1}{2}}$
17.45	$x^{-1}e^{-a/x}K_\nu(b/x)$ $-1<\mathrm{Re}\,\nu<1$	$i\left(K_\nu(ui^{\frac{1}{2}})K_\nu(vi^{\frac{1}{2}})-K_\nu\{u(-i)^{\frac{1}{2}}\}K_\nu\{v(-i)^{\frac{1}{2}}\}\right)$ $\genfrac{}{}{0pt}{}{v}{u}=y^{\frac{1}{2}}\{(a+b)^{\frac{1}{2}}\pm(a-b)^{\frac{1}{2}}\}$

2.18 Modified Bessel Functions of Argument $(ax^2+bx+c)^{\frac{1}{2}}$

	$f(x)$	$g_s(y)$
18.1	$e^{-bx}I_0(ax^{\frac{1}{2}})$	$u\exp(\tfrac{1}{4}a^2bu)\{b\sin(\tfrac{1}{4}a^2yu)+y\cos(\tfrac{1}{4}a^2yu)\}$ $u=(b^2+y^2)^{-1}$
18.2	$K_0(ax^{\frac{1}{2}})$	$-(2y)^{-1}\{\mathrm{Ci}(u)\cos u+\mathrm{si}(u)\sin u\};u=\tfrac{1}{4}a^2/y$
18.3	$x^{-\frac{1}{2}}\{K_1(ax^{\frac{1}{2}})-\tfrac{1}{2}\pi Y_1(ax^{\frac{1}{2}})\}$	$\pi a^{-1}\cos(\tfrac{1}{4}a^2/y)$
18.4	$\log(bx)\{K_0(ax^{\frac{1}{2}})$ $+\tfrac{1}{2}\pi Y_0(ax^{\frac{1}{2}})\}$	$\pi y^{-1}\sin(\tfrac{1}{4}a^2/y)\{\log(\tfrac{1}{4}a/y)+\tfrac{1}{2}\log b\}$

	$f(x)$	$g_s(y)$
18.5	$x^{-\frac{1}{2}}K_\nu(ax^{\frac{1}{2}})$ $-3/2<\mathrm{Re}\nu<3/2$	$-\frac{1}{2}\pi(\frac{1}{2}\pi/y)^{\frac{1}{2}}\{\cos(u-v)J_{\frac{1}{2}\nu}(v)$ $-\sin(u-v)Y_{\frac{1}{2}\nu}(v)\};\ u=\frac{1}{4}\pi(\nu-1),v=\frac{1}{8}a^2/y$
18.6	$x^{-\frac{1}{2}\nu}\{K_\nu(ax^{\frac{1}{2}})\cos(\pi\nu)$ $+\frac{1}{2}\pi Y_\nu(ax^{\frac{1}{2}})\},-1<\mathrm{Re}\nu<2$	$-\frac{1}{2}\pi(\frac{1}{2}a)^{-\nu}y^{\nu-1}\sin(\frac{1}{2}\pi\nu-\frac{1}{4}a^2/y)$
18.7	$x^{-\frac{1}{2}\nu}\{K_\nu(ax^{\frac{1}{2}})\sin(\pi\nu)$ $+\frac{1}{2}\pi J_\nu(ax^{\frac{1}{2}})\},-1<\mathrm{Re}\nu<2$	$\frac{1}{2}\pi(\frac{1}{2}a)^{-\nu}y^{\nu-1}\cos(\frac{1}{2}\pi\nu-\frac{1}{4}a^2/y)$
18.8	$J_0(ax^{\frac{1}{2}})K_0(ax^{\frac{1}{2}})$	$\frac{1}{2}y^{-1}K_0(\frac{1}{2}a^2/y)$
18.9	$I_0(ax^{\frac{1}{2}})K_0(ax^{\frac{1}{2}})$	$\frac{1}{4}\pi y^{-1}\{\sin u\,J_0(u)-\cos u\,Y_0(u)\},u=\frac{1}{2}a^2/y$
18.10	$xJ_1(ax^{\frac{1}{2}})K_1(ax^{\frac{1}{2}})$	$\frac{1}{4}y^{-3}a^2K_0(\frac{1}{2}a^2/y)$
18.11	$x^{-1}J_2(ax^{\frac{1}{2}})K_2(ax^{\frac{1}{2}})$	$\frac{1}{2}\pi a^{-2}y\{I_1(\frac{1}{2}a^2/y)-\mathbf{L}_1(\frac{1}{2}a^2/y)\}$
18.12	$J_\nu(ax^{\frac{1}{2}})K_\nu(ax^{\frac{1}{2}})$ $\mathrm{Re}\nu>-2$	$\frac{1}{4}\pi y^{-1}\mathrm{cosec}(\pi\nu)\{\mathbf{J}_\nu(iu)+\mathbf{J}_\nu(-iu)$ $-2\cos(\frac{1}{2}\pi\nu)I_\nu(u)\}\ ;\ u=\frac{1}{2}a^2/y$
18.13	$\{J_\nu(ax^{\frac{1}{2}})+J_{-\nu}(ax^{\frac{1}{2}})\}$ $\cdot K_\nu(ax^{\frac{1}{2}}),-1<\mathrm{Re}\nu<1$	$\cos(\frac{1}{2}\pi\nu)y^{-1}K_\nu(\frac{1}{2}a^2/y)$
18.14	$x^{-\frac{1}{2}}J_\nu(ax^{\frac{1}{2}})K_\nu(ax^{\frac{1}{2}})$ $\mathrm{Re}\nu>-3/2$	$(\frac{1}{2}\pi y)^{\frac{1}{2}}a^{-2}\Gamma(\frac{3}{4}+\frac{1}{2}\nu)\{\Gamma(1+\nu)\}^{-1}$ $W_{-\frac{1}{4},\frac{1}{2}\nu}(\frac{1}{2}a^2/y)M_{\frac{1}{4},\frac{1}{2}\nu}(\frac{1}{2}a^2/y)$
18.15	$\{Y_\nu(ax^{\frac{1}{2}})-Y_{-\nu}(ax^{\frac{1}{2}})\}$ $\cdot K_\nu(ax^{\frac{1}{2}}),-1<\mathrm{Re}\nu<1$	$-\sin(\frac{1}{2}\pi\nu)y^{-1}K_\nu(\frac{1}{2}a^2/y)$
18.16	$\{I_\nu(ax^{\frac{1}{2}})+I_{-\nu}(ax^{\frac{1}{2}})\}$ $\cdot x^{-\frac{1}{2}}K_\nu(ax^{\frac{1}{2}}),-2<\mathrm{Re}\nu<2$	$-\frac{1}{2}\pi y^{-1}\{\sin(\frac{1}{2}\pi\nu-u)J_\nu(u)$ $+\cos(\frac{1}{2}\pi\nu-u)Y_\nu(u)\};\ u=\frac{1}{2}a^2/y$

	$f(x)$	$g_s(y)$
18.17	$x^{-\frac{1}{2}}K_\nu(ax^{\frac{1}{2}})$ $\cdot\{\cos(\frac{1}{2}\pi\nu+\frac{1}{4}\pi)J_\nu(ax^{\frac{1}{2}})$ $-\cos(\frac{1}{2}\pi\nu-\frac{1}{4}\pi)Y_\nu(ax^{\frac{1}{2}})\}$ $-3/2<\mathrm{Re}\nu<3/2$	$(\frac{1}{2}\pi y)^{\frac{1}{2}}a^{-2}W_{\frac{1}{4},\frac{1}{2}\nu}(\frac{1}{2}a^2/y)W_{-\frac{1}{4},\frac{1}{2}\nu}(\frac{1}{2}a^2/y)$
18.18	$K_1\{(iax)^{\frac{1}{2}}\}K_1\{(-iax)^{\frac{1}{2}}\}$	$-\frac{1}{8}\pi^2 y^{-1}\{Y_1(\frac{1}{2}a/y)+\mathbf{H}_{-1}(\frac{1}{2}a/y)\}$
18.19	$K_\nu\{(iax)^{\frac{1}{2}}\}K_\nu\{(-iax)^{\frac{1}{2}}\}$ $-3/2<\mathrm{Re}\nu<3/2$	$\frac{1}{4}\pi\nu\mathrm{cosec}(\frac{1}{2}\pi\nu)y^{-1}S_{-1,\nu}(\frac{1}{2}a/y)$
18.20	$x^{-1}K_2\{(iax)^{\frac{1}{2}}\}K_2\{(iax)^{\frac{1}{2}}\}$	$-\frac{1}{4}\pi^2 a^{-1}y\{\mathbf{H}_1(\frac{1}{2}a/y)-Y_1(\frac{1}{2}a/y)\}$
18.21	$x^{-\frac{1}{2}}K_\nu\{(iax)^{\frac{1}{2}}\}K_\nu\{(-iax)^{\frac{1}{2}}\}$ $-3/2<\mathrm{Re}\nu<3/2$	$\frac{1}{2}a^{-1}(\frac{1}{2}\pi y)^{\frac{1}{2}}\Gamma(\frac{3}{4}+\frac{1}{2}\nu)\Gamma(\frac{3}{4}-\frac{1}{2}\nu)$ $\cdot W_{-\frac{1}{4},\frac{1}{2}\nu}(\frac{1}{2}ia/y)W_{-\frac{1}{4},\frac{1}{2}\nu}(-\frac{1}{2}ia/y)$
18.22	$x(a^2+x^2)^{-\frac{1}{2}}K_1\{b(a^2+x^2)^{\frac{1}{2}}\}$	$\frac{1}{2}\pi b^{-1}y(b^2+y^2)^{-\frac{1}{2}}\exp\{-a(b^2+y^2)^{\frac{1}{2}}\}$
18.23	$xK_0(br)$	$\frac{1}{2}\pi yaw^{-2}e^{-aw}\{1+(aw)^{-1}\}$
18.24	$xr^{-1}K_1(br)$	$\frac{1}{2}\pi b^{-1}a^{-2}ye^{-aw}$
18.25	$xr^\nu K_\nu(br)$	$(\frac{1}{2}\pi a)^{\frac{1}{2}}a(ab)^\nu w^{-\nu-3/2}K_{\nu+3/2}(aw)$
18.26	$K_0(bp)$	$\frac{1}{2}w^{-1}\Big(\cos(aw+ay)\,\mathrm{Ci}(aw+ay)-\cos(aw-ay)$ $\cdot\mathrm{Ci}(aw-ay)+\sin(aw+ay)\,\mathrm{si}(aw+ay)$ $-\sin(aw-ay)\,\mathrm{si}(aw-ay)\Big)$
18.27	$p^{-1}K_{2\nu}(bp)$ $-3/2<\mathrm{Re}\nu<3/2$	$\frac{1}{8}\pi^2\sec(\pi\nu)\Big(\cos(ay)\{J_\nu(v)Y_\nu(u)$ $-J_\nu(u)Y_\nu(v)\}-\sin(ay)\{J_\nu(u)J_\nu(v)+Y_\nu(u)Y_\nu(v)\}\Big)$ $\genfrac{}{}{0pt}{}{u}{v}=\frac{1}{2}a(w\pm y)$
18.28	$I_0(bu)\qquad x<2a$ $\qquad\qquad 0\qquad x>2a$	$2\sin(ay)s^{-1}\sinh(as)\qquad\qquad y<b$ $2\sin(ay)S^{-1}\mathrm{ain}(aS)\qquad\qquad y>b$

	$f(x)$	$g_s(y)$
18.29	$u^{-1}I_\nu(bu)$ $x<2a$ $\qquad\qquad 0$ $x>2a$ $\mathrm{Re}\,\nu>-1$	$\pi\sin(ay)J_{\frac12\nu}(\frac12 ay+\frac12 aS)J_{\frac12\nu}(\frac12 ay-\frac12 aS)$, $y>b$
18.30	$u^\nu I_\nu(bu)$ $x<2a$ $\qquad\qquad 0$ $x>2a$ $\mathrm{Re}\,\nu>-1$	$(2\pi a)^{\frac12}(ab)^\nu\sin(ay)s^{-\nu-\frac12}I_{\nu+\frac12}(as)$ $y<b$ $(2\pi a)^{\frac12}(ab)^\nu\sin(ay)S^{-\nu-\frac12}J_{\nu+\frac12}(aS)$ $y>b$
18.31	$u^{-1}K_0(bu)$ $x<2a$ $\qquad\qquad 0$ $x>2a$	$-\frac14\pi^2\sin(ay)\{J_0(u)Y_0(v)+J_0(v)Y_0(u)\}$, $y>b$ $\genfrac{}{}{0pt}{}{u}{v}=\frac12 ay\pm\frac12 aS$
18.32	$(2ax-x^2)^{-\frac12}$ $\cdot K_{2\nu}\{b(2ax-x^2)^{\frac12}\}$, $\qquad\qquad\qquad x<2a$ $\qquad\qquad 0$ $x>2a$ $-\frac12<\mathrm{Re}\,\nu<\frac12$	$\frac12\pi^2\mathrm{cosec}(2\pi\nu)\sin(ay)$ $\cdot\left(J_{-\nu}(u)J_{-\nu}(v)-J_\nu(u)J_\nu(v)\right)$ $\genfrac{}{}{0pt}{}{u}{v}=\frac12 ay\pm\frac12 aS,y>b$
18.33	$\qquad\qquad 0$ $x<a$ $K_0(bV)$ $x>a$	$\frac12 w^{-1}\Big(\cos(aw)\{Ci(aw+ay)-Ci(aw-ay)\}$ $+\sin(aw)\{si(aw+ay)-si(aw-ay)\}\Big)$
18.34	$\qquad\qquad 0$ $x<a$ $V^{-1}K_{2\nu}(bV)$ $x>a$ $-\frac12<\mathrm{Re}\,\nu<\frac12$	$\frac18\pi^2\sec(\pi\nu)\Big(J_\nu(\frac12 aw-\frac12 ay)Y_\nu(\frac12 aw+\frac12 ay)$ $-J_\nu(\frac12 aw+\frac12 ay)Y_\nu(\frac12 aw-\frac12 ay)\Big)$
18.35	$\log(\frac{a+x}{a-x})Y_0(bv)$ $x<a$ $-2\pi^{-1}\log(\frac{x+a}{x-a})K_0(bV)$ $\qquad\qquad\qquad x>a$	$-2w^{-1}\sin(aw)\log\{(w+y)/b\}$
18.36	$\log(\frac{a+x}{a-x})J_0(bV)$ $x<a$ $-2K_0(bV)$ $x>a$	$-2w^{-1}\cos(aw)\log\{(w+y)/b\}$
18.37	$xv^\nu Y_\nu(bv)$ $x<a$ $2\pi^{-1}xV^\nu K_\nu(bV)$ $x>a$ $\mathrm{Re}\,\nu>-1$	$(\frac12\pi a)^{\frac12}a(ab)^\nu w^{-\nu-3/2}Y_{\nu+3/2}(aw)$

$$v=(a^2-x^2)^{\frac12},\ V=(x^2-a^2)^{\frac12}\qquad s=(y^2-b^2)^{\frac12},\ S=(y^2-b^2)^{\frac12},\ w=(b^2+y^2)^{\frac12}$$

	$f(x)$	$g_s(y)$
18.38	$K_o\{b(a^2+x^2)^{\frac{1}{2}}\}\arctan(\frac{x}{a})$ $\cdot x^{\frac{1}{2}}I_{\frac{1}{4}}(p)K_{\frac{1}{4}}(q)$ $q\atop p$ $=\frac{1}{2}b\{(a^2+x^2)^{\frac{1}{2}}\pm a\}$	$\frac{1}{2}\pi w^{-1}e^{-aw}\log\{(y+w)/b\}$ $(\frac{1}{2}\pi/y)^{\frac{1}{2}}w^{-1}\exp(-aw)$

2.19 Modified Bessel Functions of Trigonometric and Hyperbolic Arguments

	$f(x)$	$g_s(y)$
19.1	$I_{2\nu}(2a\sin x)$ $\qquad x<\pi$ $\qquad\qquad 0 \qquad\quad x>\pi$ $\text{Re}\,\nu>-1$	$\sin(\frac{1}{2}\pi y)I_{\nu-\frac{1}{2}y}(a)I_{\nu+\frac{1}{2}y}(a)$
19.2	$\operatorname{cosec} x\, I_{2\nu}(2a\sin x),$ $\qquad\qquad\qquad x<\pi$ $\qquad\qquad 0 \qquad\quad x>\pi$ $\text{Re}\,\nu>0$	$\frac{1}{2}\pi a\nu^{-1}\sin(\frac{1}{2}\pi y)\Big[I_{\nu-\frac{1}{2}+\frac{1}{2}y}(a)I_{\nu-\frac{1}{2}-\frac{1}{2}y}(a)$ $-I_{\nu+\frac{1}{2}+\frac{1}{2}y}(a)I_{\nu+\frac{1}{2}-\frac{1}{2}y}(a)\Big]$
19.3	$(\sin x)^{-m}I_m(2a\sin x)$ $\qquad\qquad\qquad x<\pi$ $\qquad\qquad 0 \qquad\quad x>\pi$ $m=0,1,2,\ldots$	$a^m m!\,\pi\sin(\frac{1}{2}\pi y)\sum_{n=o}^{m}(-1)^n\varepsilon_n\{(m+n)!\,(m-n)!\}^{-1}$ $\cdot I_{n-\frac{1}{2}y}(a)I_{n+\frac{1}{2}y}(a)$
19.4	$K_{2\nu}(2a\sin x)$ $\qquad x<\pi$ $\qquad\qquad 0 \qquad\quad x>\pi$ $-1<\text{Re}\,\nu<1$	$\frac{1}{2}\pi^2\operatorname{cosec}(2\pi\nu)\sin(\frac{1}{2}\pi y)$ $\cdot\Big[I_{-\nu-\frac{1}{2}y}(a)I_{-\nu+\frac{1}{2}y}(a)-I_{\nu-\frac{1}{2}y}(a)I_{\nu+\frac{1}{2}y}(a)\Big]$
19.5	$K_{2\nu}(2a\sinh x)$ $-1<\text{Re}\,\nu<1$	$\frac{1}{8}i\pi^2\operatorname{cosec}(2\pi\nu)\Big[J_{\nu-\frac{1}{2}iy}(a)Y_{-\nu-\frac{1}{2}iy}(a)$ $-J_{-\nu-\frac{1}{2}iy}(a)Y_{\nu-\frac{1}{2}iy}(a)+J_{-\nu+\frac{1}{2}iy}(a)Y_{\nu+\frac{1}{2}iy}(a)$ $-J_{\nu+\frac{1}{2}iy}(a)Y_{-\nu+\frac{1}{2}iy}(a)\Big]$
19.6	$2\pi^{-1}K_o[a(2\sinh x)^{\frac{1}{2}}]$ $+Y_o[a(2\sinh x)^{\frac{1}{2}}]$	$iK_{iy}(a)[Y_{iy}(a)-Y_{-iy}(a)]$
19.7	$\sinh x\, K_{2\nu}(2a\cosh x)$	$-\frac{1}{8}ia\nu^{-1}\Big[K_{\nu+\frac{1}{2}iy}(a)\{K_{\nu-\frac{1}{2}iy-1}(a)$ $+K_{\nu-\frac{1}{2}iy+1}(a)\}-K_{\nu-\frac{1}{2}iy}(a)$ $\cdot\{K_{\nu+\frac{1}{2}iy-1}(a)+K_{\nu+\frac{1}{2}iy+1}(a)\}\Big]$

	$f(x)$	$g_s(y)$
19.8	$K_0\{(a^2+b^2-2ab\ \cos\ x)^{\frac{1}{2}}\}$ $\qquad\qquad x<\pi$ $\qquad 0 \qquad x>\pi$	$y\sum_{n=0}^{\infty}\varepsilon_n(y^2-n^2)^{-1}\{1-(-1)^n\cos(\pi y)\}I_n(b)K_n(a)$ $\qquad\qquad\qquad\qquad\qquad\qquad b<a$
19.9	$K_0\{(a^2+b^2+2ab\ \cosh\ x)^{\frac{1}{2}}\}$	$-\tfrac{1}{2}\pi\mathrm{cosech}(\pi y)\{I_{ix}(b)+I_{-ix}(b)\}K_{ix}(a)$ $+\sum_{n=0}^{\infty}(-1)^n\varepsilon_n(n^2+y^2)^{-1}I_n(b)K_n(a)\ ,\ b<a$
19.10	$J_0\{(2ab\ \cosh\ x-a^2-b^2)^{\frac{1}{2}}\}$ $\cosh\ x>\tfrac{1}{2}(a^2+b^2)/(ab)$ $\qquad 0$ $\cosh\ x<\tfrac{1}{2}(a^2+b^2)/ab)$	$2\pi^{-1}\sinh(\pi y)K_{iy}(a)K_{iy}(b)$
19.11	$\sinh\ x(a^2+b^2+2ab\ \cosh\ x)^{-\frac{1}{2}}$ $\cdot K_1\{(a^2+b^2+2ab\ \cosh\ x)^{\frac{1}{2}}\}$	$(ab)^{-\frac{1}{2}}yK_{iy}(a)K_{iy}(b)$
19.12	$\left(\left(\dfrac{a+be^x}{b+ae^x}\right)^{\nu}-\left(\dfrac{b+ae^x}{a+be^x}\right)^{\nu}\right)$ $\cdot K_{2\nu}\{(a^2+b^2+2ab\ \cosh\ x)\}\}$	$i\left(K_{\nu+iy}(a)K_{\nu-iy}(b)-K_{\nu+iy}(b)K_{\nu-iy}(a)\right)$

2.20   Modified Bessel Functions of Variable Order

	$f(x)$	$g_s(y)$
20.1	$\sin(\pi x)I_{\nu-x}(a)I_{\nu+x}(a)$	$\tfrac{1}{4}I_{2\nu}\{2a\ \sin(\tfrac{1}{2}y)\}\qquad\qquad y<2\pi$ $\qquad 0 \qquad\qquad\qquad\qquad\quad y>2\pi$
20.2	$\mathrm{sech}(\pi x)\cosh(bx)$ $\cdot\{I_{-ix}(a)-I_{ix}(a)\}$ $\qquad\underline{b\leq\tfrac{1}{2}}$	$\tfrac{1}{2}\left[\exp\{-a\ \cosh(y+ib)\}\mathrm{Erf}\{i(2a)^{\frac{1}{2}}\sinh(\tfrac{1}{2}y+\tfrac{1}{2}ib)\}\right.$ $\left.+\exp\{-a\ \cosh(y-ib)\}\mathrm{Erf}\{i(2a)^{\frac{1}{2}}\cosh(\tfrac{1}{2}y-\tfrac{1}{2}ib)\}\right]$
20.3	$\mathrm{sech}(\pi x)\sinh(bx)$ $\cdot\{I_{ix}(a)+I_{-ix}(a)\}$ $\qquad\underline{b\leq\tfrac{1}{2}\pi}$	$\tfrac{1}{2}\left[\exp\{-a\ \cosh(y+ib)\}\mathrm{Erf}\{i(2a)^{\frac{1}{2}}\cosh(\tfrac{1}{2}y+\tfrac{1}{2}ib)\}\right.$ $\left.-\exp\{-a\ \cosh(y-ib)\}\mathrm{Erf}\{i(2a)^{\frac{1}{2}}\cosh(\tfrac{1}{2}y-\tfrac{1}{2}ib)\}\right]$
20.4	$\sinh(bx)K_{ix}(a)$ $\qquad\underline{b\leq\tfrac{1}{2}\pi}$	$\tfrac{1}{2}\pi\exp(-a\ \cos b)\ \cosh y)\sin(a\ \sin b\ \sinh y)$

	$f(x)$	$g_s(y)$
20.5	$\operatorname{cosech}(\tfrac{1}{2}\pi x)K_{ix}(a)$	$\sinh y \int_0^\infty \exp(-t\cosh y)K_0\{(a^2+t^2)^{\frac{1}{2}}\}dt$
20.6	$\operatorname{cosech}(\pi x)K_{ix}(a)$	$\tfrac{1}{2}\sinh y \int_0^\infty \exp(-t\cosh y)K_0(a+t)dt$
20.7	$\tanh(\pi x)K_{ix}(a)$	$-\tfrac{1}{2}i\pi\exp(-a\cosh y)\operatorname{Erf}\{i(2a)^{\frac{1}{2}}\sinh(\tfrac{1}{2}y)\}$
20.8	$\operatorname{sech}(\pi x)\sinh(bx)$ $\cdot K_{ix}(a)$, $b\leq\tfrac{3}{2}\pi$	$\tfrac{1}{4}\pi i\Big(\exp\{a\cosh(y+ib)\}\operatorname{Erfc}\{(2a)^{\frac{1}{2}}\cosh(\tfrac{1}{2}y+\tfrac{1}{2}ib)\}$ $-\exp\{a\cosh(y-ib)\}\operatorname{Erfc}\{(2a)^{\frac{1}{2}}\cosh(\tfrac{1}{2}y-\tfrac{1}{2}ib)\}\Big)$
20.9	$K_{ix}(a)\{Y_{ix}(a)-Y_{-ix}(a)\}$	$-i\Big(\tfrac{1}{2}\pi Y_0\{a(2\sinh y)^{\frac{1}{2}}\}+K_0\{a(2\sinh y)^{\frac{1}{2}}\}\Big)$
20.10	$\tanh(\pi x)K_{ix}(a)$ $\cdot\{I_{ix}(a)+I_{-ix}(a)\}$	$\tfrac{1}{2}\pi \mathbf{H}_0[2a\sinh(\tfrac{1}{2}y)]$
20.11	$K_{ix}(a)\{J_{ix}(a)-J_{-ix}(a)\}$	$-\tfrac{1}{2}i\pi J_0\{a(2\sinh y)^{\frac{1}{2}}\}$
20.12	$\sinh(\pi x)\{K_{ix}(a)\}^2$	$\tfrac{1}{4}\pi^2 J_0\{2a\sinh(\tfrac{1}{2}y)\}$
20.13	$xK_{ix}(a)K_{ix}(b)$	$\tfrac{1}{2}\pi ab\sinh y\, u^{-1}K_1(u)$, $u=(a^2+b^2+2ab\cosh y)^{\frac{1}{2}}$
20.14	$\sinh(\pi x)K_{ix}(a)K_{ix}(b)$	$\tfrac{1}{4}\pi^2 J_0(u)$ $u>0$ 0 $u<0$, $u=2ab\cosh y-a^2-b^2$
20.15	$K_{\nu+ix}(a)K_{\nu-ix}(b)$ $-K_{\nu+ix}(b)K_{\nu-ix}(a)$	$-\tfrac{1}{2}i\pi\{(\tfrac{u}{v})^\nu-(\tfrac{v}{u})^\nu\}K_{2\nu}\{(uve^{-y})^{\frac{1}{2}}\}$; $u=a+be^y$ $v=b+ae^y$

2.21 Functions Related to Bessel Functions

21.1	$J_{-\nu}(ax)-J_\nu(ax)$ $=\tan(\tfrac{\pi}{2}\nu)\{\mathbf{E}_\nu(ax)+\mathbf{E}_{-\nu}(ax)\}$	$2\sin(\tfrac{1}{2}\pi\nu)(a^2-y^2)^{-\frac{1}{2}}\cos\{(\nu+\tfrac{1}{2})\arccos(\tfrac{y}{a})\}$, $y<a$ 0 $y>a$

	$f(x)$	$g_s(y)$
21.2	$(b^2+x^2)^{-1}\{\mathbf{J}_\nu(ax)-\mathbf{J}_{-\nu}(ax)\}$ $=-(b^2+x^2)^{-1}\tan(\tfrac{1}{2}\pi\nu)$ $\cdot\{\mathbf{E}_\nu)ax)+\mathbf{E}_{-\nu}(ax)$	$i\pi b^{-1}e^{-by}\{\mathbf{J}_{-\nu}(iab)-\mathbf{J}_\nu(iab)\}$
21.3	$\mathbf{J}_\nu(a\sinh x)-\mathbf{J}_{-\nu}(a\sinh x)$ $=-\tan(\tfrac{1}{2}\pi\nu)$ $\cdot[\mathbf{E}_\nu(a\sinh x)+\mathbf{E}_{-\nu}(a\sinh x)]$	$-\tfrac{1}{2}i\pi\sin(\tfrac{1}{2}\pi\nu)\,\mathrm{sech}(\tfrac{1}{2}\pi y)\Big(I_{-\frac{1}{2}\nu-\frac{1}{2}iy}(\tfrac{1}{2}a)$ $\cdot I_{\frac{1}{2}\nu-\frac{1}{2}iy}(\tfrac{1}{2}a)-I_{-\frac{1}{2}\nu+\frac{1}{2}iy}(\tfrac{1}{2}a)I_{\frac{1}{2}\nu+\frac{1}{2}iy}(\tfrac{1}{2}a)\Big)$
21.4	$\mathbf{J}_{-x}(a)-\mathbf{J}_x(a)$ $=-\tan(\tfrac{\pi}{2}x)\{\mathbf{E}_x(a)+\mathbf{E}_{-x}(a)\}$	$\sin(a\sin y) \qquad y<\pi$ $0 \qquad y>\pi$
21.5	$\mathbf{J}_x(a)+\mathbf{J}_{-x}(a)$ $=\cot(\tfrac{\pi}{2}x)\{\mathbf{E}_x(a)-\mathbf{E}_{-x}(a)\}$	$\cos(a\sin y) \qquad y<\pi$ $0 \qquad y>\pi$
21.6	$\mathbf{H}_o(ax)$	$(a^2-y^2)^{-\frac{1}{2}} \qquad y<a$ $0 \qquad y>a$
21.7	$\mathbf{H}_o(ax)-Y_o(ax)$	$2\pi^{-1}(a^2-y^2)^{-\frac{1}{2}}\arccos(y/a), \qquad y<a$ $0 \qquad y>a$
21.8	$x^{-\nu}\{\mathbf{H}_\nu(ax)-Y_\nu(ax)\}$ $\mathrm{Re}\,\nu<1$	$(\tfrac{1}{2}\pi)^{-\frac{1}{2}}2^{-2\nu}\Gamma(1-\nu)\{\Gamma(\tfrac{1}{2}+\nu)\}^{-1}$ $\cdot(a^2-y^2)^{\frac{1}{2}\nu-\frac{1}{4}}P_{\nu-\frac{1}{2}}^{\nu-\frac{1}{2}}(\tfrac{y}{a}) \qquad y<a$ $(\tfrac{1}{2}\pi)^{-\frac{1}{2}}2^{-2\nu}\Gamma(1-\nu)\{\Gamma(\tfrac{1}{2}+\nu)\}^{-1}$ $\cdot(y^2-a^2)^{\frac{1}{2}\nu-\frac{1}{4}}P_{\nu-\frac{1}{2}}^{\nu-\frac{1}{2}}(\tfrac{y}{a}) \qquad y>a$
21.9	$x^{1+\nu}\{\mathbf{H}_\nu(ax)-Y_\nu(ax)\}$ $-3/2<\mathrm{Re}\,\nu<0$	$(\tfrac{1}{2}\pi y)^{-\frac{1}{2}}\Gamma(3+2\nu)(a/y)^\nu(a^2-y^2)^{-\frac{1}{2}\nu-3/4}$ $\cdot P_{-\nu-3/2}^{-\nu-3/2}(y/a) \qquad y<a$ $(\tfrac{1}{2}\pi y)^{-\frac{1}{2}}\Gamma(3+2\nu)(a/y)^\nu(y^2-a^2)^{-\frac{1}{2}\nu-3/4}$ $\cdot P_{-\nu-3/2}^{-\nu-3/2}(y/a) \qquad y>a$
21.10	$I_o(ax)-\mathbf{L}_o(ax)$	$(a^2+y^2)^{-\frac{1}{2}}$
21.11	$x^{-\nu}\{I_{-\nu}(ax)-\mathbf{L}_\nu(ax)\}$	$\pi^{\frac{1}{2}}\{\Gamma(\tfrac{1}{2}+\nu)\}^{-1}(2a)^{-\nu}(a^2+y^2)^{\nu-\frac{1}{2}}$

	$f(x)$	$g_s(y)$	
21.12	$x^{-\nu}\mathbf{H}_\nu(ax)$ $\mathrm{Re}\,\nu > -\tfrac{1}{2}$	$\pi^{\frac{1}{2}}\{\Gamma(\tfrac{1}{2}+\nu)\}^{-1}(2a)^{-\nu}(a^2-y^2)^{\nu-\frac{1}{2}},$ 0	$y<a$ $y>a$
21.13	$x^{-\nu}(b^2+x^2)^{-1}\mathbf{H}_\nu(ax)$ $\mathrm{Re}\,\nu > -5/2$	$\tfrac{1}{2}\pi b^{-\nu-1}e^{-by}\mathbf{L}_\nu(ab)$	$y<a$
21.14	$x^{\frac{1}{2}}\mathbf{H}_{\frac{1}{4}}(ax^2)$	$-\tfrac{1}{2}a^{-1}(\tfrac{1}{2}\pi y)^{\frac{1}{2}}Y_{\frac{1}{4}}(\tfrac{1}{4}y^2/a)$	
21.15	$x^{3/2}\mathbf{H}_{-\frac{1}{4}}(ax^2)$	$-\tfrac{1}{4}a^{-2}y(\tfrac{1}{2}\pi y)^{\frac{1}{2}}Y_{3/4}(\tfrac{1}{4}y^2/a)$	
21.16	$\mathbf{H}_0(ax^{\frac{1}{2}})$	$y^{-1}\Big(\sin u\{C(u)+S(u)\}+\cos u\{C(u)-S(u)\}\Big)$ $u=\tfrac{1}{4}a^2/y$	
21.17	$I_0(ax^{\frac{1}{2}})-\mathbf{L}_0(ax^{\frac{1}{2}})$	$y^{-1}\Big(\sin u\{C(u)-S(u)\}+\cos u\{1-C(u)-S(u)\}\Big)$ $u=\tfrac{1}{4}a^2/y$	
2.18	$x^{\frac{1}{2}\nu}\{I_\nu(ax^{\frac{1}{2}})-\mathbf{L}_\nu(ax^{\frac{1}{2}})\}$ $-2<\mathrm{Re}\,\nu<\tfrac{1}{2}$	$y^{-1}(2y/a)^{-\nu}\Big[\sin(\tfrac{1}{2}\pi\nu+u)\{C(u)-S(u)\}$ $+\cos(\tfrac{1}{2}\pi\nu+u)\{1-C(u)-S(u)\}\Big]\quad u=\tfrac{1}{4}a^2/y$	
21.19	$x^{\frac{1}{2}\nu}\mathbf{H}_\nu(ax^{\frac{1}{2}})$ $-5/2<\mathrm{Re}\,\nu<\tfrac{1}{2}$	$2^{\frac{1}{2}}y^{-1}(2y/a)^{-\nu}\Big(\sin(\tfrac{1}{4}\pi-\tfrac{1}{2}\pi\nu+u)C(u)$ $-\cos(\tfrac{1}{4}\pi-\tfrac{1}{2}\pi\nu+u)S(u)\Big),\quad u=\tfrac{1}{4}a^2/y$	
21.20	$\mathbf{H}_0(a\sinh x)$	$\tfrac{1}{2}\tanh(\tfrac{1}{2}\pi y)K_{\frac{1}{2}iy}(\tfrac{1}{2}a)\{I_{\frac{1}{2}iy}(\tfrac{1}{2}a)+I_{-\frac{1}{2}iy}(\tfrac{1}{2}a)\}$	
21.21	$I_0(a\sinh x)$ $-\mathbf{L}_0(a\sinh x)$	$\tfrac{1}{4}\pi\tanh(\tfrac{1}{2}\pi y)\Big[\{J_{\frac{1}{2}iy}(\tfrac{1}{2}a)\}^2+\{Y_{\frac{1}{2}iy}(\tfrac{1}{2}a)\}^2\Big]$	
21.22	$x^{-\nu}\mathbf{S}_{\nu,\nu+1}(ax)$ $\mathrm{Re}\,\nu<\tfrac{1}{2}$	$2^{-\nu-3/2}\pi\Gamma(\tfrac{1}{2}-\nu)a^{-\frac{1}{2}}(a^2-y^2)^{\frac{1}{2}\nu-\frac{1}{4}}$ $\cdot P_{\nu+\frac{1}{2}}^{\nu-\frac{1}{2}}(y/a)$ $2^{-\nu-3/2}\pi\Gamma(\tfrac{1}{2}-\nu)a^{-\frac{1}{2}}(y^2-a^2)^{\frac{1}{2}\nu-\frac{1}{4}}$ $\cdot P_{\nu+\frac{1}{2}}^{\nu-\frac{1}{2}}(y/a)$	$y<a$ $y>a$

	$f(x)$	$g_s(y)$
21.23	$x^{-\mu}S_{\mu,\nu}(ax)$ $\mathrm{Re}(\pm\nu-\mu)>-1$	$2^{-\mu}(\tfrac{1}{2}\pi/a)^{\frac{1}{2}}\Gamma(1-\tfrac{1}{2}\nu-\tfrac{1}{2}\mu)\Gamma(1+\tfrac{1}{2}\nu-\tfrac{1}{2}\mu)$ $\cdot\begin{cases} \cdot(a^2-y^2)^{\frac{1}{2}\mu-\frac{1}{4}}P_{\nu-\frac{1}{2}}^{\mu-\frac{1}{2}}(y/a) & y<a \\ (y^2-a^2)^{\frac{1}{2}\mu-\frac{1}{4}}P_{\nu-\frac{1}{2}}^{\mu-\frac{1}{2}}(y/a) & y>a \end{cases}$
21.24	$x^{-1}S_{-1,\nu}(ax)$ $-1<\mathrm{Re}\nu<1$	$2\nu^{-1}\sin(\tfrac{1}{2}\pi\nu)K_\nu\{(2aiy)^{\frac{1}{2}}\}K_\nu\{(-2aiy)^{\frac{1}{2}}\}$
21.25	$\tanh(\tfrac{1}{2}\pi x)S_{0,ix}(a)$	$-\tfrac{1}{2}i\left[\exp(-a\sinh y)\overline{Ei}(a\sinh y)\right.$ $\left.-\exp(a\sinh y)Ei(-a\sinh y)\right]$

2.22 Parabolic Cylinder- and Whittaker Functions

	$f(x)$	$g_s(y)$
22.1	$\exp(-\tfrac{1}{4}x^2)\{D_\nu(x)-D_\nu(-ax)$ $\mathrm{Re}\nu>0$	$(2\pi)^{\frac{1}{2}}y^{-\nu}\exp(-\tfrac{1}{2}y^2)\sin(\tfrac{1}{2}\pi\nu)$
22.2	$\exp(\tfrac{1}{4}a^2x^2)D_\nu(ax)$ $\mathrm{Re}\nu\ 0$	$\pi^{\frac{1}{2}}\{\Gamma(\tfrac{1}{2}-\tfrac{1}{2}\nu)\}^{-1}y^{-1}(2a/y)^\nu\exp(\tfrac{1}{2}y^2/a^2)$ $\cdot\Gamma(1+\tfrac{1}{2}\nu,\tfrac{1}{2}y^2/a^2)$
22.3	$x^\mu\exp(-\tfrac{1}{4}a^2x^2)D_\nu(ax)$ $\mathrm{Re}\mu>-2$	$2^{\frac{1}{2}\nu-\frac{1}{2}\mu-1}\Gamma(2+\mu)\{\Gamma(\tfrac{3}{2}+\tfrac{1}{2}\mu-\tfrac{1}{2}\nu)\}^{-1}a^{-\mu-2}y$ $\cdot {}_2F_2(1+\tfrac{1}{2}\mu,\tfrac{3}{2}+\tfrac{1}{2}\mu;\tfrac{3}{2},\tfrac{3}{2}+\tfrac{1}{2}\mu-\tfrac{1}{2}\nu;-\tfrac{1}{2}y^2/a^2)$
22.4	$x^{-\frac{1}{2}-\frac{1}{2}\nu}\exp(\tfrac{1}{4}a^2x)$ $\cdot D_\nu(ax^{\frac{1}{2}})$, $\mathrm{Re}\nu<3$	$(\pi/y)^{\frac{1}{2}}\{y+(a+y^{\frac{1}{2}})^2\}^{\frac{1}{2}\nu}$ $\cdot\sin\left[\tfrac{1}{4}\pi-\nu\arctan\{y^{\frac{1}{2}}(a+y^{\frac{1}{2}})^{-1}\}\right]$
22.5	$x^{-\frac{1}{2}\nu-3/2}\exp(\tfrac{1}{4}a^2x)$ $\cdot D_\nu(ax^{\frac{1}{2}})$, $\mathrm{Re}\nu<1$	$(\tfrac{1}{2}\pi)^{\frac{1}{2}}(\tfrac{1}{2}+\tfrac{1}{2}\nu)^{-1}\{y+(a+y^{\frac{1}{2}})^2\}^{\frac{1}{2}\nu+\frac{1}{2}}$ $\cdot\sin\left[(\nu+1)\arctan\{y^{\frac{1}{2}}(a+y^{\frac{1}{2}})^{-1}\}\right]$
22.6	$x^{-\frac{1}{2}\nu-\frac{1}{2}}\exp(-\tfrac{1}{4}a^2/x)$ $\cdot D_\nu(a/x)$, $\mathrm{Re}\nu>-1$	$(\pi/y)^{\frac{1}{2}}(2y)^{\frac{1}{2}\nu}\exp(-ax^{\frac{1}{2}})\cos(\tfrac{1}{2}\pi\nu-ay^{\frac{1}{2}})$
22.7	$D_{-\nu-1}(u)\{D_\nu(u)-D_\nu(-u)\}$ $u=(2x)^{\frac{1}{2}}$, $\mathrm{Re}\nu>0$	$(2\pi)^{\frac{1}{2}}\sin(\tfrac{1}{2}\pi\nu)y^\nu(1+y^2)^{-\frac{1}{2}}$ $\cdot\{1+(1+y^2)^{\frac{1}{2}}\}^{-\nu-\frac{1}{2}}$
22.8	$\exp(-\tfrac{1}{4}u^2)$ $\cdot\{D_\nu(u)+D_\nu(-u)\}$ $\mathrm{Re}\nu>-1$, $u=2a\sinh x$	$2^{\frac{1}{2}\nu}(2\pi a^2)^{-\frac{1}{2}}\exp(a^2)\sin(\tfrac{1}{2}\pi\nu)\sinh(\tfrac{1}{2}\pi y)$ $\cdot\Gamma(\tfrac{1}{2}+\tfrac{1}{2}\nu+\tfrac{1}{2}iy)\Gamma(\tfrac{1}{2}+\tfrac{1}{2}\nu-\tfrac{1}{2}iy)W_{-\frac{1}{2}\nu,\frac{1}{2}iy}(2a^2)$

	$f(x)$	$g_s(y)$
22.9	$x^{2\nu-1}\exp(-\tfrac{1}{4}x^2)$ $\cdot M_{3\nu,\nu}(\tfrac{1}{2}x^2)$, $\mathrm{Re}\,\nu>-\tfrac{1}{4}$	$(\tfrac{1}{2}\pi)^{\tfrac{1}{2}}2^{\nu-1}\exp(-\tfrac{1}{4}y^2)M_{3\nu,\nu}(\tfrac{1}{2}y^2)$
22.10	$x^{-2\nu}\exp(\tfrac{1}{4}x^2)$ $\cdot W_{3\nu-1,\nu}(\tfrac{1}{2}x^2)$, $\mathrm{Re}\,\nu<\tfrac{1}{2}$	$(\tfrac{1}{2}\pi)^{\tfrac{1}{2}}y^{-2\nu}\exp(\tfrac{1}{4}y^2)W_{3\nu-1,\nu}(\tfrac{1}{2}y^2)$
22.11	$x^{-2\mu}\exp(-\tfrac{1}{4}x^2)M_{\kappa,\mu}(\tfrac{1}{2}x^2)$ $\mathrm{Re}\,(\mu+\kappa)>\tfrac{1}{2}$	$\pi^{\tfrac{1}{2}}2^{-\tfrac{1}{2}\kappa-3/2\mu}\Gamma(1+2\mu)\{\Gamma(\tfrac{1}{2}+\mu+\kappa)\}^{-1}y^{\kappa+\mu-1}$ $\cdot\exp(-\tfrac{1}{4}y^2)W_{\alpha,\beta}(\tfrac{1}{2}y^2);2\alpha=1+\kappa-3\mu,$ $2\beta=\kappa+\mu-1$
22.12	$x^{2\mu-1}\exp(-\tfrac{1}{4}x^2)M_{\kappa,\mu}(\tfrac{1}{2}x^2)$ $-\tfrac{1}{2}<\mathrm{Re}\,\mu<\mathrm{Re}\,\kappa$	$(\tfrac{1}{2}\pi)^{\tfrac{1}{2}}2^{3/2\mu-\tfrac{1}{2}\kappa}\Gamma(1+2\mu)\{\Gamma(1+\kappa-\mu)\}^{-1}y^{\kappa-\mu-1}$ $\cdot\exp(-\tfrac{1}{4}y^2)M_{\alpha,\beta}(\tfrac{1}{2}y^2);2\alpha=\kappa+3\mu,2\beta=\kappa-\mu$
22.13	$x^{-2\mu}\exp(-\tfrac{1}{4}x^2)W_{\kappa,\mu}(\tfrac{1}{2}x^2)$ $\mathrm{Re}\,\mu<3/4$	$\pi^{\tfrac{1}{2}}2^{-\tfrac{1}{2}\kappa-3/2\mu}\Gamma(\tfrac{3}{2}-2\mu)\{\Gamma(2-\kappa-\mu)\}^{-1}y^{\kappa+\mu-1}$ $\cdot\exp(-\tfrac{1}{4}y^2)M_{\alpha,\beta}(\tfrac{1}{2}y^2);2\alpha=1+\kappa-3\mu,$ $2\beta=1-\kappa-\mu$
22.14	$x^{-2\mu-1}\exp(\tfrac{1}{4}x^2)W_{\kappa,\mu}(\tfrac{1}{2}x^2)$ $\mathrm{Re}\,(\mu-\kappa)>\tfrac{1}{2}$, $\mathrm{Re}\,\mu<3/4$	$\pi^{\tfrac{1}{2}}2^{\tfrac{1}{2}\kappa-3/2\mu}\Gamma(\tfrac{3}{2}-2\mu)\{\Gamma(\tfrac{1}{2}+\mu-\kappa)\}^{-1}y^{\mu-\kappa-1}$ $\cdot\exp(\tfrac{1}{4}y^2)W_{\alpha,\beta}(\tfrac{1}{2}y^2);\ 2\alpha=\kappa+3\mu-1,$ $2\beta=\kappa-\mu+1$
22.15	$x^{2\nu-1}\exp(-\tfrac{1}{2}x^2)W_{\kappa,\mu}(x^2)$ $\mathrm{Re}\,\nu>|\mathrm{Re}\,\mu|>-1$	$\tfrac{1}{2}\Gamma(1+\mu+\nu)\Gamma(1-\mu+\nu)\{\Gamma(\tfrac{3}{2}-\kappa+\nu)\}^{-1}y$ $\cdot {}_2F_2(1+\nu+\mu,1+\nu-\mu;\tfrac{3}{2},\tfrac{3}{2}-\kappa+\nu;-\tfrac{1}{4}y^2)$
22.16	$x^{-\tfrac{1}{2}}K_{\mu-\tfrac{1}{4}}(\tfrac{1}{2}x^2)M_{\kappa,\mu}(x^2)$ $\mathrm{Re}\,\mu>-\tfrac{1}{2}$, $\mathrm{Re}\,\kappa>-\tfrac{1}{4}$	$\tfrac{1}{2}\Gamma(1+2\mu)\{\Gamma(\kappa+\tfrac{5}{4})\}^{-1}(\tfrac{1}{2}\pi/y)^{\tfrac{1}{2}}W_{\alpha,\beta}(\tfrac{1}{2}y^2)$ $\cdot M_{\gamma,\delta}(\tfrac{1}{2}y^2);2\alpha=\kappa-\mu,2\beta=\kappa-\tfrac{1}{4},$ $2\gamma=\kappa-\mu,2\delta=\kappa+\tfrac{1}{4}$
22.17	$x^{2\mu-1}W_{\kappa,\mu}(ax)M_{-\kappa,\mu}(ax)$ $\mathrm{Re}\,(\mu+\kappa)<\tfrac{1}{4}$, $\mathrm{Re}\,\mu>-\tfrac{1}{2}$	$\tfrac{1}{2}\pi^{\tfrac{1}{2}}a^{2\kappa}\Gamma(1+2\mu)\{\Gamma(1-\mu-\kappa)\}^{-1}(\tfrac{1}{2}y)^{-2\mu-2\kappa}$ $\cdot {}_3F_2(\tfrac{1}{2}-\kappa,1-\kappa,\tfrac{1}{2}+\mu-\kappa;1-2\kappa,1-\mu-\kappa;-y^2/a^2)$
22.18	$x^{\nu}W_{\kappa,\mu}(ax)W_{-\kappa,\mu}(ax)$ $\mathrm{Re}\,\nu>2|\mathrm{Re}\,\mu|-3$	$\tfrac{1}{2}a^{1-2\nu}\Gamma(\nu+\mu)\Gamma(\nu-\mu)\Gamma(2\nu)$ $\cdot y\{\Gamma(\tfrac{1}{2}+\kappa+\nu)\Gamma(\tfrac{1}{2}-\kappa+\nu)\}^{-1}$ $\cdot {}_4F_3(\nu,\tfrac{1}{2}+\nu,\nu+\mu,\nu-\mu;\tfrac{1}{2}+\kappa+\nu,\tfrac{1}{2}-\kappa+\nu,\tfrac{3}{2};-y^2/a^2)$

	$f(x)$	$g_s(y)$
22.19	$x^{-3/2}W_{\nu+\mu,\kappa+1/8}(\tfrac{1}{2}x^2)$ $\cdot M_{\mu-\nu,-\kappa+1/8}(\tfrac{1}{2}x^2)$	$(\tfrac{1}{2}\pi)^{\frac{1}{2}}\Gamma(\tfrac{5}{4}-2\kappa)\{\Gamma(\tfrac{5}{2}-2\nu)\}^{-1}$ $\cdot y^{-3/2}W_{\mu+\nu,\nu+1/8}(\tfrac{1}{2}y^2)M_{\mu-\kappa,-\nu+1/8}(\tfrac{1}{2}y^2)$
22.20	$\Gamma(\tfrac{1}{2}+\nu+ix)\Gamma(\tfrac{1}{2}+\nu-ix)$ $\cdot\sinh(\pi x)W_{-\nu,ix}(a)$ $\mathrm{Re}\,\nu>-1$	$\pi 2^{-\nu-2}(\pi a)^{\frac{1}{2}}\mathrm{cosec}(\pi\nu)\exp\{\tfrac{1}{2}a\cosh^2(\tfrac{1}{2}y)\}$ $\{D_{2\nu}(u)-D_{2\nu}(-u)\};\ u=(2a)^{\frac{1}{2}}\sinh(\tfrac{1}{2}y)$

2.23 Elliptic Integrals*

	$f(x)$	$g_s(y)$
23.1	$K\{(\tfrac{1}{2}-\tfrac{1}{2}x)^{\frac{1}{2}}\}\quad x<1$ $\qquad\qquad 0\qquad\quad x>1$	$\tfrac{9}{32}\pi^{3/2}\{\Gamma(\tfrac{7}{4})\}^{-2}s_{\frac{1}{2},0}(y)$
23.2	$a^{-1}K(x/a)\qquad x<a$ $x^{-1}K(a/x)\qquad x>a$	$\tfrac{1}{4}\pi^2\left[J_0(\tfrac{1}{2}ay)\right]^2$
23.3	$0\qquad\qquad\quad x<a$ $(a+x)^{-\frac{1}{2}}K\left[\left(\tfrac{x-a}{x+a}\right)^{\frac{1}{2}}\right]x>a$	$\tfrac{1}{4}\pi(\tfrac{1}{2}\pi/y)^{\frac{1}{2}}\{J_0(ay)-Y_0(ay)\}$
23.4	$(a+x)^{-\frac{1}{2}}K\{(\tfrac{x}{a+x})^{\frac{1}{2}}\}$	$\tfrac{1}{4}\pi(\pi/y)^{\frac{1}{2}}$ $\cdot\left[J_0(ay)\sin(\tfrac{1}{2}ay+\tfrac{1}{4}\pi)-Y_0(ay)\cos(\tfrac{1}{2}ay+\tfrac{1}{4}\pi)\right]$
23.5	$0\qquad\qquad\quad x<a$ $x^{-1}K\{(1-a^2/x^2)^{\frac{1}{2}}\}x<a$	$-\tfrac{1}{4}\pi^2 J_0(\tfrac{1}{2}ay)Y_0(\tfrac{1}{2}ay)$
23.6	$0\qquad\qquad\quad x<a$ $(b^2-x^2)^{-\frac{1}{2}}K(u^{\frac{1}{2}}),x>a>b$ $u=(x^2-a^2)/(x^2-b^2)$	$-\tfrac{1}{8}\pi^2\left[J_0(\tfrac{1}{2}ay+\tfrac{1}{2}by)Y_0(\tfrac{1}{2}ay-\tfrac{1}{2}by)\right.$ $\left.+Y_0(\tfrac{1}{2}ay+\tfrac{1}{2}ab)J_0(\tfrac{1}{2}ay-\tfrac{1}{2}ab)\right]$
23.7	$0\qquad\qquad\quad x<b$ $(a^2-b^2)^{-\frac{1}{2}}K(u^{\frac{1}{2}})b<x<a$ $(x^2-b^2)^{-\frac{1}{2}}K(u^{-\frac{1}{2}})x>a$ $u=(x^2-b^2)/(a^2-b^2)x>b$	$\tfrac{1}{4}\pi^2 J_0(\tfrac{1}{2}ay+\tfrac{1}{2}by)J_0(\tfrac{1}{2}ay-\tfrac{1}{2}by)$

*Note that the notations for the Elliptic Integrals $\mathbf{K}(k),\mathbf{E}(e)$ (in bold letters) are here replaced by the ordinary ones K and E

	$f(x)$	$g_s(y)$
23.8	$(a^2+x^2)^{-\frac{1}{2}}K\{x(a^2+x^2)^{-\frac{1}{2}}\}$	$\frac{1}{2}\pi I_0(\frac{1}{2}ay)K_0(\frac{1}{2}ay)$
23.9	$\begin{array}{ll} 0 & u>1 \\ u\,K\{(1-u^2)^{\frac{1}{2}}\} & u<1 \\ u=a\,\text{cosech}\,x \end{array}$	$-\frac{1}{4}\pi a\,\tanh(\frac{1}{2}\pi y)P_{-\frac{1}{2}+\frac{1}{2}iy}(v)$ $\cdot\left[q_{-\frac{1}{2}+\frac{1}{2}iy}(v)+q_{-\frac{1}{2}-iy}(v)\right]\ ;\ v=(1+a^2)^{\frac{1}{2}}$
23.10	$u^{-\frac{1}{2}}K(au^{-\frac{1}{2}}\sinh x)$ $a>1\ ,\ u=1+a^2\sinh^2 x$	$\frac{1}{4}\pi^2 a^{-1}\sinh(\frac{1}{2}\pi y)\,\text{sech}^2(\frac{1}{2}\pi y)$ $\cdot P_{-\frac{1}{2}+iy}(v)P_{-\frac{1}{2}+iy}(-v)\ ;\ v=(1-a^{-2})^{\frac{1}{2}}$

Part III
Exponential Fourier Transforms
(Tables III)

III. Exponential Fourier Transforms

	$f(x)$		$g_e(y)$
3.1	A	$0<x<a<b$	$iAy^{-1}(e^{-iay}-e^{iby})$
	0	otherwise	
3.2	x^n	$0<x<a$	$n!(-iy)^{-n-1}\left[1-e^{iay}\sum_{k=o}^{n}(-iay)^k/k!\right]$
	0	otherwise	
	$n=1,2,\ldots$		
3.3	x^ν	$0<x<a$	$(-iy)^{-\nu-1}\gamma(\nu+1,-iay)$
	0	otherwise	
	$\mathrm{Re}\nu>-1$		
3.4	0	$0<x<a$	$(-iy)^{-\nu-1}\Gamma(\nu+1,-iay)$
	x^ν	$x>a$	
	$\mathrm{Re}\nu<0$		
3.5	$(a-x)^\nu$	$0<x<a$	$(-iy)^{-\nu-1}e^{iay}\gamma(\nu+1,iay)$
	0	otherwise	
	$\mathrm{Re}\nu>-1$		
3.6	$(a+x)^\nu$	$x>0$	$(-iy)^{-\nu-1}e^{-iay}\Gamma(\nu+1,-iay)$
	0	$x<0$	
	$\mathrm{Re}\nu<0$		
3.7	$x^\nu(a+x)^{-1}$	$x>0$	$a^\nu e^{-iay}\Gamma(\nu+1)\Gamma(-\nu,-iay)$
	0	$x<0$	
	$-1<\mathrm{Re}\nu<1$		
3.8	$x^\mu(a+x)^\nu$	$x<0$	$a^{\frac{1}{2}\mu+\frac{1}{2}\nu}(-iy)^{-1-\frac{1}{2}\mu-\frac{1}{2}\nu}\Gamma(1+\mu)e^{-iay/2}$
	0	$x<0$	$\cdot W_{\frac{1}{2}\nu-\frac{1}{2}\mu,\,\frac{1}{2}+\frac{1}{2}\nu+\frac{1}{2}\mu}(-iay)$
	$-1<\mathrm{Re}\mu<-\mathrm{Re}\nu$		
3.9	$x^{-1}(x-a)^\nu$	$x>a>0$	$a^\nu\Gamma(\nu+1)\Gamma(-\nu,-iay)$
	0	$x<a$	
	$-1<\mathrm{Re}\nu<1$		

	$f(x)$	$g_e(y)$
3.10	$(x-a)^\mu (b-x)^\nu$ $0<a<x<b$ 0 otherwise $\mathrm{Re}(\nu,\mu)>-1$	$(b-a)^{\frac{1}{2}\mu+\frac{1}{2}\nu}B(1+\mu,1+\nu)\exp\{\frac{1}{2}iy(a+b)\}$ $\cdot(-iy)^{-\frac{1}{2}\nu-\frac{1}{2}\mu-1}M_{\frac{1}{2}\mu-\frac{1}{2}\nu,\frac{1}{2}\mu+\frac{1}{2}\nu+\frac{1}{2}}\{-iy(b-a)\}$
3.11	$(x+a)^\mu (x-b)^\nu$ $x>b>0$ 0 otherwise $-1<\mathrm{Re}\,\nu<-\mathrm{Re}\,\mu$	$(a+b)^{\frac{1}{2}\mu+\frac{1}{2}\nu}\Gamma(1+\nu)\exp\{-\frac{1}{2}iy(a-b)\}$ $\cdot(-iy)^{-\frac{1}{2}\mu-\frac{1}{2}\nu-\frac{1}{2}}W_{\frac{1}{2}\mu-\frac{1}{2}\nu,\frac{1}{2}+\frac{1}{2}\nu+\frac{1}{2}\mu}\{-iy(a+b)\}$
3.12	$(x-a)^{\nu-1}(x+a)^{-\nu-\frac{1}{2}},x>a>0$ 0 otherwise $\mathrm{Re}\,\nu>0$	$2^{\nu-\frac{1}{2}}a^{-\frac{1}{2}}\Gamma(\nu)D_{-2\nu}\{2(-iay)^{\frac{1}{2}}\}$
3.13	$x^{\nu-1}(a+x)^{-\nu-\frac{1}{2}}$ $x>0$ 0 $x<0$ $\mathrm{Re}\,\nu>0$	$2^\nu a^{-\frac{1}{2}}e^{-\frac{1}{2}iay}\Gamma(\nu)D_{-2\nu}\{(-2iay)^{\frac{1}{2}}\}$
3.14	e^{-cx} $a<x<b$ 0 otherwise	$(c-iy)^{-1}\left[\exp\{-a(c-iy)\}-\exp\{-b(c-iy)\}\right]$
3.15	$x^{\nu-1}e^{-ax}$ $x>0$ 0 $x<0$ $\mathrm{Re}\,\nu>0$	$\Gamma(\nu)(a-iy)^{-\nu}$
3.16	$x^{-1}(e^{-ax}-e^{-bx})$ $x>0$ 0 $x<0$ $a<b$	$\log\{(b-iy)/(a-iy)\}$
3.17	$e^{-\lambda x}(1-e^{-ax})^\nu$ $x>0$ 0 $x<0$ $\mathrm{Re}\,\nu>-1$, $\mathrm{Re}\,\lambda>0$	$a^{-1}B\{1+\nu,(\lambda-iy)/a\}$
3.18	$xe^{-\lambda x}(1-e^{-ax})^\nu$ $x>0$ 0 $x<0$ $\mathrm{Re}\,\nu>-1$, $\mathrm{Re}\,\lambda>0$	$a^{-2}B\{1+\nu,(\lambda-iy)/a\}$ $\cdot\left[\psi\{1+\nu+(\lambda-iy)/a\}-\psi\{(\lambda-iy)/a\}\right]$
3.19	$x^n e^{-\lambda x}(1-e^{-ax})^{-1}$, $x>0$ 0 $x<0$ $n=1,2,\ldots,\mathrm{Re}\,\lambda>0$	$(-a)^{-n-1}\psi^{(n)}\{(\lambda-iy)/a\}$

	$f(x)$	$g_e(y)$
3.20	$x^{\nu-1}e^{-\lambda x}(e^{ax}-1)^{-1}$, $x>0$ 0 $x<0$ $\mathrm{Re}\,\nu>1$, $\mathrm{Re}\,\lambda>-a$	$a^{-\nu}\Gamma(\nu)\zeta\{\nu,1+(\lambda-iy)/a\}$
3.21	$e^{-\lambda x}(1+e^{-ax})^{-1}$ $x>0$ 0 $x<0$ $\mathrm{Re}\,\lambda>0$	$\tfrac{1}{2}a^{-1}\left[\psi\{\tfrac{1}{2}+\tfrac{1}{2}(\lambda-iy)/a\}-\psi\{\tfrac{1}{2}(\lambda-iy)/a\}\right]$
3.22	$x^{\nu-1}e^{-\lambda x}(1+e^{-ax})^{-1}$, $x>0$ 0 $x<0$ $\mathrm{Re}\,\nu>0$, $\mathrm{Re}\,\lambda>0$	$-(2a)^{-\nu}$ $\cdot\left[\zeta\{\nu,\tfrac{1}{2}+\tfrac{1}{2}(\lambda-iy)/a\}-\zeta\{\nu,\tfrac{1}{2}(\lambda-iy)/a\}\right]$
3.23	$x^{\nu-1}e^{-\lambda x}e^{-a/x}$ $x>0$ 0 $x<0$ $\mathrm{Re}\,\nu>0$, $\mathrm{Re}\,\lambda>0$	$2a^{\frac{1}{2}\nu}(\lambda-iy)^{-\frac{1}{2}\nu}K_\nu\{2(a\lambda-iay)^{\frac{1}{2}}\}$
3.24	$x^{\nu-1}\exp(-ax^{\frac{1}{2}})$ $x>0$ 0 $x<0$ $\mathrm{Re}\,\nu>0$	$2(-2iy)^{-\nu}\Gamma(2\nu)\exp(\tfrac{1}{8}ia^2/y)$ $\cdot D_{-2\nu}\{(\tfrac{1}{2}ia^2/y^{\frac{1}{2}}\}$
3.25	0 $x<b$ $\exp(-ax^2)$ $x>b>0$	$\tfrac{1}{2}(\pi/a)^{\frac{1}{2}}\exp(-\tfrac{1}{4}y^2/a)\mathrm{Erfc}(-\tfrac{1}{2}iya^{-\frac{1}{2}}+ba^{\frac{1}{2}})$
3.26	$x^{\nu-1}\exp(-ax^2)$ $x>0$ 0 $x<0$ $\mathrm{Re}\,\nu>0$	$(2a)^{-\frac{1}{2}\nu}\Gamma(\nu)\exp(-\tfrac{1}{8}y^2/a)D_{-\nu}\{(-\tfrac{1}{2}y^2/a)^{\frac{1}{2}}\}$
3.27	$\exp(-a^3x^3)$ $x>0$ 0 $x<0$	$2iy^{-1}(-\tfrac{1}{3}iy/a)^{3/2}S_{0,1/3}\{2(-\tfrac{1}{3}iy/a)^{3/2}\}$
3.28	$x^\nu\exp(-ax^p)$ $x>0$ 0 $x<0$ $\mathrm{Re}\,\nu>-1$	$(iy)^{-\nu-1}\sum_{k=0}^{\infty}\Gamma(1+\nu+kp)(-iy)^{-kp}(-a)^k/k!$ $0<p<1$ $p^{-1}a^{-(1+\nu)/p}\sum_{k=0}^{\infty}\Gamma\{(1+\nu+k)/p\}a^{-k/p}(iy)^k/k!$ $p>1$
3.29	$e^{-\lambda x}\exp(-ae^{-x})$, $x>0$ 0 $x<0$ $\mathrm{Re}\,\lambda>0$	$a^{-(\lambda-iy)}\gamma(\lambda-iy,a)$

	$f(x)$		$g_e(y)$
3.30	$e^{-\lambda x}\exp(-ae^x)$,	$x>0$	$a^{\lambda-iy}\Gamma(-\lambda+iy,a)$
	0	$x<0$	
3.31	$\exp(-a\sinh x)$	$x>0$	$i\pi\,\mathrm{cosech}(\pi y)\{J_{-iy}(a)-J_{-iy}(a)\}$
	0	$x<0$	
3.32	$e^{-\lambda x}\log(a+bx)$	$x>0$	$(\lambda-iy)^{-1}[\log a-\exp\{a(\lambda-iy)/b\}$
	0	$x<0$	$\cdot\mathrm{Ei}\{-a(\lambda-iy)/b\}]$
	$\mathrm{Re}\,\lambda>0$		
3.33	$e^{-\lambda x}\log(1+e^{-ax})$	$x>0$	$(\lambda-iy)^{-1}\Big(\log 2-\tfrac{1}{2}\psi\{1+\tfrac{1}{2}(\lambda-iy)/a\}$
	0	$x<0$	$+\tfrac{1}{2}\psi\{\tfrac{1}{2}+\tfrac{1}{2}(\lambda-iy)/a\}\Big)$
	$\mathrm{Re}\,\lambda>0$		
3.34	$e^{-\lambda x}\log(1-e^{-ax})$	$x>0$	$(\lambda-iy)^{-1}\Big(\gamma+\psi\{1+(\lambda-iy)/a\}\Big)$
	0	$x<0$	
	$\mathrm{Re}\,\lambda>-a$		
3.35	$e^{-\lambda x}\mathrm{sech}(ax)$	$x>0$	$\tfrac{1}{2}a^{-1}\Big(\psi\{\tfrac{3}{3}+\tfrac{1}{4}(\lambda-iy)/a\}-\psi\{\tfrac{1}{4}+\tfrac{1}{4}(\lambda-iy)/a\}\Big)$
	0	$x<0$	
	$\mathrm{Re}\,\lambda>-a$		
3.36	$e^{-\lambda x}x^{\nu-1}\mathrm{cosech}(ax)$,	$x>0$	$2(2a)^{-\nu}\Gamma(\nu)\,\zeta\{\nu,\tfrac{1}{2}+\tfrac{1}{2}(\lambda-iy)/a\}$
	0	$x<0$	
	$\mathrm{Re}\,\nu>1,\quad \mathrm{Re}\,\lambda>-a$		
3.37	$e^{-\lambda x}x^{\nu-1}\mathrm{sech}(ax)$	$x>0$	$2^{1-2(\lambda-iy)}a^{-\nu}\Gamma(\nu)$
	0	$x<0$	$\cdot\Big(\zeta\{\nu,\tfrac{1}{4}+\tfrac{1}{4}(\lambda-iy)/a\}-\zeta\{\nu,\tfrac{3}{4}+\tfrac{1}{4}(\lambda-iy)/a\}\Big)$
	$\mathrm{Re}\,\nu>0,\quad \mathrm{Re}\,\lambda>-a$		
3.38	$e^{-\lambda x}\tanh(ax)$	$x>0$	$(2a)^{-1}\Big(\psi\{\tfrac{1}{2}+\tfrac{1}{4}(\lambda-iy)/a\}-\psi\{\tfrac{1}{4}(\lambda-iy)/a\}\Big)$
	0	$x<0$	$-(\lambda-iy)^{-1}$
	$\mathrm{Re}\,\lambda>0$		
3.39	$\exp(-\lambda x-a\sinh x)$	$x>0$	$\pi\cosec\{\pi(\lambda-iy)\}\Big(J_{\lambda-iy}(a)-J_{\lambda-iy}(a)\Big)$
	0	$x<0$	

	$f(x)$	$g_e(y)$
3.40	$x^2 \exp(-ax^2)$ $x>0$ 0 $x<0$	$2(2a)^{-3/2} \exp(-\tfrac{1}{4}y^2/a) D_{-3}\{(-\tfrac{1}{2}y^2/a)^{1/2}\}$ $=\tfrac{1}{4}a^{-3/2}\exp(-\tfrac{1}{4}y^2/a)$ $\cdot\left(\pi^{1/2}(1-\tfrac{1}{2}y^2/a)\operatorname{Erfc}(-\tfrac{1}{2}iya^{-1/2})+iya^{-1/2}\right)$
3.41	$(a+x)^\mu (a-x)^\nu$ $-a<x<a$ 0 otherwise $\operatorname{Re}(\mu,\nu)>-1$	$(2a)^{\nu+\mu+1}B(1+\nu,1+\mu)e^{-iay}$ $\cdot {}_1F_1(1+\mu;2+\nu+\mu;2iay)$
3.42	$(a+ix)^{-\nu}$ $\operatorname{Re}\nu>0$	$2\pi\{\Gamma(\nu)\}^{-1}y^{\nu-1}e^{-ay}$ $y>0$ 0 $y<0$
3.43	$(a-ix)^{-\nu}$ $\operatorname{Re}\nu>0$	$2\pi\{\Gamma(\nu)\}^{-1}(-y)^{\nu-1}e^{ay}$ $y<0$ 0 $y>0$
3.44	$(a+ix)^{-\nu}(b+ix)^{-1}$ $\operatorname{Re}\nu>-1$	$2\pi\{\Gamma(\nu)\}^{-1}(a-b)^{-\nu}e^{-by}\gamma(\nu,ay-by)$ $y>0$ 0 $y<0$
3.45	$(a+ix)^{-\nu}(b-ix)^{-1}$ $\operatorname{Re}\nu>-1$	$2\pi\{\Gamma(\nu)\}^{-1}(a+b)^{-\nu}e^{by}\Gamma(\nu,ay+by)$ $y>0$ $2\pi(a+b)^{-\nu}e^{by}$ $y<0$
3.46	$(a-ix)^{-\nu}(b+ix)^{-1}$ $\operatorname{Re}\nu>-1$	$2\pi(a+b)^{-\nu}e^{-by}$ $x<0$ $2\pi\{\Gamma(\nu)\}^{-1}(a+b)^{-\nu}e^{-by}\Gamma(\nu,ay-by),$ $y>0$
3.47	$(a-ix)^{-\nu}(b-ix)^{-1}$ $\operatorname{Re}\nu>-1$	0 $y>0$ $2\pi\{\Gamma(\nu)\}^{-1}(a-b)^{-\nu}e^{by}\gamma(\nu,by-ay)$ $y<0$
3.48	$(ix)^{-\nu}(a^2+x^2)^{-1}$ $-2<\operatorname{Re}\nu<1$ $\arg(ix)=\pm\tfrac{1}{2}\pi$ for $x\gtrless0$	$\pi\{\Gamma(\nu)\}^{-1}a^{-\nu-1}e^{ay}$ $+\pi\{\Gamma(1+\nu)\}^{-1}a^{-1}y^{\nu}e^{-ay}{}_1F_1(\nu;1+\nu;ay)$, $y>0$ $a^{-\nu-1}e^{ay}$ $y<0$
3.49	$(-ix)^{-\nu}(a^2+x^2)^{-1}$ $-2<\operatorname{Re}\nu<1$ $\arg(ix)=\pm\tfrac{1}{2}\pi$ for $x\lessgtr0$	$\pi a^{-\nu-1}e^{-ay}$ $y>0$ $\pi\{\Gamma(\nu)\}^{-1}a^{-\nu-1}e^{-ay}\Gamma(\nu,-ay)$ $+\pi\{\Gamma(1+\nu)\}^{-1}a^{-1}(-y)^{\nu}e^{ay}{}_1F_1(\nu;1+\nu;-ay)$ $y<0$

	$f(x)$	$g_e(y)$
3.50	$(a-ix)^{-\nu}(b^2+x^2)^{-1}$ $\mathrm{Re}\,\nu>-2$	$\pi b^{-1}(a+b)^{-\nu}e^{-by}$ $\qquad\qquad y>0$ $\pi b^{-1}\{\Gamma(\nu)\}^{-1}\Big((a-b)^{-\nu}e^{by}e^{by}\gamma(\nu,by-ay)$ $+(a+b)^{-\nu}e^{-by}\Gamma(\nu,-ay-by)\Big)$ $\qquad y<0$
3.51	$(a+ix)^{-\nu}(b^2+x^2)^{-1}$ $\mathrm{Re}\,\nu>-2$	$\pi b^{-1}(a+b)^{-\nu}e^{by}$ $\qquad\qquad y<0$ $\pi b^{-1}\{\Gamma(\nu)\}^{-1}\Big((a+b)^{-\nu}e^{by}\Gamma(\nu,ay+by)$ $+(a-b)^{-\nu}e^{-by}\gamma(\nu,ay-by)\Big]$ $\qquad y>0$
3.52	$(a+ix)^{-\mu}(b+ix)^{-\nu}$ $\mathrm{Re}(\mu+\nu)>0$	$2\pi\{\Gamma(\mu+\nu)\}^{-1}e^{-ay}y^{\mu+\nu-1}$ $\cdot {}_1F_1(\nu;\nu+\mu;ay-by)$ $\qquad y>0$ 0 $\qquad\qquad y<0$
3.53	$(a-ix)^{-\mu}(b-ix)^{-\nu}$ $\mathrm{Re}(\mu+\nu)>0$	$2\pi\{\Gamma(\mu+\nu)\}^{-1}e^{ay}(-y)^{\mu+\nu-1}$ $\qquad y<0$ $\cdot {}_1F_1(\nu;\nu+\mu;by-ay)$ 0 $\qquad\qquad y>0$
3.54	$(a+ix)^{-\mu}(b-ix)^{-\nu}$ $\mathrm{Re}(\nu+\mu)>0$	$2\pi\{\Gamma(\nu)\}^{-1}(a+b)^{-\frac12\nu-\frac12\mu}y^{\frac12\nu+\frac12\mu-1}$ $\cdot\exp(\tfrac12 by-\tfrac12 ay)W_{\alpha,\beta}(ay+by)$ $\qquad y>0$ $2\pi\{\Gamma(\mu)\}^{-1}(a+b)^{-\frac12\nu-\frac12\mu}(-y)^{\frac12\nu+\frac12\mu-1}$ $\cdot\exp(\tfrac12 ay-\tfrac12 by)W_{-\alpha,\beta}(-ay-by)$ $\qquad y<0$ $2\alpha=\nu-\mu$, $\quad 2\beta=1-\nu-\mu$
3.55	$(a+ix)^{-\nu}\exp\{-b(a+ix)^{-1}\}$	$2\pi e^{-ay}(y/b)^{\frac12\nu-\frac12}J_{\nu-1}\{2(by)^{\frac12}\}$ $\qquad y>0$ 0 $\qquad\qquad y<0$
3.56	$e^{-\lambda x}(b+ae^{-x})^{-1}$ $0<\mathrm{Re}\,\lambda<1$	$\pi(b/a)^{\nu-iy}b^{-1}\mathrm{cosec}\{\pi(\lambda-iy)\}$
3.57	$\exp(ia^3x^3)$	$2/(3a)(y/a)^{\frac12}K_{1/3}(u)$; $u=2(\tfrac{y}{3a})^{3/2}$
3.58	$\exp(-ia^3x^3)$	$\frac{2\pi}{3a}(\tfrac13 y/a)^{\frac12}\{J_{1/3}(u)+J_{-1/3}(u)\}$, u as before
3.59	$e^{-\lambda x}(b+ae^{-x})^{-\nu}$ $0<\mathrm{Re}\,\lambda<\mathrm{Re}\,\nu$	$(b/a)^{\lambda-iy}b^{-\nu}B(\lambda-iy,\nu-\lambda+iy)$
3.60	$e^{-\lambda x}(c+be^{-x})^{\nu}(d+ae^{-x})^{\mu}$ $0<\mathrm{Re}\,\lambda<-\mathrm{Re}(\nu+\mu)$	$c^{\nu}a^{-\lambda+iy}d^{\mu+\lambda-iy}B(\lambda-iy,-\mu-\nu-\lambda+iy)$ $\cdot {}_2F_1(-\nu,\lambda-iy;-\mu-\nu;\ 1-\tfrac{bd}{ac})$

	$f(x)$	$g_e(y)$
3.61	$e^{-\lambda x}\exp(-ae^{-bx})$ $\mathrm{Re}\,\lambda>0$	$b^{-1}a^{-(\lambda-iy)/b}\Gamma\{(\lambda-iy)/b\}$
3.62	$e^{-\lambda x}\exp(-ae^{bx})$ $\mathrm{Re}\,\lambda<0$	$b^{-1}a^{(\lambda-iy)/b}\Gamma\{-(\lambda-iy)/b\}$
3.63	$e^{-\lambda x}\{\exp(ae^{-x})-1\}^{-1}$ $\mathrm{Re}\,\lambda>0$	$a^{-\lambda+iy}\Gamma(\lambda-iy)\,\zeta(\lambda-iy)$
3.64	$e^{-\lambda x}\{\exp(ae^{-x})+1\}^{-1}$ $\mathrm{Re}\,\lambda>0$	$a^{-\lambda+iy}\Gamma(\lambda-iy)(1-2^{1-\lambda+iy})\,\zeta(\lambda-iy)$
3.65	$e^{-\lambda x}\exp(-ae^{-px}-be^{px})$ $-p<\mathrm{Re}\,\lambda<p$	$2p^{-1}(b/a)^{\frac12(\lambda-iy)/p}K_{(\lambda-iy)/p}\{2(ab)^{\frac12}\}$
3.66	$e^{-\lambda x}\exp(-a\cosh x-b\sinh x)$ $\quad\quad a>b$ (For $b=a$ see 3.62)	$2\{(a+b)/(a-b)\}^{\frac12(\lambda-iy)}K_{\lambda-iy}\{(a^2-b^2)^{\frac12}\}$
3.67	$e^{-\lambda x}\log(1+ae^{-bx})$ $-b<\mathrm{Re}\,\lambda<0$	$\pi b^{-1}\{(\lambda-iy)/b\}^{-1}e^{-(\lambda-iy)/b}\operatorname{cosec}\{\pi(\lambda-iy)/b\}$
3.68	$e^{-\lambda x}(a+e^{-x})^{-\nu}\log(a+e^{-x})$ $0<\mathrm{Re}\,\lambda<\mathrm{Re}\,\nu$	$a^{\lambda-\nu+iy}B(\lambda+iy,\nu-\lambda-iy)$ $\cdot\{\psi(\nu)-\psi(\nu-\lambda-iy)+\log a\}$
3.69	$(\cos x)^{\mu}(a^2e^{-ix}+b^2e^{ix})^{\nu}$ $-\tfrac12\pi<x<\tfrac12\pi$ $\quad\quad 0\quad\quad |x|>\tfrac12\pi$	$\pi b^{2\nu}2^{-\mu}\Gamma(1+\mu)\left[\Gamma\!\left(1-\tfrac{\nu}{2}+\tfrac{\mu}{2}-\tfrac{y}{2}\right)\Gamma\!\left(1+\tfrac{\nu}{2}+\tfrac{\mu}{2}+\tfrac{y}{2}\right)\right]^{-1}$ $\cdot {}_2F_1(-\nu,-\tfrac12\nu-\tfrac12\mu-\tfrac12 y;1+\tfrac12\mu-\tfrac12\nu-\tfrac12 y;a^2/b^2)$, $\quad\quad\quad\quad\quad\quad\quad\quad\quad a<b$ $\pi a^{2\nu}2^{-\mu}\Gamma(1+\mu)\left[\Gamma\!\left(1+\tfrac12+\tfrac{\mu}{2}-\tfrac{y}{2}\right)\Gamma\!\left(1+\tfrac{\mu}{2}+\tfrac{\nu}{2}-\tfrac{y}{2}\right)\right]^{-1}$ $\cdot {}_2F_1(-\nu,\tfrac12 y-\tfrac12\nu-\tfrac12\mu;1+\tfrac12\mu-\tfrac12\nu+\tfrac12 y;b^2/a^2)\quad a>b$
3.70	$e^{-\lambda x}\{\operatorname{sech}(ax-b)\}^{\nu}$ $\mathrm{Re}\,\nu>0,\,-a\mathrm{Re}\,\nu<\mathrm{Re}\,\lambda<a\mathrm{Re}\,\nu$	$2^{\nu-1}a^{-1}\{\Gamma(\nu)\}^{-1}e^{-b\lambda/a}e^{iby/a}$ $\cdot\Gamma(\tfrac12\nu+\tfrac12\lambda/a-\tfrac12 iy/a)\Gamma(\tfrac12\nu-\tfrac12\lambda/a+iy/a)$
3.71	$e^{-\lambda x}\operatorname{sech}(ae^{-x})$ $\mathrm{Re}\,\lambda>0$	$a^{-\lambda+iy}2^{1-2\lambda+2iy}\Gamma(\lambda-iy)$ $\cdot\left[\zeta(\lambda-iy,\tfrac14)-\zeta(\lambda-iy,3/4)\right]$

	$f(x)$	$g_e(y)$
3.72	$e^{-\lambda x}\mathrm{cosech}(ae^{-x})$ $\mathrm{Re}\,\lambda > 1$	$2a^{-\lambda+y}(1-2^{-\lambda+iy})\,\Gamma(\lambda-iy)\,\zeta(\lambda-iy)$
3.73	$e^{-\lambda x}\tanh(ae^{-x})$ $-1 < \mathrm{Re}\,\lambda < 0$	$2(2a)^{-\lambda+iy}\{2^{1-\lambda+iy}-1\}\Gamma(\lambda-iy)\,\zeta(\lambda-iy)$
3.74	$e^{-\lambda x}\exp(a^2 e^x + b^2 e^{-x})$ $\cdot\mathrm{Erfc}(ae^{\frac{1}{2}x}+be^{-\frac{1}{2}x})$	$2(b/a)^{-\lambda+iy}\mathrm{sech}\{\pi(y+i\lambda)\}K_{-\lambda+iy}(2ab)$
3.75	$2^{-\lambda x}P_\nu(x) \qquad |x|<1$ $\qquad\qquad 0 \qquad\quad |x|>1$	$2\pi\nu^{-1}(1+\nu^2)^{-1}\sin(\pi\nu)e^{\lambda-iy}$ $\cdot\,{}_2F_2(1,1;-\nu,2+\nu;2iy-2\lambda)$
3.76	$x^{-\frac{1}{2}}J_{n+\frac{1}{2}}(x)$ $n=0,1,2,\ldots$	$(-i)^n(2\pi)^{\frac{1}{2}}P_n(y) \qquad |y|<1$ $\qquad\qquad 0 \qquad\qquad\quad |y|>1$
3.77	$x^{-\nu-\frac{1}{2}}J_{n+\nu+\frac{1}{2}}(x)$ $n=0,1,2,\ldots;\ \mathrm{Re}\,\nu>-1$	$(-i)^n n!\,(2\pi)^{\frac{1}{2}}2^{-\nu}$ $\cdot(1-y^2)^\nu P_n^{(\nu,\nu)}(y)\{\Gamma(n+\nu+1)\}^{-1}\ |y|<1$ $\qquad\qquad 0 \qquad\qquad\qquad\qquad |y|>1$
3.78	$u^{-\frac{1}{2}\nu}(\cos x)^{\frac{1}{2}\nu}$ $\cdot J_\nu\{c(2u\cos x)^{\frac{1}{2}}\}|x|<\frac{1}{2}\pi$ $\qquad\qquad 0 \qquad\quad |x|>\frac{1}{2}\pi$ $u=a^2 e^{-ix}+b^2 e^{ix}$ $\mathrm{Re}(\mu+\nu)>-1$	$\pi(2ab)^{-\frac{1}{2}\nu}(a/b)^{\frac{1}{2}y}J_{\frac{1}{2}\nu-\frac{1}{2}y}(ac)J_{\frac{1}{2}\nu+\frac{1}{2}y}(bc)$
3.79	$x^{\frac{1}{2}}\{J_{1/3}(u)+J_{-1/3}(u)\}$ $u=ax^{3/2}$	$2a^{-1}\exp(iv^3 y^3)$ $v=1/3(\frac{1}{2}a)^{-2/3}$
3.80	$x^{\frac{1}{2}}K_{1/3}(ax^{3/2})$	$2\pi a^{-1}3^{-\frac{1}{2}}\exp(-iv^3 y^3\ ,\ u,v\ \text{as before}$
3.81	$e^{-\lambda x}(u/v)^\nu J_{2\nu}(uv)$ $u=ae^x+be^{-x}, v=ae^{-x}+be^x$ $-3/2 < \mathrm{Re}\,\lambda < 3/2$	$-\frac{1}{2}\pi\Big(J_{\nu-\frac{1}{2}\lambda+\frac{1}{2}iy}(b)Y_{\nu+\frac{1}{2}\lambda-\frac{1}{2}iy}(a)$ $+Y_{\nu-\frac{1}{2}\lambda+\frac{1}{2}iy}(b)J_{\nu+\frac{1}{2}\lambda-\frac{1}{2}iy}(a)\Big)$
3.82	$e^{-\lambda x}(u/v)^\nu Y_{2\nu}(uv)$ $u,v\ \text{as before},-\frac{3}{2}<\mathrm{Re}\,\nu<\frac{3}{2}$	$\frac{1}{2}\pi\Big(J_{\nu-\frac{1}{2}\lambda+\frac{1}{2}iy}(b)J_{\nu+\frac{1}{2}\lambda-\frac{1}{2}iy}(a)$ $-Y_{\nu-\frac{1}{2}\lambda+\frac{1}{2}iy}(b)Y_{\nu+\frac{1}{2}\lambda-\frac{1}{2}iy}(a)\Big)$

	$f(x)$	$g_e(y)$
3.83	$e^{-\lambda x}(u/v)^{\nu}K_{2\nu}(uv)$ u,v as before	$K_{\nu-\frac{1}{2}\lambda+\frac{1}{2}iy}(b)K_{\nu+\frac{1}{2}\lambda-\frac{1}{2}iy}(a)$
3.84	$J_{\mu+x}(a)J_{\nu-x}(a)$ $\mathrm{Re}(\nu+\mu)>-1$	$\exp\{-\frac{1}{2}iy(\mu-\nu)\}J_{\nu+\mu}\{2a\,\cos(\frac{1}{2}y)\}$ $|y|<\pi$ 0 $|y|>\pi$
3.85	$a^{-x-\mu}J_{\mu+x}(a)b^{x-\nu}J_{\nu-x}(b)$ $\mathrm{Re}(\nu+\mu)>-1$ $z=a^2e^{-\frac{1}{2}iy}+b^2e^{\frac{1}{2}iy}$	$(2\cos\frac{1}{2}y)^{\frac{1}{2}\nu+\frac{1}{2}\mu}z^{-\frac{1}{2}\nu-\frac{1}{2}\mu}\exp\{-\frac{1}{2}iy(\mu-\nu)\}$ $\cdot J_{\mu+\nu}\{(2z\,\cos\frac{1}{2}y)^{\frac{1}{2}}\}$ $|y|<\pi$ 0 $|y|>\pi$
3.86	$J_{\nu+ix}(a)J_{\nu-ix}(b)$ $-Y_{\nu+ix}(a)Y_{\nu-ix}(b)$	$2(u/v)^{\nu}Y_{2\nu}(uv); u=ae^{\frac{1}{2}x}+be^{-\frac{1}{2}x},$ $v=ae^{-\frac{1}{2}x}+be^{\frac{1}{2}x}$
3.87	$Y_{\nu+ix}(a)J_{\nu-ix}(b)$ $+J_{\nu+ix}(a)Y_{\nu-ix}(b)$	$-2(u/v)^{\nu}J_{2\nu}(uv);$ u,v as before
3.88	$K_{\nu-ix}(a)K_{\nu+ix}(b)$	$\pi(u/v)^{\nu}K_{2\nu}(uv);$ u,v as before
3.89	$\Gamma(\frac{1}{2}+\nu+ix)\Gamma(\frac{1}{2}+\nu-ix)$ $\cdot M_{-ix,\nu}(a)$	$\pi 2^{-2\nu}\Gamma(1+2\nu)a^{\nu+\frac{1}{2}}\exp\{-\frac{1}{2}a\,\tanh(\frac{1}{2}y)\}$ $\cdot\{\cosh(\frac{1}{2}y)\}^{-2\nu-1}$
3.90	$\Gamma(\frac{1}{2}+\nu+ix)\Gamma(\frac{1}{2}+\nu-ix)$ $\cdot M_{-ix,\nu}(a)M_{-ix,\nu}(b)$ $\mathrm{Re}\,\nu>-\frac{1}{2}$	$2\pi(ab)^{\frac{1}{2}}\mathrm{sech}(\frac{1}{2}y)\exp\{-(a+b)\tanh(\frac{1}{2}y)\}$ $\cdot J_{2\nu}\{2(ab)^{\frac{1}{2}}\mathrm{sech}(\frac{1}{2}y)\}$
3.91	$\exp(-a^2\cosh^2 x)$ $\cdot D_{\nu}(2a\sinh x);\ \mathrm{Re}\,\nu>-1$	$2^{\frac{1}{2}\nu}(2\pi a^2)^{-\frac{1}{2}}\Gamma(\frac{1}{2}+\frac{1}{2}\nu+iy)\Gamma(\frac{1}{2}+\frac{1}{2}\nu-\frac{1}{2}iy)$ $\cdot\cos(\frac{1}{2}\pi\nu-\frac{1}{2}i\pi y)W_{-\frac{1}{2}\nu,\frac{1}{2}iy}(2a^2)$
3.92	$\exp(-a\cosh x)$	$2\,K_{iy}(a)$

Part IV
Fourier Transforms of Distributions
(Tables IV and V)

IV. Fourier Transforms of Distributions

This part is concerned with the previous parts I-III in as much as
those results are singled out involving such functions f(x) which are
nonnegative and integrable over the range of integration under considera-
tion. Because of the importance of such functions in probability theory a short explanation of some basic facts may be in order.

<u>Definitions</u>. Let X be a random variable. Suppose that R_X, the
range space of X, is an interval or a collection of intervals. Then
it is said that X is a continuous variable.

Furthermore, let X be a continuous variable. Then the probability
density function (denoted by pdf, $\phi(x)$ is a function satisfying the
condition:

(1) $\phi(x) \geq 0$ for all $x \epsilon R_X$

(2) $\int_{R_X} \phi(x)\,dx = 1$

Furthermore, we define for any   a<b   (in R_X)

(3) $P(a<X<b) = \int_a^b \phi(t)\,dt$ and

(4) $F(x) = P(X<x) = \int_{-\infty}^x \phi(t)\,dt.$

The function F(x) is often denoted as the cumulative distribution
function (cdf) or sometimes just the distribution funtion of the ran-
dom variable X. It represents the probability that X is in the interval $(-\infty, x)$. Also:

$$\lim_{x \to -\infty} F(x) = 0 \quad \text{and} \quad \lim_{x \to \infty} F(x) = 1$$

A number of definitions are derived from $\phi(x)$. A short list of the
most frequently encountered is:

(a) $\phi(x)$, Probability density function pdf.

(b) $\Phi(x) = \int_{R_X} \phi(t)e^{itx}\,dt$, Characteristic function of X.

(c) $F(x) = P(X<x) = \int_{-\infty}^x \phi(t)\,dt$, Distribution function of X.

(d) $\mu_k = \int\limits_{R_X} x^k \phi(x)\,dx$, kth moment of $\phi(x)$.

(e) $E(X) = \mu_1$, Expectation of X.

(f) $E(X^2) = \mu_2$

(g) $V(X) = \mu_2 - \mu_1^2$, Variance of X.

(h) $\sigma = \{V(X)\}^{\frac{1}{2}}$, Standard deviation (dispersion).

Description of the Tables IV and V.

We return to the Tables I-III for $g_c(y)$, $g_s(y)$ and $g_e(y)$ and ex-
tract from these the results involving functions f(x) with the conditions:

(5) $f(x) \geq 0$ and $\int\limits_{R_X} f(x)\,dx = N < \infty$.

With this result we define a function $\phi(x) = f(x)/N$. Then $\phi(x)$ is a
probability density function as defined in (1), (2). It should be
pointed out that in some instances the function f(x) in the Tables
I-III is negative over the interval of integration. This is for instance the case in the entries 4.1 and 4.9 of the Tables IV. In this
case an asterisk has been provided to indicate that the corresponding
$\phi(x)$ has to be provided with a negative sign.

In the Tables IV to follow, we list in the first two columns the location of a function of the type (5), together with their transforms
$g_c(y)$, $g_s(y)$ according to their numeration in the Tables I and II.
The third column gives the (normalization) constant $N=g_c(0)$. The
fourth column lists a possible restriction (which may be different from
those in I, II, because of the absolute convergence of g_c and g_s).
The range space here is $R_X(0,\infty)$. With respect to the Tables V, the
same is done for the exponential Fourier Transforms as given in the
Tables III. The range space is here $R_X(-\infty,\infty)$. Take for instance
$f(x) = x^\nu \exp(-ax^2)$, $0 < x < \infty$.

Then its transforms g_c, g_s are listed in I, II under the numbers (3.21)
and (3.31) respectively. We find under the same numbers in the Tables
IV $N = \frac{1}{2} a^{-\frac{1}{2} - \frac{1}{2}\nu} \Gamma(\frac{1}{2} + \frac{1}{2}\nu)$ and $\nu > -1$. The pdf is

$$\phi(x) = 2a^{\frac{1}{2} + \frac{1}{2}\nu} x^\nu \exp(-ax^2)/\Gamma(\frac{1}{2} + \frac{1}{2}\nu), 0 < x < \infty , \nu > -1.$$

We consider three cases for $\phi(x)$.
A $\phi(x)$ is an even function of x , $R_X(-\infty,\infty)$.

B $\phi(x)=0$ for $x<0$, $R_X(0,\infty)$.

C $\phi(x)$ is neither   , $R_X(-\infty,\infty)$.

The characteristic function (b) is for the case A:

$$\phi(x) = \int_{-\infty}^{\infty} \phi(t)e^{ixt}dt = 2 \int_{0}^{\infty} \phi(t)\cos(xt)dt.$$

Since $N=2g_c(0)$ we obtain in the terminology of the Tables I, II:

(6) $\phi(x)=\{g_c(0)\}^{-1} \int_{0}^{\infty} f(t)\cos(xt)dt = g_c(x)/g_c(0)$.

In the case B the characteristic function becomes:

(7) $\phi(x) = \int_{0}^{\infty} \phi(t)e^{ixt}dt = \{g_c(x)+ig_s(x)\}/g_c(0)$.

On this case both, g_c and g_s are needed as listed in the Tables I,II. The integral (7) can also be written in the form

(8) $\phi(x) = g_e(x)/g_e(0)$.

Here g_e is the exponential Fourier transform of $f(x)$ for the case

$f(x)=0$ for $x<0$. Here the Tables III and V have to be used. Note that

(8) represents also a Laplace integral $g(s) = \int_{0}^{\infty} F(t)e^{-st}dt$ with $s=-iy$.

Thus tables of such integrals can be used for the evaluation of integral of the form (7). Finally for the case C.

(9) $\phi(x) = \int_{-\infty}^{\infty} \phi(t)e^{it}dt = g_e(x)/g_e(0)$.

Again, the Tables III and V are to be used. A few examples are:

Example I.   The Maxwell-Boltzmann Distribution.

Let $f(x)=x^2\exp(-ax^2)$. This is the case. B and (8) is used. We obtain then (3.40) Tables III and V for the Pdf:

(10) $\phi(x)=4\pi^{-\frac{1}{2}}a^{3/2}x^2\exp(-ax^2)$, $0<x<\infty$, $a>0$.   Then (b)-(h) become:

$\phi(x)=\exp(-\frac{1}{4}x^2/a)\left[(1-\frac{1}{2}x^2/a)\text{Erfc}(-\frac{1}{2}ixa^{-\frac{1}{2}})+i\pi^{-\frac{1}{2}}xa^{-\frac{1}{2}}\right]$,   by (3.40)

$P(X<x)=4\pi^{-\frac{1}{2}}a^{3/2}\int_{0}^{x} t^2\exp(-at^2)dt=2\pi^{-\frac{1}{2}}\gamma(3/2,ax^2)$.   Then (p.250),

$P(X<x)=\text{Erf}(xa^{\frac{1}{2}})-2(a/\pi)^{\frac{1}{2}}x\exp(-ax^2)$. Furthermore,

$\mu_k=4\pi^{-\frac{1}{2}}a^{3/2}\int_{0}^{\infty} x^{k+2}\exp(-ax^2)dx=2\pi^{-\frac{1}{2}}a^{-\frac{1}{2}k}\Gamma(\frac{1}{2}k+3/2)$.   Finally,

$E(X)=\mu_1=2(\pi a)^{-\frac{1}{2}}$, $E(X^2)=\mu_2 = \frac{3}{2}a^{-1}$, $V(x) = \frac{3}{2}a^{-1}-4(\pi a)^{-1}$.

Example II.   The Generalized Gamma Distribution and special cases.

A number of results involving the Gamma- and the incomplete Gamma function is needed here. These are listed on Pages 244, 250, 251, 307.
Also needed is the integral.

$$K = \int_{0}^{x} (a+t)^{\nu-1}e^{-st}dt = s^{-\nu}e^{as}\{\Gamma(\nu,as)-\Gamma(\nu,as+xs)\}.$$

This follows easily from the definition of $\Gamma(\nu,z)$ P. 250.

We choose $f(x) = (a+x)^{\nu-1}e^{-sx}$, $0<x<\infty$, $s>0$.
The case B applies here. This leads to 3.6, Tables III. As a one-sided Fourier transform $g_e(y)$ is an analytic function of y in an upper half plane, in this case $\text{Im}(y)>0$. We replace y by $y+is$ with

$s>0$. From 3.6, Tables III, the following pair of transform follows.

$f(x)=(a+x)^{\nu-1}$, $g_e(y)=(s-iy)^{-\nu}\exp(as-iay)\Gamma(\nu,as-iay)$. The pdf is:

(11) $\phi(x) = \dfrac{s^{\nu}e^{-as}}{\Gamma(\nu,as)} (a+x)^{\nu-1}e^{-sx}=\phi(x,s,\nu,a)$, $0<x<\infty$, $s>0$.

With this by (8),

(12) $\Phi(x) = \dfrac{s^{\nu}e^{-iax}}{\Gamma(\nu,as)} (s-ix)^{-\nu}\Gamma(\nu,as-iax)=A(x)+iB(x)$. Also,

$P(X<x) = \dfrac{s^{\nu}e^{-as}}{\Gamma(\nu,as)} \int_{0}^{x} (a+t)^{\nu-1}e^{-st}dt=1-\Gamma(\nu,as+xs)/\Gamma(\nu,as)$, $x>0$.

$\mu_k = \dfrac{s^{\nu}e^{-as}}{\Gamma(\nu,as)} \int_{0}^{\infty} x^k(a+x)^{\nu-1}e^{-sx}dx$

$\mu_k = \dfrac{k!e^{-\frac{1}{2}as}}{\Gamma(\nu,as)} s^{\frac{1}{2}\nu-\frac{1}{2}k-\frac{1}{2}}a^{\frac{1}{2}\nu+\frac{1}{2}k-\frac{1}{2}}W_{\frac{1}{2}\nu-\frac{1}{2}k-\frac{1}{2},\frac{1}{2}\nu+\frac{1}{2}k}(as)$.

With this result the remaining parameters (e)-(h) are established. If $\nu=\pm n$ or $\nu=\frac{1}{2}\pm n$ the previous results simplify (pp. 250, 251).

The Gamma Distribution, (a=0).

(13) $\phi(x)=\phi(x,s,\nu,0) = \dfrac{s^{\nu}}{\Gamma(\nu)} x^{\nu-1}e^{-sx}$, $x>0$, $\nu>0$.

(14) $\Phi(x)=s^{\nu}(s-ix)^{-\nu}$

$P(X<x)=\gamma(\nu,xs)/\Gamma(\nu)$, $\mu_k = \dfrac{s^{\nu}}{\Gamma(\nu)} \int_{0}^{\infty} x^{k+\nu-1}e^{-sx}dx = s^{-k} \dfrac{\Gamma(\nu+k)}{\Gamma(\nu)}$.

(This result follows also from the general case for $\lim a\to 0$). Then,

$E(X)=\nu s^{-1}$, $E(X^2)=\nu(\nu+1)s^{-2}$, $V(X)=\nu s^{-2}$, $\sigma=\nu^{\frac{1}{2}}s^{-1}$.

The Poisson Distribution, (a=0, $\nu=n+1$, $n=0,1,2,...$) .

(15) $\Phi(x)=\phi(x,s,n+1,0)=(n!)^{-1}s^{n+1}x^ne^{-sx}$, $x>0$

(16) $\Phi(x)=s^{n+1}(s-ix)^{-n-1}$.

$$P(X<x)=(n!)^{-1}\gamma(n+1,xs)=1-e^{-sx}\sum_{n=o}^{n}(sx)^{k}/k! \ , \ \mu_{k}=s^{-k}(k+n)!/k!$$

The Exponential Distribution,   ($a=0,\nu=1$).

(17) $\phi(x)=\phi(x,s,1,0)=se^{-sx}$, $x>0$.

$\Phi(x)=s(s-ix)^{-1}, P(X<x)=1-e^{-sx}$, $\mu_{k}=s^{-k}k!$.

The Chi-Square Distribution,   ($a=0,\nu=\frac{1}{2}n,s=\frac{1}{2},n=1,2,3,...$)

(18) $\phi(x)=\phi(x,\frac{1}{2},\frac{1}{2}n,0)=2^{-\frac{1}{2}n}x^{\frac{1}{2}n-1}e^{-\frac{1}{2}x}/\Gamma(\frac{1}{2}n)$, $x>0$.

$\Phi(x)=2^{-\frac{1}{2}n}(\frac{1}{2}-ix)^{-\frac{1}{2}n}$, $P(X<x)=\gamma(\frac{1}{2}n,\frac{1}{2}x)/\Gamma(\frac{1}{2}n)$, $\mu_{k}=\dfrac{2^{k}\Gamma(k+\frac{1}{2}n)}{\Gamma(\frac{1}{2}n)}$.

$P(X<x)=\gamma(m,\frac{1}{2}x)/\Gamma(m)$ if n is even, equal to   2m.

$P(X<x)=\gamma(\frac{1}{2}+m,\frac{1}{2}x)/\Gamma(\frac{1}{2}+m)$ if n is odd, equal to   2m+1.

Then by the formulas pp. 244, 250, 251

$$P(X<x) = 1-e^{-\frac{1}{2}x}\sum_{k=1}^{m-1}(\frac{1}{2}x)^{k}/k! \ , \qquad n=2m.$$

$$P(X<x)=\mathrm{Erf}\{(\frac{1}{2}x)^{\frac{1}{2}}\}-(\frac{1}{2}\pi x)^{-\frac{1}{2}}e^{-\frac{1}{2}x}\sum_{k=1}^{m}(2x)^{k}k!/(2k)! \ , \qquad n=2m+1. \ \ \text{Also,}$$

$\mu_{k}=2^{k}(k+m-1)!/(m-1)! \ , \ n=2m; \ \mu_{k}=2^{-k}m!(2k+2m)!/(2m)!(m+k)! \ , \ n=2m+1.$

Example III.   Distribution connected with Example II.

We consider instead of (12) the distribution

(19) $\phi(x) = \dfrac{s^{\nu}e^{-as}}{2\Gamma(\nu,as)}(a+|x|)^{\nu-1}e^{-s|x|}$, $-\infty<x<\infty$, $s>0$.

This is an even function (case A) and it follows

(20) $\phi(x) = \dfrac{s^{\nu}e^{-as}}{\Gamma(\nu,as)}\int_{o}^{\infty}(a+t)^{\nu-1}e^{-st}\cos(xt)dt=A(x)$,

where A(x) is defined by (12). Also

$$P(X<x) = \frac{s^{\nu}e^{-as}}{2\Gamma(\nu,as)}\left(\int_{o}^{\infty}(a+t)^{\nu-1}e^{-st}dt + \int_{o}^{x}(a+t)^{\nu-1}e^{-st}dt\right).$$

These integrals are obtained from the integral K in Example II.

$P(X<x)=\frac{1}{2}\left[1\pm\{1 - \dfrac{\Gamma(\nu,as+|x|s)}{\Gamma(\nu,as)}\}\right]$,   the $\pm$ sign according as $x\gtrless 0$.

Since $\phi(x)$ is even $\mu_{2k+1}=0, \mu_{2k} = \dfrac{s^{\nu}e^{-as}}{\Gamma(\nu,as)}\int_{o}^{\infty}x^{2k}(a+x)^{\nu-1}e^{-sx}dx.$

This value is the same as in the Example II with k replaced by 2k.
Again, it may be pointed out that specialization of the parameters as
demonstrated in the Example II can be considered. Especially the
choice of particular values for ν lead to cases of the so-called
"Sargan Distribution".

<u>Example IV.</u>   The Gauss Distribution.

Here we choose $f(x) = \exp\{-a(x-c)^2\}$ $-\infty < x < \infty$, $a > 0$.

This represents the case   C   with equation (9).   At first

$$g_e(y) = \int_{-\infty}^{\infty} \exp\{-a(x-c)^2\}e^{ixy}dx = 2e^{icy}\int_0^{\infty}\exp(-at^2)\cos(yt)dt.$$

By 3.17 (Tables I) $g_e(y) = e^{icy}(\pi/a)^{\frac{1}{2}}\exp(-\tfrac{1}{4}y^2/a)$ and by (9)

(21) $\phi(x) = (a/\pi)^{\frac{1}{2}}\exp\{-a(x-c)^2\}$, $\Phi(x) = e^{icx}\exp(-\tfrac{1}{4}x^2/a)$.   Furthermore,

$$P(X<x) = (a/\pi)^{\frac{1}{2}}\int_{-\infty}^{x}\exp\{-a(t-c)^2\}dt = (a/\pi)^{\frac{1}{2}}\int_{-\infty}^{x-c}\exp(-au^2)du$$

$$P(X<x) = (a/\pi)^{\frac{1}{2}}\left[\int_{-\infty}^{0}(\ )du + \int_{0}^{x-c}(\ )du\right] = \tfrac{1}{2}\left[1+\text{Erf}\{(x-c)a^{\frac{1}{2}}\}\right].$$ Also

$$\mu_k = (a/\pi)^{\frac{1}{2}}\int_{-\infty}^{\infty}x^k\exp\{-a(x-c)^2\}dx = (a/\pi)^{\frac{1}{2}}\int_{-\infty}^{\infty}(c+t)^k\exp(-at^2)dt.$$

With the binomial theorem $(c+t)^k = \sum_{n=0}^{k}\binom{k}{n}c^{k-n}t^n$ one obtains

$$\mu_k = (a/\pi)^{\frac{1}{2}}\sum_{n=0}^{k}\binom{k}{n}c^{k-n}\int_{-\infty}^{\infty}t^n e^{-at^2}dt.$$ The integral is zero for   n   odd.

$$\mu_k = (a/\pi)^{\frac{1}{2}}\sum_{n=0}^{<\frac{1}{2}k}\binom{k}{2n}c^{k-2n}2\int_0^{\infty}t^{2n}e^{-at^2}dt = (a/\pi)^{\frac{1}{2}}\sum_{n=0}^{<\frac{1}{2}k}\binom{k}{2n}c^{k-2n}\int_0^{\infty}x^{n-\frac{1}{2}}e^{-ax}dx.$$

The integral is equal to $\Gamma(\tfrac{1}{2}+n)a^{-n-\frac{1}{2}} = a^{-n-\frac{1}{2}}\pi^{\frac{1}{2}}2^{-2n}(2n)!/n!$.   Then

$$\mu_k = \sum_{n=0}^{<\frac{1}{2}k}\binom{k}{2n}c^{k-2n}(4a)^{-n}(2n)!/n! .$$

With this result the parameters (e)-(h) are established.

$\mu_1 = \mu = E(X) = c$; $\mu_2 = E(X^2) = c^2 + (2a)^{-1}$; $V(X) = (2a)^{-1} = \sigma^2$.

If instead of   a   and   c   new parameters $c = \mu$ and $a = \tfrac{1}{2}\sigma^{-2}$ are used, then the former properties become

(21a) $\phi(x) = N(\mu,\sigma^2) = (2\pi\sigma^2)^{-\frac{1}{2}}\exp\left[-\tfrac{1}{2}(\tfrac{x-\mu}{\sigma})^2\right]$,   normal distribution

$\quad\quad P(X<x) = \tfrac{1}{2}\left[1+\text{Erf}\{2^{-\frac{1}{2}}(\tfrac{x-\mu}{\sigma})\}\right]$; $\Phi(x) = e^{i\mu x}\exp(-\tfrac{1}{2}\sigma^2 x^2)$

$\quad\quad E(X) = \mu$; $E(X^2) = \mu^2 + \sigma^2$; $V(X) = \sigma^2$; $\mu_k = \mu^k \sum_{n=0}^{<\frac{1}{2}k}\binom{k}{2n}2^{-n}(\sigma/\mu)^{2n}$.

The special case   N(0,1)   of   (21a) is called a standardized normal distribution.   We have also the theorem: If   X   has the distribution

$N(\mu,\sigma^2)$ and if   Y=aX+b   then   Y   has the distribution $N(a\mu+b, a^2\sigma^2)$.

<u>Example V.</u> $f(x) = \text{sech}(ax-b)$, $-\infty < x < \infty$

We find from 3.70 (Tables III) with $\lambda = 0$, $\nu = 1$

$$g_e(y)=\pi a^{-1}\text{sech}(\tfrac{1}{2}\pi y/a)\exp(iby/a),\ N=g_e(0)=\pi a^{-1}.\quad\text{Then}$$

(22) $\phi(x)=a\pi^{-1}\text{sech}(ax-b)$, $\Phi(x)=\text{sech}(\tfrac{1}{2}\pi x)\exp(ibx/a)$

$$P(X)=\pi^{-1}\left[\arctan\{\sinh(ax-b)\}+\tfrac{1}{2}\pi\right],\int_0^\infty\text{sech }z\ dz=\arctan(\sinh z)\quad.$$

For the remaining properties (c)-(h) we evaluate first

$$\mu_k=a\pi^{-1}\int_{-\infty}^\infty x^k\text{sech}(ax-b)dx=\pi^{-1}a^{-k}\int_{-\infty}^\infty(b+t)^k\text{sech }t\ dt\ .$$

With the binomial theorem $(b+t)^k=\sum\limits_{n=0}^k\binom{k}{n}b^{k-n}t^n$ we obtain:

$$\mu_k=\pi^{-1}a^{-k}\sum_{n=0}^k\binom{k}{n}b^{k-n}\int_{-\infty}^\infty t^n\text{sech }t\ dt.\quad\text{Finally,}$$

$$\mu_k=2\pi^{-1}a^{-k}\sum_{n=0}^{<\frac{1}{2}k}\binom{k}{2n}b^{k-2n}\int_0^\infty t^{2n}\text{sech }t\ dt,\ \text{since all terms }n=2m+1\text{ vanish.}$$

But $\int_0^\infty t^{2n}\text{sech }t\ dt=(-1)^n(\tfrac{1}{2}\pi)^{2n+1}E_{2n}$, where E_{2n} are Euler's numbers*.

Then $\mu_k=(b/a)^k\sum\limits_{n=0}^{<\frac{1}{2}k}\binom{k}{2n}(\tfrac{1}{2}\pi/b)^{2n+1}E_{2n}$ and with $E_0=1,\ E_2=-1$ we have

$$E(X)=\mu_1=b/a,E(X^2)=\mu_2=a^{-2}(b^2+\tfrac{1}{4}\pi^2),\ V(X)=\tfrac{1}{4}\pi^2/a^2),\ \sigma=\tfrac{1}{2}\pi/a.$$

<u>Example VI.</u> $f(x)=(b^2+x^2)^{-m-1}\exp(-ax^2)$, $-\infty<x<\infty$, $m=0,1,2,\ldots$

Here the case (A) and hence equation (6) applies. By (3.23), Tables IV:

(23) $g_c(0)=\tfrac{1}{2}\pi^{\frac{1}{2}m-\frac{1}{4}}b^{-m-3/2}\exp(\tfrac{1}{2}ab^2W_{-\frac{1}{4}-\frac{1}{2}m,\frac{1}{4}+\frac{1}{2}m}(ab^2)$ and by (3.23),Table I

(24) $g_c(0)\Phi(x)=2^{-\frac{1}{2}m-3/2}\pi^{\frac{1}{2}}a^{\frac{1}{2}m}b^{-m-1}(m!)^{-1}\exp(\tfrac{1}{2}ab^2-\tfrac{1}{8}y^2/a)$

$\cdot\sum\limits_{k=0}^m(m+k)!(8b^2/a)^{-\frac{1}{2}k}\left(\exp(\tfrac{1}{2}by)D_{-\nu}(u)+\exp(-\tfrac{1}{2}by)D_{-\nu})v)\right),\ \genfrac{}{}{0pt}{}{u}{v}=(2a)^{\frac{1}{2}}(b\pm\tfrac{1}{2}y/a)$

$\hspace{11cm}\nu=m-k+1$

It may be desirable to give a short outline of the derivation of the formula (3.23). At first we use the identity (inversion of (1.57)Table I).

$$(b^2+x^2)^{-m-1}=2(2b)^{-m-1}(m!)^{-1}\sum_{k=0}^m\frac{(m+k)!}{(m-k)!}(2b)^{-k}\int_0^\infty t^{m-k}e^{-bt}\cos(xt)dt$$

Inserting this into (3.23) and interchanging the order of the (now repeated) integral, the occuring inner integral is given by (3.17). The remaining integral is of the form {A. Erdelyi et al.   (Tables of Integral Transform, Vol. 1, p. 146 (24)}.

$$\int_0^\infty t^{\nu-1}\exp(-\tfrac{1}{4}a^{-1}t^2)e^{-bt}\ dt=(2a)^{\frac{1}{2}\nu}\Gamma(\nu)\exp(\tfrac{1}{2}ap^2)D_{-\nu}\{p(2a)^{\frac{1}{2}}\}$$

---

* (Erdelyi, A. Higher Transcendental Functions. Vol. 1, p. 42)

IV. Fourier Transforms of Distributions (Tables IV)

$g_c(y)$	$g_s(y)$	$g_c(0)$	Restrictions
1.1	1.1	a	
1.2	1.2	$\frac{1}{2}b^2 - \frac{1}{2}a^2$	
1.3	1.3	a^2	
1.6	1.7	$2a^{\frac{1}{2}}$	
1.8	1.10	$\log(1+b/a)$	
1.10	1.12	$-\log(1-b/a)$	
1.11	1.16	$a^{-1}\log(1+a/b)$	
1.13	1.67	$(n-1)^{-1}(a+b)^{1-n}$	
1.15	1.18	$\pi a^{-\frac{1}{2}}$	
1.16	1.19	$2(a+b)^{\frac{1}{2}} - 2a^{\frac{1}{2}}$	
1.18	1.21	$2a^{-\frac{1}{2}}$	
1.22	1.25	$\pi a^{-\frac{1}{2}}$	
1.23	1.24	$2a^{\frac{1}{2}}$	
1.24	1.28	$\pi(a+b)^{-\frac{1}{2}}$	
1.26	1.31	$\frac{1}{2}\pi a^{-1}$	
1.28	1.34	$\frac{1}{2}a^{-1}\log\{(a+b)/(a-b)\}$	$a>b$
1.29	1.35	$\frac{1}{2}a^{-1}\log\{(b+a)/(b-a)\}$	$a<b$
1.33	1.46	$\pi(\pi/a)^{\frac{1}{2}}\Gamma^{-2}(3/4)$	
1.34		$\frac{1}{2}\pi\{ab(a+b)^{-1}\}$	
1.39	1.49	$\pi(2a)^{-\frac{1}{2}}$	

$g_c(y)$	$g_s(y)$	$g_c(0)$	Restrictions
1.40		$\pi 2^{-\frac{1}{2}} a^{-1}$	
1.41		$2^{-\frac{1}{2}} a^{-2}$	
1.42		$\pi(2a)^{-\frac{1}{2}}$	
1.43	1.43	$\frac{1}{2}\pi$	
1.44		$\arccos(b/a)$	
1.45	1.44	a	
1.46	1.47	$\pi(\frac{1}{2}\pi/a)^{\frac{1}{2}} \Gamma^{-2}(3/4)$	
1.48		$\frac{1}{2}\pi a^{-1}$	
1.49	1.66	$n^{-1} a^{-n}$	
1.50	1.48	$\pi(\frac{1}{2}\pi/a)^{\frac{1}{2}} \Gamma^{-2}(3/4)$	
1.51		$\pi 2^{-3/2} a^{-3}$	
1.52		$\frac{1}{4}\pi a^{-3} \sec(\frac{1}{2}b)$	$-\pi < b < \pi$
1.53		$\frac{1}{4}\pi a^{-1} \sec(\frac{1}{2}b)$	$-\pi < b < \pi$
1.56		$\frac{1}{2}\pi n^{-1} a^{2m-2n+1} \operatorname{cosec}\{(m+\frac{1}{2})\pi/n\}$	
1.57		$\pi(2a)^{-2m-1} (2m)!/m!$	$m = 0,1,2,\ldots$
2.1	2.1	$-(1+\nu)^{-1} a^{1+\nu}$	$\nu < -1$
2.2	2.2	$(1+\nu)^{-1}\{(a+b)^{1+\nu} - a^{1+\nu}\}$	
2.3	2.3	$-\pi a^{\nu} \operatorname{cosec}(\pi\nu)$	$-1 < \nu < 0$
2.4	2.4	$(\nu+1)^{-1} a^{\nu+1}$	$\nu > -1$
2.5	2.5	$B(\nu+1, \mu+1) a^{\nu+\mu+1}$	$(\nu,\mu) > -1$

$g_c(y)$	$g_s(y)$	$N=g_c(0)$	Restrictions
2.6	2.6	$-\pi a^{\nu} \operatorname{cosec}(\pi \nu)$	$-1 < \nu < 0$
2.7	2.8	$\tfrac{1}{2}\pi^{\frac{1}{2}} a^{-2\nu} \Gamma(\nu)/\Gamma(\tfrac{1}{2}+\nu)$	$\nu > 0$
2.8	2.22	$\tfrac{1}{2}\pi^{\frac{1}{2}} a^{2\nu} \Gamma(\tfrac{1}{2}+\nu)/\Gamma(1+\nu)$	$\nu > -\tfrac{1}{2}$
2.9	2.31	$\tfrac{1}{2}\pi^{-\frac{1}{2}} a^{-2\nu} \Gamma(\nu)\,\Gamma(\tfrac{1}{2}-\nu)$	$0 < \nu < \tfrac{1}{2}$
2.10	2.23	$a^{1+2\nu}(1+2\nu)^{-1}$	$\nu > -\tfrac{1}{2}$
2.11	2.33	$\tfrac{1}{2} a^{-2\nu-1}\sec(\pi\nu)$	$-\tfrac{1}{2} < \nu < \tfrac{1}{2}$
2.12	2.34	$\tfrac{1}{2}\pi^{-\frac{1}{2}} a^{-2\nu} \Gamma(\nu)\,\Gamma(\tfrac{1}{2}-\nu)$	$0 < \nu < \tfrac{1}{2}$
2.13	2.11	$\tfrac{1}{2}\pi^{-\frac{1}{2}} a^{-2\nu} \Gamma(\nu)\,\Gamma(\tfrac{1}{2}-\nu)$	$0 < \nu < \tfrac{1}{2}$
2.14	2.24	$\pi^{\frac{1}{2}} a^{2\nu} \Gamma(\tfrac{1}{2}+\nu)/\Gamma(1+\nu)$	$\nu > -\tfrac{1}{2}$
2.15	2.26	$\tfrac{1}{2}\pi^{\frac{1}{2}} a^{2\nu} \Gamma(\tfrac{1}{2}+\nu)/\Gamma(1+\nu)$	$\nu > -\tfrac{1}{2}$
2.16	2.14	$(\tfrac{1}{2}\pi)^{\frac{1}{2}} a^{\nu-\frac{1}{2}} \Gamma(\tfrac{1}{4}-\tfrac{1}{2}\nu)/\Gamma(3/4-\tfrac{1}{2}\nu)$	$\nu < \tfrac{1}{2}$
2.17	2.15	$(\tfrac{1}{2}\pi/a)^{\frac{1}{2}} \Gamma(\tfrac{1}{4}-\tfrac{1}{2}\nu)/\Gamma(3/4-\tfrac{1}{2}\nu)$	$\nu < \tfrac{1}{2}$
2.18	2.12	$\nu a^{1-\nu}(\nu^{2}-1)^{-1}$	$\nu > -1$
2.19	2.13	$a^{-\nu}\nu^{-1}$	$\nu > -1$
2.21	2.21	$\tfrac{1}{2} a^{\nu-2\mu-1} B(\tfrac{1}{2}+\tfrac{1}{2}\nu,\, \tfrac{1}{2}-\tfrac{1}{2}\nu+\mu)$	$\nu > -1$
2.22	2.30	$(2a)^{-\frac{1}{2}} B(\tfrac{1}{4}+\tfrac{1}{2}\nu,\, \tfrac{1}{4}-\tfrac{1}{2}\nu)$	$-\tfrac{1}{2} < \nu < \tfrac{1}{2}$
2.23	2.27	$\tfrac{1}{2} a^{1+\nu+2\mu} B(1+\mu,\, \tfrac{1}{2}+\tfrac{1}{2}\nu)$	$(\nu,\mu) > -1$
2.25	2.36	$(2\pi)^{-\frac{1}{2}} a^{\nu-\frac{1}{2}} \Gamma(\tfrac{1}{4}-\tfrac{1}{2}\nu)\,\Gamma(\tfrac{1}{4}+\tfrac{1}{2}\nu)$	$-\tfrac{1}{2} < \nu < \tfrac{1}{2}$
3.1	3.1	a^{-1}	
3.2		$\log(b/a)$	$b > a$

$g_c(y)$	$g_s(y)$	$N=g_c(0)$	Restrictions
3.3	3.4	$\frac{1}{2}\pi^{\frac{1}{2}}a^{-3/2}$	
3.4	3.5	$(\pi/a)^{\frac{1}{2}}$	
3.5	3.7	$n!a^{-n-1}$	
3.6	3.8	$\pi^{\frac{1}{2}}2^{-2n}(2n)!/n!$	
3.7	3.9	$a^{-\nu}\Gamma(\pi)$	$\nu>0$
3.8	3.10	$a^{-\nu-1}e^{-ab}\Gamma(1+\nu)$	$\nu>-1$
3.9	3.11	$a^{-1}\log 2$	
3.12		$1/6(\pi/a)^2$	
3.13	3.13	$a^{-\nu}\Gamma(\nu)\zeta(\nu)$	$\nu>-1$
3.14	3.14	$(1-2^{1-\nu})\zeta(\nu)\Gamma(\nu)$	$\nu>0$
3.15		$2a\log 2$	
3.16	3.21	$b^{-1}B(\nu,a/b)$	$\nu>0$
3.17	3.23	$\frac{1}{2}(\pi/a)^{\frac{1}{2}}$	
3.18	3.26	$\frac{1}{2}\Gamma(3/4)a^{-3/4}$	
3.19	3.27	$\frac{1}{2}\Gamma(1/4)a^{-1/4}$	
3.20		$\pi^{\frac{1}{2}}(2a)^{-2n-1}(2n)!/n!$	
3.21	3.31	$\frac{1}{2}a^{-\frac{1}{2}\nu-\frac{1}{2}}\Gamma(\frac{1}{2}+\frac{1}{2}\nu)$	$\nu>-1$
3.22		$\frac{1}{2}\pi b^{-1}\exp(a^2b^2)\mathrm{Erfc}(ab)$	
3.23		$\frac{1}{2}\pi^{\frac{1}{2}}a^{\frac{1}{2}m-\frac{1}{4}}b^{-\frac{3}{2}-m}\exp(\frac{1}{2}ab^2)W_{-\frac{1}{4}-\frac{1}{2}m,\frac{1}{4}+\frac{1}{2}m}(ab^2)$	$m=0,1,2,\ldots$

$g_c(y)$	$g_s(y)$	$N=g_c(0)$	Restrictions
3.24		$(\pi/a)^{\frac{1}{2}}(2a)^{-\frac{1}{2}\nu}\exp(\frac{1}{2}ab^2)D_\nu\{b(2a)^{\frac{1}{2}}\}$	
3.25	3.30	$\frac{1}{2}(\pi/b)^{\frac{1}{2}}\exp(\frac{1}{2}a^2/b)\mathrm{Erfc}(\frac{1}{2}ab^{-\frac{1}{2}})$	
3.27	3.32	$(2b)^{-\frac{1}{2}\nu}\exp(\frac{a^2}{8b})\Gamma(\nu)D_{-\nu}\{a(2b)^{-\frac{1}{2}}\}$	$\nu>0$
3.29	3.27	$(\pi/a)^{\frac{1}{2}}$	
3.30	3.38	$\Gamma(\nu)a^{-\nu}$	
3.31	3.39	$(\pi/a)^{\frac{1}{2}}\exp(-2ba^{\frac{1}{2}})$	
3.32	3.40	$\pi^{\frac{1}{2}}b^{-1}\exp(-2ba^{\frac{1}{2}})$	
3.33	3.41	$2b^\nu a^{-\frac{1}{2}\nu}K_\nu(2ba^{\frac{1}{2}})$	
3.34	3.42	$\frac{1}{2}a^{-1}\pi^{\frac{1}{2}}$	
3.35	3.43	$2a^{-2}$	
3.36	3.45	$2a^{-1}$	
3.37	3.46	$2(\pi/a)^{\frac{1}{2}}$	
3.38	3.47	$2a^{-2\nu}\Gamma(2\nu)$	$\nu>0$
3.39		$aK_1(ab)$	
3.40		$\int_b^\infty K_0(at)dt$	
3.41		$K_0(ab)$	
3.42		$\pi^{\frac{1}{2}}(ab/a)^{\frac{1}{4}}K_{\frac{1}{4}}(ab)[\Gamma(\frac{3}{4})]^{-1}$	
3.43		$(\frac{1}{2}b/\pi)^{\frac{1}{2}}\{K_{\frac{1}{4}}(\frac{1}{2}ab)\}^2$	
3.44	3.49	$\pi(2a)^{-\frac{1}{2}}\exp(ab)\mathrm{Erfc}\{(2ab)^{\frac{1}{2}}\}$	
3.45		$(\pi/b)^{\frac{1}{2}}\exp(-ab)$	

$g_c(y)$	$g_s(y)$	$N=g_c(0)$	Restrictions
3.46		$2a^\nu K_\nu(ab)$	
3.47	3.53	$2^{\frac{1}{2}\nu-\frac{1}{2}}a^{-\frac{1}{2}}\Gamma(\frac{1}{2}+\frac{1}{2}\nu)D_{-\nu-\frac{1}{2}}\{2(ab)^{\frac{1}{2}}\}$	$\nu>-\frac{1}{2}$
3.48		$\frac{1}{2}a^{\frac{1}{2}}\exp(\frac{1}{2}ba^2)K_{\frac{1}{4}}(\frac{1}{2}ba^2)$	
3.50		$\frac{1}{2}(\pi b)^{\frac{1}{2}}K_{\frac{1}{4}}(\frac{1}{2}ab)\{I_{\frac{1}{4}}(\frac{1}{2}ab)+I_{-\frac{1}{4}}(ab)\}$	
3.51		$\frac{1}{2}\pi(\frac{1}{2}\pi b)^{\frac{1}{2}}\{I^2_{-\frac{1}{4}}(\frac{1}{2}ab)+I^2_{\frac{1}{4}}(\frac{1}{2}ab)\}$	
3.52		$(\pi b)^{\frac{1}{2}}K_{\frac{1}{4}}(\frac{1}{2}ab)\{I_{\frac{1}{4}}(\frac{1}{2}ab)+I_{-\frac{1}{4}}(\frac{1}{2}ab)\}$	
3.53		$\pi(\frac{1}{2}\pi b)^{\frac{1}{2}}\{I^2_{\frac{1}{4}}(\frac{1}{2}ab)+I^2_{-\frac{1}{4}}(ab)\}$	
3.54	3.54	$2^{-\nu-\frac{1}{4}}\pi a^{-\frac{1}{2}}\sec(2\pi\nu)\{\Gamma(3/4+\nu)\}^{-1}$ $\left[D_{2\nu-\frac{1}{2}}\{2(ab)^{\frac{1}{2}}\}+D_{2\nu-\frac{1}{2}}\{-2(ab)^{\frac{1}{2}}\}\right]$	$-\frac{1}{4}<\nu<\frac{1}{4}$
3.55	3.33	$c^{-1}a^{-(1+\nu)/c}\Gamma\{(1+\nu)/c\}$	$c<1,\ \nu>-1$
3.56	3.34	same as 3.55	$c>1,\ \nu>-1$
3.57	3.35	$\frac{1}{3}a^{-1}\ (\frac{1}{3})$	
4.1*	4.1	$-a(1-\log a)$	
4.3		$\frac{1}{4}\pi^2 a^{-1}$	
4.9*	4.5	$-a(1-\log a)$	$a\geq1$
4.10*	4.6	$-\{b+(a-b)\log(a-b)-a\log a\}$	$b\leq a\leq1$
4.11	4.7	$(a+b)\log(a+b)-a\log a-b$	$a\geq1$
4.14		$\pi a^{-1}\log(2a)$	$a\geq1$
4.15*		$-\frac{1}{2}\pi\log(2a)$	
4.17*	4.9	$-\{(a-b)\log(a-b)-(a+b)\log(a+b)+2b\}$	$b\leq a\leq1$

$g_c(y)$	$g_s(y)$	$N=g_c(0)$	Restrictions
4.18	4.23	$\pi(a-b)$	$a>b$
4.23	4.25	πa	
4.26*	4.8	$-2a\{1-\log(2a)\}$	$a\leq 1$
4.27*		$-\pi\log(2/a)$	$a\leq 1$
4.29*	4.35	$-2\pi\log(2/a)$	$a\leq 1$
4.31	4.41	a	
4.32		$\tfrac{1}{4}\pi^2$	
4.42		$a^{-1}\{\tfrac{1}{6}\pi^2+(\gamma+\log a)^2\}$	
4.43	4.46	$(12a)^{-1}\pi^2$	
4.44*	4.47	$-(6a)^{-1}\pi^3$	
4.45		$(\tfrac{1}{2}a)^{-2n}{}_n{}^{-1}(n!)/(2n)!$ $\cdot\left[\log(\tfrac{1}{2}a)-\sum\limits_{k=1}^{2n-1}(-1)^k/k\right]$	$a\geq 1$
4.46		$n!\sum\limits_{k=0}^{n-1}a^k(\tfrac{1}{2}b)^{k-n}K_k(ab)/\{k!(n-k)!\}$	
4.47*		$-\pi(\tfrac{1}{2}a)^{2n}(2n)!(n!)^{-2}$ $\cdot\left[\log(2/a)+\sum\limits_{k=1}^{2n}(-1)^k/k\right]$	$a\leq 1$
4.48*		The same expression as before multiplied by two	$a\leq 1$
5.10	5.11	$\tfrac{1}{2}\pi a$	
5.12	5.15	$(-1)^m 2^{1-2m}{}_m\pi\sum\limits_{k=1}^{m}(-1)^k(2ak)^{2m-1}$ $\cdot\{(m+k)!(m-k)!\}^{-1}$	

$g_c(y)$	$g_s(y)$	$N = g_c(0)$	Restrictions
5.14	5.17	$i(-1)^n \dfrac{2^{-2n}}{4n+2}\left(A_n^{-1} - B_n^{-1}\right), \quad \begin{matrix} A_n \\ B_n \end{matrix} = \begin{pmatrix} n \pm \frac{1}{2} i a \\ 2n+1 \end{pmatrix}$	
5.16	5.20	$2^{1-\nu}\Gamma(\nu)\{\Gamma(\frac{1}{2}+\frac{1}{2}\nu)\}^{-2}$	
5.25		$2^{1-\nu}\Gamma(\nu)\{\Gamma(\frac{1}{2}+\frac{1}{2}\nu)\}^{-2}$	$\nu > 0$
5.26	5.28	$\pi\,\mathrm{cosech}(a)$	
5.27		$\pi(\sinh a)^{-\nu}P_{-\nu}(\coth a)$	
5.28		$(\frac{1}{2}\pi)^{\frac{1}{2}}\sin^\nu a\,\Gamma(\frac{1}{2}+\nu)\,P_{-\frac{1}{2}}^{-\nu}(\cos a)$	$0 < a < \pi, \nu > -\frac{1}{2}$
5.29		$\frac{1}{2}\pi a^{-1}(e^a+b)(e^a-b)^{-1}(1-b^2)^{-1}$	$b < 1$
5.30		same as before	$b < 1$
5.31		$\frac{1}{2}\pi a^{-1}(e^a-b)^{-1}$	
5.98		$\frac{1}{2}\pi a J_1(ab)$	$ab \le \pi$
5.99		$\frac{1}{2}\pi J_0(ab)$	$ab \le \frac{1}{2}\pi$
5.100	5.86	$2^{-3/4}\pi^{3/2}(ab)^{-\frac{1}{4}}\{\Gamma(\frac{3}{4})\}^{-1}J_1(ab)$	$ab \le \frac{1}{2}\pi$
5.101		$(\frac{1}{2}\pi)^{3/2}\{J_{\frac{1}{4}}(\frac{1}{2}ab)\}^2 b^{\frac{1}{2}}$	$ab \le \pi$
5.102		$\frac{1}{2}\pi(\frac{1}{2}\pi b)^{\frac{1}{2}}\{J_{-\frac{1}{4}}(\frac{1}{2}ab)\}^2$	$ab \le \frac{1}{2}\pi$
5.103	5.88	$\pi J_0(ab)$	$ab \le \frac{1}{2}\pi$
5.104	5.99	$\pi(\frac{1}{2}\pi b)^{\frac{1}{2}}\{J_{-\frac{1}{4}}(\frac{1}{2}ab)\}^2$	$ab \le \frac{1}{2}\pi$
5.105	5.90	$\pi(\frac{1}{2}\pi b)^{\frac{1}{2}}\{J_{\frac{1}{4}}(\frac{1}{2}ab)\}^2$	$ab \le \pi$
5.106	5.91	$\pi a J_1(ab)$	$ab \le \pi$
5.115		$I_0(a)$	

$g_c(y)$	$g_s(y)$	$N=g_c(0)$	Restrictions
5.116	5.107	$(2\pi a)^{\frac{1}{2}} K_{\frac{1}{4}}(a)\{I_{\frac{1}{4}}(a)+I_{-\frac{1}{4}}(a)\}$	
5.117	5.108	$\pi(\pi a)^{\frac{1}{2}}\{I_{\frac{1}{4}}^2(a)+I_{-\frac{1}{2}}^2(a)\}$	
5.118		$(\frac{1}{2}\pi a)^{\frac{1}{2}} K_{\frac{1}{4}}(a)\{I_{\frac{1}{4}}(a)+I_{-\frac{1}{4}}(a)\}$	
5.119		$(\frac{1}{2}\pi a)^{\frac{1}{2}}\{I_{-\frac{1}{4}}^2(a)+I_{\frac{1}{4}}^2(a)\}$	
5.120		$(\frac{1}{2}a)^{\frac{1}{2}}\pi^{-1} K_{\frac{1}{4}}^2(\frac{1}{2}a)$	
5.121		$\frac{1}{2}\pi(\pi a)^{\frac{1}{2}} J_{\frac{1}{4}}^2(a)$	$a\leq\frac{1}{2}\pi$
5.122		$\frac{1}{2}\pi(\pi a)^{\frac{1}{2}} J_{-\frac{1}{4}}^2(a)$	$a\leq\frac{1}{2}\pi$
5.123	5.109	$\pi(\pi a)^{\frac{1}{2}} J_{\frac{1}{4}}^2(a)$	$a\leq\frac{1}{2}\pi$
5.124	5.110	$\pi(\pi a)^{\frac{1}{2}} J_{-\frac{1}{4}}^2(a)$	$a\leq\frac{1}{2}\pi$
5.125	5.105	$\pi H_0(a)$	$a\leq\pi$
5.126	5.106	$\pi J_0(a)$	$a\leq\frac{1}{2}\pi$
5.127		$\frac{1}{2}\pi H_0(a)$	$a\leq\pi$
5.128		$\frac{1}{2}\pi J_0(a)$	$a\leq\frac{1}{2}\pi$
5.129	5.114	$2(\pi a)^{3/2}\{J_{-\frac{1}{4}}^2(a)+J_{3/4}^2(a)\}$	$a\leq\frac{1}{2}\pi$
5.130*		same expression as before	$a\leq\frac{1}{2}\pi$
5.131*		$-\log 2$	
5.132		$-\log 2$	
5.133		$-2^{1-\nu}\Gamma(\nu)\{\Gamma(\frac{1}{2}+\frac{1}{2}\nu)\}^{-2}$ $\{\psi(\frac{1}{2}+\frac{1}{2}\nu)-\psi(\nu)+\log 2\}$	$\nu>0$

$g_c(y)$	$g_s(y)$	$N=g_c(0)$	Restrictions
5.134		Same expression as before	$\nu>0$
5.135		$\pi 2^{-\frac{1}{2}}\{2\pi^{-1}\log(\sin a)\mathbf{K}(\sin\frac{1}{2}a)$ $-\mathbf{K}(\cos\frac{1}{2}a)\}$	$a\leq\pi$
6.1	6.1	$\frac{1}{2}\pi-1$	
6.2	6.2	1	
6.10		$\pi(\frac{1}{2}a)^{\frac{1}{2}}$	
6.11		$\frac{1}{2}\pi a \sum\limits_{k=1}^{n}(-1)^{k+1}\cos\{(k+\frac{1}{2})\pi/n\}$	
6.13		$\frac{1}{2}\pi\log 2$	
7.1	7.1	$\frac{1}{2}\pi a^{-1}$	
7.2	7.2	a^{-1}	
7.3		$2^{2n-2}\{a(2n-1)!\}^{-1}\{(n-1)!\}^{2}$	
7.4		$2^{-2n-1}\pi a^{-1}(2n)!(n!)^{-2}$	
7.5		$\frac{1}{2}\pi^{\frac{1}{2}}a^{-1}\Gamma(\frac{1}{2}\nu)\{\Gamma(\frac{1}{2}+\frac{1}{2}\nu)\}^{-1}$	$\nu>0$
7.6	7.38	$\frac{1}{2}a^{-1}\pi^{-\frac{1}{2}}\Gamma(\frac{1}{2}\nu)\{\Gamma(\frac{1}{2}-\frac{1}{2}\nu)\}^{-1}$	$0<\nu<1$
7.7	7.18	$ba^{-1}\mathrm{cosec}\ b$	$0<b<\pi$
7.8	7.17	$ba^{-1}\mathrm{cosec}\ b$	$0<b<\pi$
7.10		$2^{\frac{1}{2}}\mathbf{K}(\sin\frac{1}{2}b)$	$0<b<\pi$
7.11		$2^{\frac{1}{2}}\mathrm{sech}(\frac{1}{2}b)\mathbf{K}(\tanh b)$	
7.12		$2^{\frac{1}{2}}\mathbf{K}(\cos\frac{1}{2}b)$	$0<b<\pi$
7.14		$\frac{1}{3}(\frac{1}{2}\pi)^{\frac{1}{2}}$	

$g_c(y)$	$g_s(y)$	$N=g_c(0)$	Restrictions
7.15		$(\sinh a)^{-\nu} q_{\nu-1}(\coth a)$	$\nu>0$
7.16		$(\tfrac{1}{2}\pi)^{\frac{1}{2}} \Gamma(\nu)(\sin b)^{\frac{1}{2}-\nu} P_{-\frac{1}{2}}^{\frac{1}{2}-\nu}(\cos b)$	$0<b<\pi, \nu>0$
7.17		$(\sinh a)^{-\nu} q_{-\nu}(\coth a)$	$\nu<1$
7.18	7.22	$(\tfrac{1}{2}\pi)^{-\frac{1}{2}} \Gamma(\tfrac{1}{2}+\nu)(\sinh a)^{\nu} e^{i\pi\nu} q_{-\frac{1}{2}}^{-\nu}(\cosh a)$	$0<\nu<1$
7.19		$\tfrac{1}{2}b^{-1}\sec(\tfrac{1}{2}a/b)$	
7.20	7.26	$\tfrac{1}{2}b^{-1}\tanh(\tfrac{1}{2}a/b)$	
7.21		$\tfrac{1}{4}a^{-2}$	
722*	7.8	$-\log 2$	
7.24	7.25	$\tfrac{1}{2}b^{-1}\left[\pi\tan(\tfrac{1}{2}\pi a/b)+\psi(3/4-\tfrac{1}{4}a/b) -\psi(3/4+\tfrac{1}{4}a/b)\right]$	$a<b$
7.30	7.30	$\log\{\cos(\tfrac{1}{2}b/c)\sec(\tfrac{1}{2}a/c)\}$	$(a,b)<\pi c$
7.26		$-\tfrac{1}{2}\log\{\cos(a/b)\}$	$a\leq\tfrac{1}{2}\pi b$
7.27*	7.14	$-2\Gamma(\nu)(2^{-\nu}-1)\zeta(\nu)$	$-1<\nu<1$
7.28		$\tfrac{1}{2}\pi a^{-1}\operatorname{sech}(\tfrac{1}{2}b)$	
7.29		$\tfrac{1}{2}\pi a^{-1}\sec(\tfrac{1}{2}b)$	$0<b<\pi$
7.25		$\tfrac{1}{2}\log\left[\{1+\sin(\tfrac{1}{2}a/b)\}/\{1-\sin(\tfrac{1}{2}a/b)\}\right]$	
7.31		$2a^3$	
7.32		$\tfrac{3}{8}a^4$	
7.33		$2-\tfrac{1}{2}\pi$	
7.34		$\log 2$	

$g_c(y)$	$g_s(y)$	$N=g_c(0)$	Restrictions
7.35		$\frac{1}{2}\log\{(1-ab)/(a-b)\}$	$a \leq b$
7.36	7.10	$a^{-1}\log 2$	
7.40		$\frac{1}{2}b^{-1}\mathrm{cosec}(a/b)-\frac{1}{2}a^{-1}$	
7.41		$\frac{1}{2}a^{-1}+\frac{1}{2}b^{-1}\cot(a/b)$	
7.42		$\frac{1}{2}v\pi c^{-1}\mathrm{cosec}(bv)$ $+\pi b^{-1}\sum\limits_{n=1}^{\infty}(-1)^{n}v_{m}^{-1}(1-v^2/c_n^2)^{-1}$	
7.43		$\frac{1}{2}\pi c^{-1}\sec(bv)$ $-\pi b^{-1}\sum\limits_{n=o}^{\infty}(-1)^{n}(v_n c_n)^{-1}(1-v^2/c_n^2)^{-1}$	
7.44		$\frac{1}{2}\pi c^{-1}\sin(av)\mathrm{cosec}(bv)$ $+\pi b^{-1}\sum\limits_{n=1}^{\infty}(-1)^{n}(1-v^2/c_n^2)^{-1}$ $\cdot(v_n c_n)^{-1}\sin(ac_n)$	
7.45		$\frac{1}{2}\pi c^{-1}\cos(av)\sec(bv)$ $-\pi b^{-1}\sum\limits_{n=o}^{\infty}(-1)^{n}(1-v^2/c_n^2)^{-1}$ $\cdot(v_n c_n)^{-1}\cos(ac_n)$	
7.46		$\frac{1}{2}\pi c^{-1}\sin(ac)\mathrm{cosec}(bc)$ $-\pi b^{-1}\sum\limits_{n=o}^{\infty}(-1)^{n}(c^2-c_n^2)^{-1}\cdot\sin(ac_n)$	
7.47		$\frac{1}{2}\pi c^{-1}\cos(ac)\sec(bc)$ $+\pi b^{-1}\sum\limits_{n=o}^{\infty}(-1)^{n}(c^2-c_n^2)^{-1}\cos(ac_n)$	
7.48		$\pi b^{-1}\sum\limits_{n=o}^{\infty}(-1)^{n+1}c_n v_n^{-1}\sin(ac_n)$	

$g_c(y)$	$g_s(y)$	$N = g_c(0)$
7.49		$b^{-1} \sum\limits_{n=0}^{\infty} (-1)^n c_n v_n^{-1} \cos(ac_n)$
7.50		$\pi b^{-1} \sum\limits_{n=0}^{\infty} (-1)^n v_n^{-1} \sin(ac_n)$
7.51		$\tfrac{1}{2}\pi b^{-1} \sum\limits_{n=0}^{\infty} (-1)^n \varepsilon_n v_n^{-1} \cos(ac_n)$
7.52		$\tfrac{1}{2}\pi a(bk)^{-1} + \pi b^{-1} \sum\limits_{n=1}^{\infty} (-1)^n (c_n v_n)^{-1} \sin(ac_n)$
7.53		$\tfrac{1}{2}\pi k^{-1} - \pi b^{-1} \sum\limits_{n=0}^{\infty} (-1)^n (c_n v_n)^{-1} \cos(ac_n)$
7.57		$\tfrac{1}{2}\pi I_0(ab)$
7.58		$\tfrac{1}{2}\pi a I_1(ab)$
7.59		$\tfrac{1}{2}\pi (\tfrac{1}{2}\pi b)^{\frac{1}{2}} \{ I_{\frac{1}{4}}(\tfrac{1}{2}ab) \}^2$
7.60		$\tfrac{1}{2}\pi (\tfrac{1}{2}\pi b)^{\frac{1}{2}} \{ I_{-\frac{1}{4}}(\tfrac{1}{2}ab) \}^2$
7.63	7.51	$2^{-3/4}(b/a)^{\frac{1}{2}}\pi^{3/2} \{ \Gamma(3/4) \}^{-1} I_{-\frac{1}{4}}(ab)$
7.64	7.55	$\pi (\tfrac{1}{2}\pi b)^{\frac{1}{2}} \{ I_{-\frac{1}{4}}(\tfrac{1}{2}ab) \}^2$
7.65	7.54	$\pi (\tfrac{1}{2}\pi b)^{\frac{1}{2}} \{ I_{\frac{1}{4}}(\tfrac{1}{2}ab) \}^2$
7.67		$2^{-\nu-1}b^{-1}\Gamma(1+\nu)\,\Gamma(\tfrac{1}{2}a/b-\tfrac{1}{2}\nu)/\Gamma(1+\tfrac{1}{2}a/b+\tfrac{1}{2}\nu),\quad -1<\nu<a/b$
7.68		$\tfrac{1}{2}(\pi/b)^{\frac{1}{2}}\exp(\tfrac{1}{4}a^2/b)$
7.69		$(\tfrac{1}{2}\pi)^{\frac{1}{2}}$
7.76	7.57	$K_0(ab)$

$g_c(y)$	$g_s(y)$	$N=g_c(0)$
7.77	7.56	$\frac{1}{2}\pi\{\mathbf{H}_o(a)-Y_o(a)\}$
7.78		$K_o\{(a^2-b^2)^{\frac{1}{2}}\},a>b,$ (Replace in 7.78 b by ib)
7.79		$(a/\pi)^{\frac{1}{2}}\{K_{\frac{1}{2}}(a)\}^2$
7.80	7.61	$\frac{1}{2}\pi(a\pi)^{\frac{1}{2}}\{J_{\frac{1}{4}}(a)Y_{-\frac{1}{4}}(a)-J_{-\frac{1}{4}}(a)Y_{\frac{1}{4}}(a)\}$
7.81	7.60	$\frac{1}{2}\pi(\frac{1}{2}\pi)^{\frac{1}{2}}a\left(\{J_{\frac{1}{4}}(\frac{1}{2}a)\}^2+\{Y_{\frac{1}{4}}(\frac{1}{2}a)\}^2\right)$
7.82	7.72	$2\pi a(\pi a)^{\frac{1}{2}}\left(\{I_{-\frac{1}{4}}(a)\}^2+\{I_{3/4}(a)\}^2\right)$
7.83		$\pi a(\pi a)^{\frac{1}{2}}\left(\{I_{-\frac{1}{4}}(a)\}^2-\{I_{3/4}(a)\}^2\right)$
7.84	7.70	$\pi(\pi a)^{\frac{1}{2}}\{I_{\frac{1}{4}}(a)\}^2$
7.85	7.71	$\pi(\pi a)^{\frac{1}{2}}\{I_{-\frac{1}{4}}(a)\}^2$
7.86		$\frac{1}{2}\pi(\pi a)^{\frac{1}{2}}\{I_{\frac{1}{4}}(a)\}^2$
7.87		$\frac{1}{2}\pi(\pi a)^{\frac{1}{2}}\{I_{-\frac{1}{4}}(a)\}^2$
7.88		$\pi\mathbf{L}_o(a)$
7.89		$\pi I_o(a)$
7.90		$\frac{1}{2}\pi\mathbf{L}_o(a)$
7.91		$\frac{1}{2}\pi I_o(a)$
7.102*		-1
7.103		$2^{-\frac{1}{2}}\log(1-a^2)\mathbf{K}|\{[(\frac{1}{2}-\frac{1}{2}a)^{\frac{1}{2}}]\}+2^{-\frac{1}{2}}\pi\mathbf{K}\{(\frac{1}{2}+\frac{1}{2}a)^{\frac{1}{2}}\}|$
7.104		$\pi(1+z)^{-\frac{1}{2}}\left(\mathbf{K}\{(\frac{1}{2}+\frac{1}{2}z)^{-\frac{1}{2}}\}+\log(z^2-1)\mathbf{K}[\{(z-1)/(z+1)\}^{\frac{1}{2}}]\right)$

$g_c(y)$	$g_s(y)$	$N=g_c(0)$	Restrictions
9.1		$\pi 2^{-2a}\Gamma(2a)$	$a>0$
9.3		$(\pi b)^{-1}$	
9.5		$2^{2a-3}\{\Gamma(2a-1)\}^{-1}b^{-1}$	$a>\tfrac{1}{2}$
9.7		$\pi^3 b^{-1}\operatorname{cosec}(2\pi a)$	$0<a<\tfrac{1}{2}$
9.8		$2^{1-2b-2c}\pi^{3/2}a^{-1}\Gamma(2b)\Gamma(2c)$ $\cdot\Gamma(b+c)\{\Gamma(\tfrac{1}{2}+b+c)\}^{-1}$	$a,b,c>0$
9.9		$2^{2b-2a-2}c^{-1}\Gamma(2a)$ $\cdot B(b-a-\tfrac{1}{2},\tfrac{1}{2})\{\Gamma(2b-1)\}^{-1}$	$b-a>-\tfrac{1}{2}$
10.1		$\pi^{-\frac{1}{2}}$	
10.3	10.13	$\tfrac{1}{4}$	
10.4	10.8	$\nu^{-1}\pi^{-\frac{1}{2}}\Gamma(\tfrac{1}{2}+\tfrac{1}{2}\nu)$	$\nu>0$
10.5		$\log(b/a)$	$a<b$
10.8*		$-\tfrac{1}{2}\pi$	
10.9*	10.9	$-\tfrac{1}{2}\pi^{-\frac{1}{2}}a^{-1}\operatorname{Ei}(-b^2)$	
10.10	10.11	$\tfrac{1}{2}$	
10.18		$\tfrac{1}{2}a^{-2}\exp(-2a^2 b)$	
10.19		$\tfrac{1}{2}\pi\{\operatorname{Erfc}(a)\}^2$	
10.20		$a^{-1}b\pi^{\frac{1}{2}}{}_2F_2(\tfrac{1}{2},\tfrac{1}{2};3/2,1;-b^2)$	
10.21		$-\tfrac{1}{2}\operatorname{Ei}(-a^2)$	

$g_c(y)$	$g_s(y)$	$N=g_c(0)$
10.22		$\frac{1}{2}K_0(\frac{1}{2}a^2)$
10.23		$\frac{1}{2}i\pi I_0(\frac{1}{2}a^2)$
10.24		$2\pi^{-\frac{1}{2}}a\{K_{\frac{1}{4}}(\frac{1}{2}a)\}^2$
10.29	10.22	$(2\pi a)^{-\frac{1}{2}}$
10.39	10.31	$\frac{1}{4}(\frac{1}{2}\pi/a)^{\frac{1}{2}}$
10.42		a
10.54		$\frac{1}{4}\pi^{-1}\Gamma^2(\frac{1}{4})S_{\frac{1}{2},0}(a)$
10.55		$\frac{1}{2}\pi^{-1}\Gamma^2(3/4)S_{-\frac{1}{2},0}(a)$
11.1*		$-a^{-1}$
11.2*	11.2	$-\{b^{-1}-aEi(-ab)\}$
11.4*	11.3	$-a^{-1}\log(1+a/b)$
11.7		$\frac{1}{2}\pi^2$
11.9*		$-(\pi/a)^{\frac{1}{2}}$
11.10*		$-\frac{1}{2}\pi(\pi/a)^{\frac{1}{2}}$
11.12*		$-\frac{1}{4}\pi^2\{I_0(a)-\mathbf{L}_0(a)\}+\{S_{-1,0}(ia)+S_{-1,0}(-ia)\}/2$
11.13		$\frac{1}{4}\pi^2\{I_0(a)-\mathbf{L}_0(a)\}-\frac{1}{2}\{S_{-1,0}(ia)+S_{-1,0}(ia)\}$
11.14*		$-\frac{1}{2}\pi(\pi/a)^{-\frac{1}{2}}Erfc(b^2)$
11.15*		$-2a^{-1}K_0(ab)$

$g_c(y)$	$g_s(y)$	$N=g_c(0)$	Restrictions
11.18*	11.11	$\frac{-1}{2b}\left[\pi-\arctan\{2ab(b^2-a^2)^{-1}\}\right]$	
11.19	11.10	$a\mathrm{Si}(ab)$	
11.27*		$-\tfrac{1}{2}\pi a^{-1}$	
11.29		$\tfrac{1}{2}\pi a^{-1}$	
11.30		$\tfrac{1}{2}\pi a^{-1}$	
11.31		$(\tfrac{1}{2}\pi/a)^{\frac{1}{2}}$	
11.33*		$-\tfrac{1}{2}\pi(\tfrac{1}{2}\pi/a)^{\frac{1}{2}}$	
11.34*		$-\tfrac{1}{2}\pi(\tfrac{1}{2}\pi/a)^{\frac{1}{2}}$	
11.37		$\tfrac{1}{2}\pi^2\{\mathbf{H}_0(a)-Y_0(a)\}$	
11.38		$S_{-1,0}(a)$	
12.11		$\pi 2^{-\frac{1}{2}}(2\nu+1)^{-1}$	$\nu>-\tfrac{1}{2}$
12.14		$\pi a^{-\nu}b^{\nu+1}(1+2\nu)^{-1}$	$\nu>-\tfrac{1}{2}$ $a>b$
12.27		$-\tfrac{1}{4}\pi^{-2}\sin(\pi\nu)\{\Gamma(-\tfrac{1}{2}\nu)\Gamma(\tfrac{1}{2}+\tfrac{1}{2}\nu)\}^2$	$-1<\nu<0$
12.28		$-2^{\nu-3/2}(\pi a)^{-\frac{1}{2}}\cot(\tfrac{1}{2}\pi\nu)\{\Gamma(\tfrac{1}{2}+\tfrac{1}{2}\nu)\}^2$ $P_{-\frac{1}{2}}^{-\nu-\frac{1}{2}}(s)P_{-\frac{1}{2}}^{-\nu-\frac{1}{2}}(-s)\,;\ s=(1-a^{-2})^{\frac{1}{2}}$	$-1<\nu<0,a\geq1$
12.29		$2^{\nu-1}(\tfrac{1}{2}\pi/a)^{\frac{1}{2}}\{\Gamma(\tfrac{1}{2}+\tfrac{1}{2}\nu)\}^2 P_{-\frac{1}{2}}^{-\nu-\frac{1}{2}}(s)$ s as before	$\nu>-1,\ a\geq1$
12.30		$\tfrac{1}{4}\pi^{-\frac{1}{2}}2^{-\mu}\{\Gamma(\tfrac{1}{2}+\tfrac{1}{2}\nu-\tfrac{1}{2}\mu)\Gamma(-\tfrac{1}{2}\nu-\tfrac{1}{2}\mu)\}^2$ $\{\Gamma(-\nu-\mu)\Gamma(1+\nu-\mu)\Gamma(\tfrac{1}{2}-\mu)\}^{-1}$	$\mu+\nu<0,\mu-\nu<1$

$g_c(y)$	$g_s(y)$	$N = g_c(0)$	Restrictions
12.31		$\frac{1}{4}e^{i\pi\mu}2^{\nu}\{\Gamma(1+\nu)\,\Gamma(1+\nu-\mu)\}^{-1}$ $\{\Gamma(\frac{1}{2}+\frac{1}{2}\nu-\frac{1}{2}\mu)\,\Gamma(\frac{1}{2}+\frac{1}{2}\nu+\frac{1}{2}\mu)\}^{2}$	$-1<\nu+\mu<0$
12.34		$\frac{1}{2}\pi\left[\rho_{\nu}\{(1+a^{2})^{\frac{1}{2}}\}\right]^{2}$	
12.35	12.9	$\frac{1}{2}\left(P_{\nu}(s)\,\{Q_{\nu}(s)+Q_{-\nu-1}(s)\}\right),\ s=(1-a^{2})^{\frac{1}{2}}$	$-1<\nu<0,\,a<1$
12.36	12.10	$\frac{1}{2}a^{-1}\sec(\pi\nu)\left(P_{-\frac{1}{2}}^{\frac{1}{2}+\nu}(s)Q_{-\frac{1}{2}}^{-\frac{1}{2}-\nu}(s)\right.$ $\left.-P_{-\frac{1}{2}}^{-\frac{1}{2}-\nu}(s)Q_{-\frac{1}{2}}^{\frac{1}{2}+\nu}(s)\right.\quad s=(1-a^{-2})^{\frac{1}{2}}$	$-1<\nu<0,\,a>1$
12.38		$-\frac{1}{2}\pi a^{-1}\mathrm{cosec}(\pi\nu)P_{-\frac{1}{2}}^{-\frac{1}{2}-\nu}(s)Q_{-\frac{1}{2}}^{\frac{1}{2}+\nu}(s)$ $s\ \text{as before}$	$-\frac{1}{2}<\nu<0,\,a>1$
12.39		$\frac{1}{2}\pi^{-1}\left(\{Q_{\nu}(as)\}^{2}-\{Q_{-\nu-1}(as)\}^{2}\right)\tan(\pi\nu)$ $s=(1+a^{-2})^{\frac{1}{2}}$	$-1<\nu<0$
12.40		$\frac{1}{2}\{q_{\nu}(as)\}^{2}\ ,\ s=(1+a^{-2})^{\frac{1}{2}}$	$\nu>-1$
12.42		$\frac{1}{2}a^{-1}\left(P_{-\frac{1}{2}}^{-\nu-\frac{1}{2}}(s)Q_{-\frac{1}{2}}^{\nu+\frac{1}{2}}(s)\right.$ $\left.+P_{-\frac{1}{2}}^{\nu+\frac{1}{2}}(s)q_{-\frac{1}{2}}^{-\nu-\frac{1}{2}}(s)\right),\quad s=(1-a^{-2})^{\frac{1}{2}}$	$-1<\nu<0,\,a>1$
12.43		$\frac{1}{4}\pi a^{-1}\left(\Gamma(1+\nu)\,P_{-\frac{1}{2}}^{-\nu-\frac{1}{2}}(s)\right)^{2},\ s=(1-a^{-2})^{\frac{1}{2}}$	$\nu>-1,\,a>1$
12.44		$-\pi^{-1}\sin(\pi\nu)q_{\nu}(as)q_{-\nu-1}(as)$ $s=(1+a^{-2})^{\frac{1}{2}}$	
12.45	12.11	$-\frac{1}{2}\pi\,\mathrm{cosec}(\pi\nu)p_{\nu}(as)p_{-\nu-1}(as)$ $s\ \text{as before}$	$-1<\nu<0$
12.47		$e^{i\pi\mu}\pi^{\frac{1}{2}}2^{\mu-2}\Gamma(\frac{1}{2}-\mu)$ $\cdot\left(\Gamma(\frac{1}{2}+\frac{1}{2}\nu+\frac{1}{2}\mu)/\Gamma(1+\frac{1}{2}\nu-\frac{1}{2}\mu)\right)^{2}$	$\mu<\frac{1}{2},\,\mu+\nu>-1$

$g_c(y)$	$g_s(y)$	$N=g_c(0)$	Restrictions
12.48		$\frac{1}{4}\pi^{\frac{1}{2}}e^{-i\pi\mu}2^{-\mu}a^{-\nu+\mu-1}\{\Gamma(3/2+\nu)\}^{-1}$ $\Gamma^2(\frac{1}{2}+\frac{1}{2}\nu-\frac{1}{2}\mu)\,_2F_1(\frac{1}{2}+\frac{1}{2}\nu-\frac{1}{2}\mu,\frac{1}{2}+\frac{1}{2}\nu-\frac{1}{2}\mu;$ $3/2+\nu;a^{-2})$	$\mu-\nu<1$
12.50		$\frac{1}{4}\pi\left[\Gamma(\frac{1}{2}+\frac{1}{2}\nu)/\Gamma(1+\frac{1}{2}\nu)\right]^2$ $\cdot\{-\gamma-\log 4+\psi(\nu+1)-\psi(\frac{1}{2}+\frac{1}{2}\nu)-\psi(1+\frac{1}{2}\nu)\}$	$\nu>-1$
12.58		$(2+2a)^{-\frac{1}{2}}$	$-1<a<1$
12.59		$b^{-1}\sum_{n=o}^{\infty}(-1)^n\varepsilon_n\cos(n\pi y/b)Q_{\frac{1}{2}n\pi-\frac{1}{2}}(a)$ $-b\leq y\leq b$	$-1<a<1$
12.60		$2^{-\mu}(2+2a)^{-\frac{1}{2}}p_\mu^\mu\{(\frac{1}{2}+\frac{1}{2}a)^{-\frac{1}{2}}\}$	$-1<a<1,\mu<\frac{1}{2}$
12.61		$\pi^{-1}(2-2a)^{-\frac{1}{2}}\log(u/v)$ $\frac{u}{v}=(1+\cosh)^{\frac{1}{2}}\pm(\cosh y-a)^{\frac{1}{2}}$	$-1<a<1$
12.63		$2^{3/2}\pi^{\frac{1}{2}}\{1+(1-a^2)^{\frac{1}{2}}\}^{-\frac{1}{2}}\left[\{\frac{1}{2}+\frac{1}{2}(1-a^2)^{-\frac{1}{2}}\}^{-\frac{1}{2}}\right]$	$a<1$
12.65		$\pi^{\frac{1}{2}}2^{1+\mu}(1-a^2)^{-\frac{1}{4}}q_{-\mu-\frac{1}{2}}\{(1-a^2)^{-\frac{1}{2}}\}$	$a<1,\mu<\frac{1}{2}$
12.66		$(\frac{1}{2}\pi)^{\frac{1}{2}}\Gamma(\mu)(1-a)^{\frac{1}{2}\mu-\frac{1}{4}}(1+a)^{-\frac{1}{2}\mu-\frac{1}{4}}$	$-1<a<1,\mu>0$
12.67		$2^{1-\pi}(1-a^2)^{\frac{1}{2}\mu-\frac{1}{4}}\Gamma(2\mu)(1-a)^{-\frac{1}{2}\mu-\frac{1}{4}}$ $\cdot e^{i\pi(\mu-\frac{1}{4})}q_{\mu-\frac{1}{2}}^{\frac{1}{2}-\mu}(z)\ ,\quad z=(\frac{1}{2}-\frac{1}{2}a)^{-\frac{1}{2}}$	$0<a<1,\mu>0$
12.69		$\pi^{3/2}2^{1+\mu}(1-a^2)^{-\frac{1}{4}}\sec(\pi\mu)p_{\mu-\frac{1}{2}}\{(1-a^2)^{-\frac{1}{2}}\}$	$-1<n<1$ $-\frac{1}{2}<\mu<\frac{1}{2}$
12.72		$\pi^{-1}(|a|)$	$-1<a<1$
12.73		$(1-a^2)^{-\frac{1}{4}}q_{-\mu-\frac{1}{2}}\{(1+a^2)/(1-a^2)\}$	$0<a<1,\mu<\frac{1}{2}$

$g_c(y)$	$g_s(y)$	$N=g_c(0)$	Restrictions
12.78		$(a^2-1)^{-\frac{1}{2}}q_{-\mu-\frac{1}{2}}\{(a^2+1)/(a^2-1)\}$	$0<a<1,\mu<\frac{1}{2}$
12.82		$(\pi a)^{-1}\{(1-a^{-2})^{\frac{1}{2}}\}$	$a>1$
13.30	13.31	$(2a/\pi)^{-\frac{1}{2}}\Gamma(\frac{1}{4})\{\Gamma(3/4)\}^{-3}$	
13.32	13.35	$(2a/\pi)^{-\frac{1}{2}}\Gamma(\frac{1}{4}+\nu)\{\Gamma^2(3/4)\Gamma(3/4+\nu)\}^{-1}$	$\nu>-\frac{1}{4}$
13.33		$\frac{1}{2}a^{2\nu-1}\Gamma(\nu)\{\Gamma(\frac{1}{2}+\nu)\Gamma(\frac{1}{2}+2\nu)\}^{-1}$	$\nu>0$
13.43		$\frac{1}{2}\Gamma(-\nu)\Gamma(\frac{1}{2}+2\nu)\{\pi\Gamma(\frac{1}{4}-\nu)\}^{-1}$	$-\frac{1}{4}<\nu<0$
13.84		$\frac{1}{2}2^{-\frac{1}{4}}\pi^{-\frac{1}{2}}\Gamma(1/8)/\Gamma(7/8)$	
14.43		$b^{-1}\sin(ab)$	$ab\leq\tau_{o,1}$
14.44		$2(\pi b)^{-1}\{\sin(ab)Ci(ab)-\cos(ab)Si(ab)\}$	$ab\leq\lambda_{o,1}$
14.45		$b^{-1}\Big(\sin(ab)\{Ci(2ab)-Ci(ab)+\log a\}$ $-\cos(ab)\{Si(2ab)-Si(ab)\}\Big)$	$ab\leq\tau_{o,1}$ $a\leq1$
14.46		$\frac{1}{2}\pi\{J_{\frac{1}{2}\nu}(\frac{1}{2}ab)\}^2$	$ab\leq\tau_{\nu,1}$ $\nu>-1$
14.47		$(\frac{1}{2}\pi a/b)^{\frac{1}{2}}a^\nu J_{\nu+\frac{1}{2}}(ab)$	$ab\leq\tau_{\nu,1}$ $\nu>-1$
14.48		$\frac{1}{4}\pi\Big(2J_{\frac{1}{2}\nu}(\frac{1}{2}ab)Y_{\frac{1}{2}\nu}(\frac{1}{2}ab)-\tan(\pi\nu)\{J^2_{\frac{1}{2}\nu}(\frac{1}{2}ab)$ $+Y^2_{\frac{1}{2}\nu}(\frac{1}{2}ab)\}\Big)$	$ab\leq\lambda_{\nu,1}$ $-1<\nu<1$
14.51	14.33	$2b^{-1}\sin(\frac{1}{2}ab)$	$ab\leq2\tau_{o,1}$
14.52		$4(\pi b)^{-1}\{\sin(\frac{1}{2}ab)Ci(\frac{1}{2}ab)-\cos(\frac{1}{2}ab)Si(\frac{1}{2}ab)\}$	$ab\leq2\lambda_{o,1}$

$g_c(y)$	$g_s(y)$	$N=g_c(0)$	Restrictions
14.53	14.25	$(\tfrac{1}{2}a)^{\nu}(\pi a/b)^{\tfrac{1}{2}}J_{\nu+\tfrac{1}{2}}(\tfrac{1}{2}ab)$	$ab\leq 2\tau_{o,1}$ $\nu>-1$
14.54		$\tfrac{1}{2}\pi\Big[2J_{\nu}(\tfrac{1}{4}ab)Y_{\nu}(\tfrac{1}{4}ab)-\tan(\pi\nu)$ $\cdot\{J_{\nu}^{2}(\tfrac{1}{4}ab)+Y_{\nu}^{2}(\tfrac{1}{4}ab)\}\Big]$	$-\tfrac{1}{2}<\nu<\tfrac{1}{2}$ $ab<2\lambda_{\nu,1}$
15.1		$\pi J_{\nu}^{2}(a)$	$\nu>-\tfrac{1}{2},a\leq\tfrac{1}{2}\tau_{2\nu,1}$
15.2		$\pi J_{\nu}(a)Y_{\nu}(a)$	$-\tfrac{1}{2}<\nu<\tfrac{1}{2},a\leq\tfrac{1}{2}\lambda_{2\nu,1}$
15.3	15.1	$\pi J_{\tfrac{1}{2}\nu}^{2}(a)$	$\nu>-1,a\leq\tfrac{1}{2}\tau_{\nu,1}$
15.5		$\tfrac{1}{2}(\tfrac{1}{2}\pi a)^{\tfrac{1}{2}}m^{-1}\Big[\mathbf{H}_{m-3/2}(a)+\mathbf{H}_{m+\tfrac{1}{2}}(a)\Big]$	$a\leq\tau_{m,1}$
15.6	15.3	Twice the result before	$a\leq\tau_{m,1}$
15.7	15.2	$\tfrac{1}{2}\pi a\nu^{-1}\{J_{\nu-\tfrac{1}{2}}^{2}(a)+J_{\nu+\tfrac{1}{2}}^{2}(a)\}$	$\nu>0,a\leq\tau_{2\nu,1}$
15.8		as before	as before
15.9		$\pi J_{o}(a)J_{o}(b)$	$a+b\leq\tau_{o,1}$
15.10		$\pi J_{o}(a)Y_{o}(b)$	$b>a,a+b<\pi$ $a+b\leq\lambda_{o,1}$
15.28		$2^{\mu-2}a^{\nu}\{\Gamma(1+\nu)\}^{-1}B(\tfrac{1}{2}\mu+\tfrac{1}{2}\nu),\tfrac{1}{2}\mu+\tfrac{1}{2}\nu)$ ${}_{1}F_{2}(\tfrac{1}{2}\mu+\tfrac{1}{2}\nu;\tfrac{1}{2}+\tfrac{1}{2}\mu+\tfrac{1}{2}\nu),1+\nu;-\tfrac{1}{4}a^{2})$	$\nu+\mu>0$
17.2	17.2	$(b^{2}-a^{2})^{-\tfrac{1}{2}}$	$b>a$
17.3	17.3	$(2a)^{-\nu}\pi^{-\tfrac{1}{2}}\Gamma(\tfrac{1}{2}+\nu)(b^{2}-a^{2})^{-\nu-\tfrac{1}{2}}$	$b>a,\nu>-\tfrac{1}{2}$
17.4	17.4	$\tfrac{1}{2}\pi/a$	

$g_c(y)$	$g_s(y)$	$N=g_c(0)$	Restrictions
17.6	17.6	$\frac{1}{2}(2a)^{-\frac{1}{2}}\Gamma^2(\frac{1}{4})$	
17.7	17.7	$(2a)^{-\frac{1}{2}}a^{-1}\Gamma^2(3/4)$	
17.8	17.10	$\frac{1}{2}\pi a^{-1}(2a)^{-2n}\{(2n)!/n!\}^2$	
17.9	17.11	$2^{2n}a^{-2n-2}(n!)^2$	
17.12		$(a+b)^{-1}K\{2(ab)^{\frac{1}{2}}(a+b)^{-1}\}$	$a>b$
17.13		$\pi(a+b)^{-1}K\{(a-b)/(a+b)\}$	
17.14	17.13	$\frac{1}{2}\pi a^{-1}\sec(\frac{1}{2}\pi\nu)$	$-1<\nu<1$
17.15		$\frac{1}{2}\pi^{\frac{1}{2}}2^{\nu}a^{-\nu-1}\Gamma(\frac{1}{2}+\nu)$	$\nu>-\frac{1}{2}$
17.16	17.15	$2^{\mu-1}a^{-\mu-1}\Gamma(\frac{1}{2}+\frac{1}{2}\mu+\frac{1}{2}\nu)\,\Gamma(\frac{1}{2}+\frac{1}{2}\mu-\frac{1}{2}\nu)$	$\mu+\nu>-1$ $\mu-\nu>0$
17.20		$\frac{1}{4}a^{2\nu-1}\Gamma(\nu)\Gamma(\frac{1}{2}-\nu)\{\Gamma(\frac{1}{2}+2\nu)\}^{-1}$	$0<\nu<\frac{1}{2}$
17.22		$\frac{1}{2}(ab)^{-\frac{1}{2}}q_{\nu-\frac{1}{2}}\{(a^2+b^2)/2ab\}$	$-\frac{1}{2}<\nu<\frac{1}{2}$
17.23		$\frac{1}{4}\pi^2\sec(\pi\nu)(ab)^{-\frac{1}{2}}p_{\nu-\frac{1}{2}}\{(a^2+b^2)/2ab\}$	
17.24		$\frac{1}{8}(\frac{1}{2}a)^{-\lambda}\{\Gamma(\lambda)\}^{-1}\Gamma(\frac{1}{2}\lambda+\frac{1}{2}\mu+\frac{1}{2}\nu)$ $\cdot\Gamma(\frac{1}{2}\lambda+\frac{1}{2}\mu-\frac{1}{2}\nu)\Gamma(\frac{1}{2}\lambda-\frac{1}{2}\mu+\frac{1}{2}\nu)\Gamma(\frac{1}{2}\lambda-\frac{1}{2}\mu-\frac{1}{2}\nu)$	$\lambda>|\mu|+|\nu|$
17.25		$2^{\nu-\mu-1}b^{\mu}a^{-\nu-1}\pi^{\frac{1}{2}}\Gamma(\frac{1}{2}+\nu)\{\Gamma(1+\mu)\}^{-1}$ $\cdot{}_2F_1(\frac{1}{2}+\nu,\frac{1}{2};1+\mu;b^2/a^2)$	$\mu>-\frac{1}{2},b<a$
17.27		$\frac{1}{4}\pi^2 b^{-\nu-1}\sec(\pi\nu)\{\mathbf{H}_{\nu}(ab)-Y_{\nu}(ab)\}$	$\nu>\frac{1}{2}$
17.29		$(\frac{1}{2}\pi)^{\frac{1}{2}}(2ab)^{-\frac{1}{4}}\{\Gamma(3/4)\}^{-1}$	$a>b$
17.32		$\frac{1}{4}(2a)^{-\frac{1}{2}}\Gamma^2(\frac{1}{4})$	
17.34		$\frac{1}{2}\pi(\frac{1}{2}\pi/a)^{\frac{1}{2}}$	

$g_c(y)$	$g_s(y)$	$N = g_c(0)$	Restrictions
17.35	17.34	$2^{-7/4}(a/\pi)^{-3/2}\{\Gamma(3/4)\}^{-1}$	
17.36		$\pi^{3/2}(2a)^{-5/4}\{\Gamma(\tfrac{1}{4})\}^{-1}$	
17.37		$2^{\nu-3/2}\Gamma(\nu)\{\Gamma(\tfrac{1}{2}+2\nu)\}^{-1}$	$\nu>0$
17.38	17.32	$\pi^{-1}2^{\nu-3/2}\Gamma(-\nu)\Gamma(\tfrac{1}{2}+2\nu)$	$-\tfrac{1}{4}<\nu<0$
17.40		$\tfrac{1}{2}\pi a^{-\nu-\frac{1}{2}}\Gamma(\tfrac{1}{2}+2\nu)\{\Gamma(1+\nu)\}^{-1}$	$\nu>-\tfrac{1}{4}$
17.41		$\tfrac{1}{4}2^{\nu}\pi a^{-\nu-\frac{1}{2}}\Gamma(\tfrac{1}{4}+\nu)\{\Gamma(3/4)\}^{-1}$	$\nu>-\tfrac{1}{4}$
17.42		$2^{\nu-3/2}a^{2\nu-1}\cos(\pi\nu)\Gamma(\nu)\Gamma(\tfrac{1}{2}-2\nu)$	$0<\nu<\tfrac{1}{4}$
17.43		$\tfrac{1}{2}\pi^{\frac{1}{2}}(2a)^{-\lambda}\Gamma(\lambda+\nu)\Gamma(\lambda-\nu)\{\Gamma(\tfrac{1}{2}+\lambda)\}^{-1}$	$-\nu<\lambda<\nu$
17.44		$\pi^{\frac{1}{2}}2^{-9/4}a^{-3/4}\Gamma(1/8/\Gamma(7/8)$	
17.45		$2^{-5/2}a^{-1}\pi^2$	
17.48	17.43	a^{-2}	
18.1	18.1	b^{-2}	
18.2	18.2	$2a^{-2}$	
18.10	18.5	$\pi a^{-1}\sec(\pi\nu)$	$-\tfrac{1}{2}<\nu<\tfrac{1}{2}$
18.21		$\tfrac{1}{2}\pi b^{-1}e^{-ab}$	
18.22		$\tfrac{1}{2}\pi(ab)^{-1}e^{-ab}$	
18.23		$-\tfrac{1}{2}\pi a^{-1}\text{Ei}(-ab)$	
18.24		$-\tfrac{1}{2}\pi b^{-1}\{e^{ab}\text{Ei}(-2ab)-e^{-ab}\log a\}$	$a>1$
18.25	18.26	$b^{-1}\{\sin(ab)\,\text{Ci}(ab)-\cos(ab)\,\text{si}(ab)\}$	

$g_c(y)$	$g_s(y)$	$N=g_c(0)$	Restrictions
18.26		$\pi^{\frac{1}{2}}b^{-1}n \sum\limits_{k=0}^{n} (n+k-1!\{k!(n-k)!\}^{-1}$ $\cdot(\tfrac{1}{2}ab)^{\frac{1}{2}-k}K_{k-\frac{1}{2}}(ab)$	
18.27		$\tfrac{1}{2}K_\nu^2(\tfrac{1}{2}ab)$	
18.28		$\tfrac{1}{8}b\nu^{-1}\{K_{\nu+\frac{1}{2}}^2(\tfrac{1}{2}ab)-K_{\nu-\frac{1}{2}}^2(\tfrac{1}{2}ab)\}$	
18.29		$\tfrac{1}{2}(2\pi a/b)^{\frac{1}{2}}a^{-\nu}K_{\nu-\frac{1}{2}}(ab)$	
18.30	18.27	$\tfrac{1}{8}\pi^2\sec(\pi\nu)\{J_\nu^2(\tfrac{1}{2}ab)+Y_\nu^2(\tfrac{1}{2}ab)\}$	$-\tfrac{1}{2}<\nu<\tfrac{1}{2}$
18.31		$b^{-1}\sinh(ab)$	
18.33		$\tfrac{1}{2}\pi I_\nu^2(\tfrac{1}{2}ab)$	$\nu>-\tfrac{1}{2}$
18.34		$\tfrac{1}{4}\pi\sec(\pi\nu)K_\nu(\tfrac{1}{2}ab)\{I_\nu(\tfrac{1}{2}ab)+I_{-\nu}(\tfrac{1}{2}ab)\}$	$-\tfrac{1}{2}<\nu<\tfrac{1}{2}$
18.35		$(\tfrac{1}{2}\pi a/b)^{\frac{1}{2}}a^\nu I_{\nu+\frac{1}{2}}(ab)$	
18.36	18.28	$2b^{-1}\sinh(ab)$	$\nu>-1$
18.37	18.29	$\pi I_{\frac{1}{2}\nu}^2(\tfrac{1}{2}ab)$	$\nu>-1$
18.38	18.30	$(2\pi a/b)^{\frac{1}{2}}a^\nu I_{\nu+\frac{1}{2}}(ab)$	$\nu>-1$
18.39	18.32	$\tfrac{1}{2}\pi\sec(\pi\nu)K_\nu(\tfrac{1}{2}ab)\{I_\nu(\tfrac{1}{2}ab)+I_{-\nu}(\tfrac{1}{2}ab)\}$	$-\tfrac{1}{2}<\pi<\tfrac{1}{2}$
18.40	18.33	$b^{-1}\{\sin(ab)\,\text{Ci}(ab)-\cos(av)\,\text{si}(ab)\}$	
18.41	18.34	$\tfrac{1}{8}\pi^2\sec(\pi\nu)\{J_\nu^2(\tfrac{1}{2}ab)+Y_\nu^2(\tfrac{1}{2}ab)\}$	$-\tfrac{1}{2}<\nu<\tfrac{1}{2}$
18.42		$(\tfrac{1}{2}\pi a/b)^{\frac{1}{2}}a^\nu Y_{\nu+\frac{1}{2}}(ab)$	$ab<\lambda_{\nu,1}, \nu>-\tfrac{1}{2}$
18.45		$\tfrac{1}{2}\pi^{\frac{1}{2}}b^{-1}\{H_0(ab)-Y_0(ab)\}$	

$g_c(y)$	$g_s(y)$	$N=g_c(0)$	Restrictions
18.46		$\pi b^{-1}S_{0,2\nu}(ab)$	
18.47		$\pi b^{-1}K_{2\nu}(ab)$	
19.1	19.8	$\pi I_o(b)K_o(a)$	$a>b$
19.2		$\pi I_\nu^2(a)$	$\nu>-\tfrac{1}{2}$
19.3	19.1	$\pi I_\nu^2(a)$	$\nu>-\tfrac{1}{2}$
19.4		$\tfrac{1}{2}\pi^2\sec(\pi\nu)K_\nu(a)\{I_{-\nu}(a)+I_\nu(a)\}$	$-\tfrac{1}{2}<\pi<\tfrac{1}{2}$
19.5	19.4	$\tfrac{1}{2}\pi\sec(\pi\nu)K_\nu(a)\{I_\nu(a)+I_{-\nu}(a)\}$	$-\tfrac{1}{2}<\pi<\tfrac{1}{2}$
19.6		$\tfrac{1}{2}\pi a\nu^{-1}\{I_{\nu-\frac{1}{2}}^2(a)-I_{\nu+\frac{1}{2}}^2(a)\}$	$\nu>0$
19.7	19.2	same as before	
19.8		$\pi a^m m!\displaystyle\sum_{n=o}^{m}(-1)^n\varepsilon_n\{(m-n)!(m+n)!\}^{-1}\cdot I_n^2(a)$	
19.9		same as before	
19.10		$\pi(2a)^{-\frac{1}{2}}\exp(-a^2)\{D_{-3/4}(2a)\}^2$	
19.11		$\tfrac{1}{4}\pi^2\{J_o^2(a)+Y_o^2(a)\}$	
19.12		$K_o\{a(i)^{\frac{1}{2}}\}K_o\{a(-i)^{\frac{1}{2}}\}$	
19.14		$K_o(a)K_o(b)$	
19.15		$\tfrac{1}{2}K_o\left[\tfrac{1}{2}\{b+(b^2-a^2)^{\frac{1}{2}}\}\right]K_o\left[\tfrac{1}{2}\{b-(b^2-a^2)^{\frac{1}{2}}\}\right]$	
19.16		$\tfrac{1}{2}K_o\left[\tfrac{1}{2}\{(a^2+b^2)^{\frac{1}{2}}+b\}\right]K_o\left[\tfrac{1}{2}\{(a^2+b^2)^{\frac{1}{2}}-b\}\right]$	

$g_c(y)$	$g_s(y)$	$N = g_c(0)$	Restrictions
19.17		$-\pi K_o(a) Y_o(a)$	
19.18		$\frac{1}{2}\pi (2a)^{-3/4} W_{-\frac{1}{4},o}(2a)$	
19.19		$(2a)^{-3/4}\Gamma^2(\frac{1}{4}) W_{\frac{1}{4},o}(2a)$	
19.20		$K_\nu^2(a)$	
19.21		$\frac{1}{4}\pi^2 \sec(\pi\nu)\{J_\nu^2(a) + Y_\nu^2(a)$	$-\frac{1}{2} < \pi < \frac{1}{2}$
19.22		$-\frac{1}{2}\pi\nu^{-1}\{K_{\frac{1}{2}+\frac{1}{2}\nu}^2(a) - K_{\frac{1}{2}+\frac{1}{2}\nu}^2(a)\}$	
19.23		$(-1)^m \frac{1}{2} a^m m! \sum_{n=1}^{m} (-1)^n \varepsilon_n \{(m+n)!(m-n)!\}^{-1}$ $\cdot K_n^2(a)$	
19.24		$2K_{\frac{1}{2}\nu}^2(a) K_{\frac{1}{2}\nu}^2(b)$	
20.19		$\frac{1}{2}\pi K_o(2a)$	
21.20	21.10	$-2^\nu \pi^{\frac{1}{2}} \cosec(\pi\nu) a^{-\nu-1} \{\Gamma(\frac{1}{2}-\nu)\}^{-1}$	$-\frac{1}{2} < \nu < 0$
21.26		$\frac{1}{2}(2a)^{-\frac{1}{2}}\Gamma(\frac{1}{4})\{\Gamma(3/4)\}^{-1}$	
21.29		$\frac{1}{2}\pi^{-1}(2a)^{-\frac{1}{2}}\{\Gamma(\frac{1}{4})\}^2$	
21.35		$\frac{1}{2}\pi\{J_o^2(\frac{1}{2}ab) + Y_o^2(\frac{1}{2}ab)\}$	
21.36		$J_o(\frac{1}{2}ab) K_o(\frac{1}{2}ab)$	
21.38		$\frac{1}{2}\pi\{J_o^2(a) + Y_o^2(a)\}$	
21.39		$2I_o(a) K_o(a)$	
21.45		$\frac{1}{8}\pi^2\{J_{\frac{1}{2}\nu}^2(\frac{1}{2}ab) + Y_{\frac{1}{2}\nu}^2(\frac{1}{2}ab)\}$	

$g_c(y)$	$g_s(y)$	$N = g_c(0)$	Restrictions
21.47		$\frac{1}{2}(2a)^{\mu+1}\{\Gamma(\frac{1}{2}-\frac{1}{2}\mu-\frac{1}{2}\nu)\,\Gamma(\frac{1}{2}-\frac{1}{2}\mu+\frac{1}{2}\nu)\}^{-1}$ $\cdot \int_0^\infty (a^2+t^2)^{-1} K_{\frac{1}{2}\nu}^2(\frac{1}{2}t)\,dt$	$\mu\pm\nu<1$
21.48		$2^{-\mu-1}(2a)^{-\frac{1}{2}}\Gamma^2(\frac{1}{4}-\frac{1}{2}\mu)\{\Gamma(\frac{1}{2}-\mu)\}^{-1}$ $\cdot S_{\mu+\frac{1}{2},0}(a)$	$\mu<\frac{1}{2}$
21.46		$\frac{1}{2}\pi^{\frac{1}{2}}a^{-3/2}\Gamma(2+\mu)\{\Gamma(\frac{1}{2}-\mu)\}^{-1}$	$\mu<0$
21.43		$\frac{1}{4}\pi a^{\mu}\Gamma(-\frac{1}{2}\mu-\frac{1}{2}\nu)\,\Gamma(-\frac{1}{2}\mu+\frac{1}{2}\nu)$ $\cdot\{\Gamma(\frac{1}{2}-\frac{1}{2}\mu+\frac{1}{2}\nu)\,\Gamma(\frac{1}{2}-\frac{1}{2}\mu-\frac{1}{2}\nu)\}^{-1}$	$\mu\pm<1$
22.1		$(\frac{1}{2}\pi)^{\frac{1}{2}}$	
22.3	22.2	$\pi^{\frac{1}{2}}2^{\frac{1}{2}\nu-\frac{1}{2}}a^{-1}\Gamma(-\frac{1}{2}-\frac{1}{2}\nu)\,\Gamma(-\frac{1}{2}\nu)\{\Gamma(\frac{1}{2}-\frac{1}{2}\nu)\}^{-1}$	$\nu<-1$
22.4		$\pi^{\frac{1}{2}}2^{\frac{1}{2}\nu-\frac{1}{2}}a^{-1}\{\Gamma(1-\frac{1}{2}\nu)\}^{-1}$	$\nu<1$
22.5	22.3	$2^{\frac{1}{2}(\nu-\mu-1)}\pi^{\frac{1}{2}}\Gamma(1+\mu)\{\Gamma(1+\frac{1}{2}\mu-\frac{1}{2}\nu)\}^{-1}$	$\mu>-1,\nu<1$
22.11		$-(\frac{1}{2}\pi)^{\frac{1}{2}}\nu^{-1}a^{2\nu}$	$\nu<0$
22.12		$\pi^{\frac{1}{2}}2^{\frac{1}{2}\nu}\exp(-a^2)\{D_{\frac{1}{2}\nu-\frac{1}{2}}(2a)\}^2$	$\nu<1$
22.13		$\pi^{\frac{1}{2}}2^{\frac{1}{2}\nu-3/2}a^{-1}\Gamma(-\frac{1}{2}\nu)\{\Gamma(\frac{1}{2}-\frac{1}{2}\nu)\}^{-1}$ $\cdot W_{\frac{1}{2}\nu+\frac{1}{2},0}(2a^2)$	$\nu<0$
22.14		$\pi^{\frac{1}{2}}2^{\frac{1}{2}\nu-3/2}a^{-1}W_{\frac{1}{2}\nu,0}(2a^2)$	$\nu<1$
22.15		$2^{\frac{1}{2}\nu}(2\pi a^2)^{-\frac{1}{2}}\cos(\frac{1}{2}\pi\nu)\,\Gamma^2(\frac{1}{2}+\frac{1}{2}\nu)\exp(a^2)$ $W_{-\frac{1}{2}\nu,0}(2a^2)$	$-1<\nu<1$

$g_c(y)$	$g_s(y$	$g_c(0)$
23.4	23.1	$\frac{1}{2}\pi^{3/2}\{\Gamma(5/4)\}^{-2}$
23.5		$\frac{1}{4}\pi^2 a$
23.11		$\frac{1}{4}\pi^2$
23.12		$\frac{1}{4}\pi\Gamma^2(\frac{1}{4})\{\Gamma(3/4)\}^{-2}$
23.13		$a^{-1}(z+1)^{-1}K^2\{(\frac{1}{2}+\frac{1}{2}z)^{-\frac{1}{2}}\}$, $z=(1+a^{-2})^{\frac{1}{2}}$
23.14		$\frac{1}{8}\pi\Gamma^2(\frac{1}{4})\{\Gamma(3/4)\}^{-2}$
23.15		$\frac{1}{16}(\pi a)^{-1}\Gamma^4(\frac{1}{4})$
23.16		$\frac{1}{4}(2\pi a)^{-\frac{1}{2}}\Gamma^2(\frac{1}{4})\left[K\{(\frac{1}{2}+\frac{1}{2}u)^{\frac{1}{2}}\}+K\{(\frac{1}{2}-\frac{1}{2}u)^{\frac{1}{2}}\}\right]$, $u=(1-a^{-2})^{\frac{1}{2}}$
23.17		$\frac{1}{2}(2\pi a)^{-\frac{1}{2}}\Gamma^2(\frac{1}{4})K\{(\frac{1}{2}-\frac{1}{2}u)^{\frac{1}{2}}\}$, $u=(1-a^{-2})^{\frac{1}{2}}$
23.18		$a^{-1}\{(\frac{1}{2}-\frac{1}{2}u)^{\frac{1}{2}}\}\{(\frac{1}{2}+\frac{1}{2}u)^{\frac{1}{2}}\}$, $u=(1-a^{-2})^{\frac{1}{2}}$
23.19		$2a^{-1}(1+u)^{-1}K\{(\frac{u-1}{u+1})^{\frac{1}{2}}\}K\{(\frac{1}{2}+\frac{1}{2}u)^{-\frac{1}{2}}\}$, $u=(1+a^{-2})^{\frac{1}{2}}$
23.20		$K\{(\frac{1}{2}+\frac{1}{2}u)^{\frac{1}{2}}\}K\{(\frac{1}{2}-\frac{1}{2}u)^{\frac{1}{2}}\}$, $u=(1-a^2)^{\frac{1}{2}}$
23.21		$(1+u)^{-1}\left[K\{(\frac{1}{2}+\frac{1}{2}u)^{-\frac{1}{2}}\}\right]^2$, $u=(1+a^{-2})^{\frac{1}{2}}$
23.22		$\left[K\{(\frac{1}{2}-\frac{1}{2}u)^{\frac{1}{2}}\}\right]^2$, $u=(1-a^2)^{\frac{1}{2}}$
23.23		$K\{(\frac{1}{2}-\frac{1}{2}u)^{\frac{1}{2}}\}K\{(\frac{1}{2}+\frac{1}{2}u)^{\frac{1}{2}}\}$, $u=(1-a^2)^{\frac{1}{2}}$

$g_e(y)$	$N=g_e(0)$	Restrictions
1	$A(b-a)$	
2	$(n+1)^{-1}a^{n+1}$	$n=1,2,3,\ldots$
3	$(\nu+1)^{-1}a^{\nu+1}$	$\nu>-1$

IV.   Fourier Transforms of Distributions (Tables V)

$g_e(y)$	$N = g_e(0)$	Restrictions
4	$-(\nu+1)^{-1}a^{\nu+1}$	$\nu < -1$
5	$(\nu+1)^{-1}a^{\nu+1}$	$\nu > -1$
6	$-(\nu+1)^{-1}a^{\nu+1}$	$\nu < -1$
7	$-\pi a^{\nu}\operatorname{cosec}(\pi\nu)$	$-1 < \nu < 0$
8	$a^{1+\mu+\nu}B(1+\mu,-1-\nu-\mu)$	$-1 < \mu < -1-\nu$
9	$-\pi a^{\nu}\operatorname{cosec}(\pi\nu)$	$-1 < \nu < 0$
10	$(b-a)^{\nu+\mu+1}B(1+\nu,1+\mu)$	$(\nu,\mu) > -1$
11	$(a+b)^{\mu+\nu+1}B(1+\nu,-1-\mu-\nu)$	$-1 < \nu < -1-\mu$
12	$(2a/\pi)^{-\frac{1}{2}}\Gamma(\nu)\{\Gamma(\tfrac{1}{2}+\nu)\}^{-1}$	$\nu > 0$
13	$(\pi/a^{-\frac{1}{2}}\Gamma(\nu)\{\Gamma(\tfrac{1}{2}+\nu)\}^{-1}$	$\nu > 0$
14	$c^{-1}(e^{-ac}-e^{-bc})$	$a < 0$
15	$a^{-\nu}\Gamma(\nu)$	$\nu > 0$
16	$\log(b/a)$	$b > a$
17	$a^{-1}B(1+\nu,\lambda/a)$	$\nu > -1, \lambda > 0$
18	$a^{-2}B(1+\nu,\lambda/a)\{\psi(1+\nu+\lambda/a)-\psi(\lambda/a)$	$\nu > -1, \lambda > 0$
19	$(-a)^{-n-1}\psi^{(n)}(\lambda/a)$	$n = 1,2,\ldots,\lambda > 0$
20	$a^{-\nu}\Gamma(\nu)\zeta(\nu,1+\lambda/a)$	$\nu > -1, \lambda > -a$
21	$\tfrac{1}{2}a^{-1}\{\psi(\tfrac{1}{2}+\tfrac{1}{2}\lambda/a)-\psi(\tfrac{1}{2}\lambda/a)\}$	$\lambda > 0$
22	$(2a)^{-\nu}\{\zeta(\nu,\tfrac{1}{2}\lambda/a)-\zeta(\nu,\tfrac{1}{2}+\tfrac{1}{2}\lambda/a)\}$	$\nu > 0,\ \lambda > 0$

$g_e(y)$	$N=g_e(0)$	Restrictions
23	$2a^{\frac{1}{2}\nu}\lambda^{-\frac{1}{2}\nu}K_\nu\{2(a\lambda)^{\frac{1}{2}}\}$	$\nu>0,\lambda>0$
24	$2\Gamma(2\nu)a^{-2\nu}$	$\nu>0$
25	$\frac{1}{2}(\pi/a)^{\frac{1}{2}}\mathrm{Erfc}(ba^{\frac{1}{2}})$	
26	$\frac{1}{2}a^{-\frac{1}{2}\nu}\Gamma(\frac{1}{2}\nu)$	$\nu>0$
27	$\frac{1}{3}a^{-1}\Gamma(1/3)$	
28	$p^{-1}a^{-(1+\nu)/p}\Gamma\{(1+\nu)/p\}$	$\nu>0$
29	$a^{-\lambda}\gamma(\lambda,a)$	$\lambda>0$
30	$a^{\lambda}\Gamma(-\lambda,a)$	
31	$\frac{1}{2}\pi\{\mathbf{H}_0(a)-\mathbf{Y}_0(a)\}$	
32	$\lambda^{-1}\{\log a-e^{a\lambda/b}\}\mathrm{Ei}(-a\lambda/b)$	$\lambda>0$, $a>1$
33	$\lambda^{-1}\{\log 2-\frac{1}{2}\psi(1+\frac{1}{2}\lambda/a)+\frac{1}{2}\psi(\frac{1}{2}+\frac{1}{2}\lambda/a)\}$	$\lambda>0$
34	$\lambda^{-1}\{\gamma+\psi(1+\lambda/a)\}$	$\lambda>-a$
35	$\frac{1}{2}a^{-1}\{\psi(\frac{3}{4}+\frac{1}{4}\lambda/a)-\psi(\frac{1}{4}+\frac{1}{4}\lambda/a\}$	$\lambda>-a$
36	$2(2a)^{-\nu}\Gamma(\nu)\zeta(\nu,\frac{1}{2}+\frac{1}{2}\lambda/a)$	$\nu>1,\lambda>-a$
37	$2^{1-2\lambda}a^{-\nu}\Gamma(\nu)\{\zeta(\nu,\frac{1}{4}+\frac{1}{2}\lambda/a)-\zeta(\nu,\frac{1}{4}\lambda/a)\}$	$\nu>0,\lambda>-a$
38	$(2a)^{-1}\{\psi(\frac{1}{2}+\frac{1}{2}\lambda/a)-\psi(\frac{1}{2}\lambda/a)\}$	$\lambda>0$
39	$\pi\mathrm{cosec}(\pi\lambda)\{\mathbf{J}_\lambda(a)-J_\lambda(a)\}$	$\lambda>0$
40	$\frac{1}{4}\pi^{\frac{1}{2}}a^{-3/2}$	
41	$(2a)^{\nu+\mu+1}B(1+\nu,1+\mu)$	$(\nu,\mu)>-1$

$g_e(y)$	$N=g_e(0)$	Restrictions
56	$\pi(b/a)^{\lambda}b^{-1}\operatorname{cosec}(\pi\lambda)$	$0<\lambda<1$
59	$(b/a)^{\lambda}b^{-\nu}B(\lambda,\nu-\mu)$	$0<\lambda<\nu$
60	$c^{\nu}a^{-\lambda}d^{\mu+\lambda}B(\lambda,-\mu-\nu-\lambda)\,_2F_1(-\nu,\lambda;-\mu-\nu;1-\dfrac{bd}{ac}$	$0<\lambda<-\nu-\mu$
61	$b^{-1}a^{-\lambda/b}\Gamma(\lambda/b)$	$\lambda>0$
62	$b^{-1}a^{\lambda/b}\Gamma(-\lambda/b)$	$\lambda>0$
64	$a^{-\lambda}\Gamma(\lambda)(1-2^{1-\lambda})\zeta(\lambda)$	$\lambda>0$
65	$2p^{-1}(b/a)^{\frac{1}{2}\lambda/p}K_{\lambda/p}\{2(ab)^{\frac{1}{2}}\}$	$-p<\lambda<p$
66	$2\{(a+b)/(a-b)\}^{\frac{1}{2}\lambda}K_{\lambda}\{(a^2-b^2)^{\frac{1}{2}}\}$	$a>b$
67	$\pi\lambda^{-1}e^{-\lambda/b}\operatorname{cosec}(\pi\lambda/b)$	$-b<\lambda<p$
68	$d^{\lambda-\nu}B(\lambda,\nu-\lambda)\{\psi(\nu)-\psi(\nu-\lambda)+\log a\}$	$0<\lambda<\nu,a>1$
70	$2^{\nu-1}a^{-1}\{\Gamma(\nu)\}^{-1}\Gamma(\frac{1}{2}\nu+\frac{1}{2}\lambda/a)\Gamma(\frac{1}{2}\nu-\frac{1}{2}\lambda/a)\cdot e^{-b\lambda/a}$	$-a\nu<\lambda<a\nu$ $\nu>0$
71	$a^{-\lambda}2^{1-2\lambda}\Gamma(\nu)\{\zeta(\lambda,\frac{1}{4})-\zeta(\lambda,3/4)\}$	$\lambda>0$
72	$2\alpha^{-\lambda}(1-2^{-\lambda})\Gamma(\lambda)\zeta(\lambda)$	$\lambda>0$
73	$2(2a)^{-\lambda}(2^{1-\lambda}-1)\Gamma(\lambda)\zeta(\lambda)$	$\lambda>1$
74	$2(b/a)^{-\lambda}\sec(\pi\nu)K_{\lambda}(2ab)$	$-1<\lambda<0$
83	$K_{\nu-\frac{1}{2}\lambda}(b)K_{\nu+\frac{1}{2}\lambda}(a)$	
91	$2^{\frac{1}{2}\nu}(2\pi a^2)^{-\frac{1}{2}}\Gamma^2(\frac{1}{2}+\frac{1}{2}\nu)\cos(\frac{1}{2}\pi\nu)W_{-\frac{1}{2}\nu,0}(2a^2)$	$0>\nu>-1$

APPENDIX

List of Notations and Definitions

Abbreviations: ε_n = Neumann's number; $\varepsilon_0=1$, $\varepsilon_n=2$, $n=1,2,3,\ldots$

Pochhammer's symbol; $(\alpha)_n=\alpha(\alpha+1)\ldots(\alpha+n-1)=\Gamma(\alpha+n)/\Gamma(\alpha)$

Binomial coefficient; $\binom{\alpha}{n}=\alpha(\alpha-1)\ldots(\alpha-n+1)/n!$

$$= \frac{\Gamma(1+\alpha)}{n!\,\Gamma(1+\alpha-n)} = (-1)^n \frac{\Gamma(n-a)}{n!\,\Gamma(-\alpha)}$$

$\gamma=0.57721\ldots$ = Euler's constant.

1. Elementary Functions

Trigonometric and inverse trigonometric functions:

$\sin x$, $\cos x$, $\tan x=\sin x/\cos x$, $\cot x=\cos x/\sin x$

$\sec x=1/\cos x$, $\operatorname{cosec} x=1/\sin x$, $\arcsin x$, $\arccos x$

Hyperbolic and inverse hyperbolic functions:

$\sinh x = \tfrac{1}{2}(e^x-e^{-x})$, $\cosh x=\tfrac{1}{2}(e^x+e^{-x})$, $\tanh x=\dfrac{\sinh x}{\cosh x}$, $\coth x=\dfrac{\cosh x}{\sinh x}$

$\operatorname{sech} x = 1/\cosh x \quad \operatorname{cosech} x = 1/\sinh x$

$\sinh^{-1}x= \tfrac{1}{2}\log\left[x+(x^2+1)^{\frac{1}{2}}\right]$, $\cosh^{-1}x=\tfrac{1}{2}\log\left[x+(x^2-1)^{\frac{1}{2}}\right]$

$\tanh^{-1}x = \tfrac{1}{2}\log\left[(1+x)/(1-x)\right]$, $\coth^{-1}x=\tfrac{1}{2}\log\left[(x+1)/(x-1)\right]$

2. The Gamma Function and Related Functions

$$\Gamma(z) = \int_0^\infty t^{z-1}e^{-t}dt \, , \quad \operatorname{Re} z>0$$

ψ-function

$$\psi(z) = \frac{d}{dz}\log\Gamma(z) = \frac{\Gamma'(z)}{\Gamma(z)}$$

$\Gamma(1+n)=n!$, $\Gamma(\tfrac{1}{2})=\pi^{\frac{1}{2}}$, $\Gamma(-\tfrac{1}{2}) = -2\pi^{\frac{1}{2}}$

$\Gamma(\tfrac{1}{2}+n)=\pi^{\frac{1}{2}}2^{-2n}(2n)!/n!$, $\Gamma(\tfrac{1}{2}-n)=(-1)^n\pi^{\frac{1}{2}}2^{2n}n!/(2n)!$, $n=1,2,3,\ldots$

$\psi(1)=-\gamma$, $\psi(\tfrac{1}{2})=-\gamma-\log 4$, $\psi(-\tfrac{1}{2})=2-\gamma-\log 4$, $\psi(1+n) = \sum\limits_{k=0}^{n-1} k^{-1}-\gamma$

$\psi(\tfrac{1}{2}+n)=\psi(\tfrac{1}{2}-n)=-\gamma-\log 4+2\sum\limits_{k=0}^{n-1}(2k+1)^{-1}$, $n=1,2,3,\ldots$

$\Gamma(2z)=\pi^{-\frac{1}{2}}2^{2z-1}\Gamma(z)\Gamma(\tfrac{1}{2}+z)$, $\psi(2z)=\log 2+\tfrac{1}{2}\psi(z)+\tfrac{1}{2}\psi(\tfrac{1}{2}+z)$

The Beta Function: $B(x,y)=\Gamma(x)\Gamma(y)/\Gamma(x+y)$

3. Orthogonal Polynomials

Legendre Polynomials $P_n(x)$

$$P_n(x)=2^{-n}(n!)^{-1}\frac{d^n}{dx^n}(x^2-1)^n = {}_2F_1(-n,n+1;1;\tfrac{1}{2}-\tfrac{1}{2}x)$$

Gegenbauer Polynomials $C_n^\nu(x)$

$$C_n^\nu(x)=\{n!\,\Gamma(2\nu)\}^{-1}\Gamma(2\nu+n)\,{}_2F_1(-n,2\nu+n;\tfrac{1}{2}+\nu;\tfrac{1}{2}-\tfrac{1}{2}x)$$

$$= (-2)^{-n}(1-x^2)^{\frac{1}{2}-\nu}(2\nu)_n\{n!\,(\tfrac{1}{2}+\nu)_n\}^{-1}\frac{d^n}{dx^n}\left[(1-x^2)^{n+\nu-\frac{1}{2}}\right]$$

$$C_n^{\frac{1}{2}}(x) = P_n(x),\quad C_\alpha^1(x) = \sin\{(\alpha+1)x\}\sin\alpha,\quad x = \cos\alpha$$

$$\lim_{\nu\to0}\Gamma(\nu)C_\alpha^\nu(\cos x)=2\alpha^{-1}\cos(\alpha x)$$

Chebyshev Polynomials $T_n(x)$, $U_n(x)$

$$T_n(x)=\cos(n\arccos x)=\tfrac{1}{2}\left({\{x+i(1-x^2)^{\frac{1}{2}}\}}^n+{\{x-i(1-x^2)^{\frac{1}{2}}\}}^n\right)=\tfrac{1}{2}n\lim_{\nu\to0}\Gamma(\nu)C_n^\nu(x)$$

$$U_n(x)=(1-x^2)^{-\frac{1}{2}}\sin\{(n+1)\arccos x\}=C_n^1(x)=(n+1)x\,{}_2F_1(-n,n+1;\tfrac{3}{2};\tfrac{1}{2}-\tfrac{1}{2}x)$$

$$U_n(x)=-\tfrac{1}{2}i(1-x^2)^{-\frac{1}{2}}\left({\{x+i(1-x^2)^{\frac{1}{2}}\}}^{n+1}-{\{x-i(1-x^2)^{\frac{1}{2}}\}}^{n+1}\right)$$

Jacobi Polynomials $P_n^{(\alpha,\beta)}(x)$

$$P_n^{(\alpha,\beta)}(x)=\{n!\,\Gamma(1+\alpha)\}^{-1}\Gamma(1+\alpha+n)\,{}_2F_1(-n,1+n+\alpha+\beta;n+1;\tfrac{1}{2}-\tfrac{1}{2}x)$$

$$= (-1)^n2^{-n}(n!)^{-1}(1-x)^{-\alpha}(1+x)^{-\beta}\frac{d^n}{dx^n}\left[(1-x)^{\alpha+n}(1+x)^{\beta+n}\right]$$

Laguerre Polynomials $L_n^\alpha(x)$

$$L_n^\alpha(x)=(n!)^{-1}x^{-\alpha}e^x\frac{d^n}{dx^n}(x^{n+\alpha}e^{-x})=\{n!\,\Gamma(1+\alpha)\}^{-1}\Gamma(1+\alpha+n)\,{}_1F_1(-n;1+\alpha;x)$$

Hermite Polynomials $He_n(x)$, $H_n(x)$

$$He_n(x)=(-1)^n\exp(\tfrac{1}{2}x^2)\frac{d^n}{dx^n}\exp(-\tfrac{1}{2}x^2);\quad H_n(x)=2^{\frac{1}{2}n}He_n(2^{\frac{1}{2}}x)$$

$$He_{2n}(x)=(-1)^n2^{-n}(n!)^{-1}(2n)!\,{}_1F_1(-n;\tfrac{1}{2};\tfrac{1}{2}x^2)$$

$$He_{2n+1}(x)=(-1)^n2^{-n}(n!)^{-1}(2n+1)!\,x\,{}_1F_1(-n;3/2;\tfrac{1}{2}x^2)$$

4. Legendre Functions. (Definitions according to Hobson)

$$p_\nu^\mu(z)=\{\Gamma(1-\mu)\}^{-1}\left((z+1)/(z-1)\right)^{\frac{1}{2}\mu}{}_2F_1(-\nu,1+\nu;1-\mu;\tfrac{1}{2}-\tfrac{1}{2}z)$$

$$e^{-i\pi\mu}q_\nu^\mu(z)=2^{-\nu-1}\{\Gamma(\tfrac{3}{2}+\nu)\}^{-1}\pi^{\frac{1}{2}}\Gamma(1+\nu+\mu)z^{-\nu-\mu-1}$$

$$\cdot(z^2-1)^{\frac{1}{2}\mu}{}_2F_1(\tfrac{1}{2}+\tfrac{1}{2}\nu+\tfrac{1}{2}\mu,1+\tfrac{1}{2}\nu+\tfrac{1}{2}\mu;\tfrac{3}{2}+\nu;z^{-2})$$

z is a point in the complex z-plane cut along the real z-axis from $-\infty$ to +1. Also:

$$(z^2-1)^{\frac{1}{2}\mu}=(z-1)^{\frac{1}{2}\mu}(z+1)^{\frac{1}{2}\mu},\quad\text{with}\quad-\pi<\arg z<\pi,\ -\pi<\arg(z\pm1)<\pi.$$

For real x with $-1<x<+1$ we have the definitions

$$P_\nu^\mu(x)=\{\Gamma(1-\mu)\}^{-1}\left((1+x)/(1-x)\right)^{\frac{1}{2}\mu}{}_2F_1(-\nu,1+\nu;1-\mu;\tfrac{1}{2}-\tfrac{1}{2}x)$$

$$Q_\nu^\mu(x)=\tfrac{1}{2}\pi\cosec(\pi\mu)\left[\cos(\pi\mu)P_\nu^\mu(x)-\frac{\Gamma(\nu+\mu+1)}{\Gamma(\nu-\mu+1)}P_\nu^{-\mu}(x)\right];\quad\text{or also}$$

$$P_\nu^\mu(x)=e^{\frac{1}{2}i\pi\nu}p_\nu^\mu(x+i0)=e^{-\frac{1}{2}i\pi\nu}p_\nu^\mu(x-i0)$$

$$Q_\nu^\mu(x)=\tfrac{1}{2}e^{-i\pi\mu}\left[e^{-\frac{1}{2}i\pi\mu}q_\nu(x+i0)+e^{\frac{1}{2}i\pi\mu}q_\nu^\mu(x-i0)\right]$$

$$p_\nu^0(z)=p_\nu(z);\quad P_\nu^0(x)=P_\nu(x);\quad q_\nu^0(z)=q_\nu(z);\quad q_\nu^0(x)=q_\nu(x)$$

Special cases

$$p_\nu^{\frac{1}{2}}(z)=(2\pi)^{-\frac{1}{2}}(z^2-1)^{-\frac{1}{4}}\left[\{z+(z^2-1)^{\frac{1}{2}}\}^{\frac{1}{2}+\nu}+\{z+(z^2-1)^{\frac{1}{2}}\}^{-\frac{1}{2}-\nu}\right]$$

$$q_\nu^{\frac{1}{2}})z)=i(\tfrac{1}{2}\pi)^{\frac{1}{2}}(z^2-1)^{-\frac{1}{4}}\{z+(z^2-1)^{\frac{1}{2}}\}^{-\frac{1}{2}-\nu}$$

$$p_\nu^{-\frac{1}{2}}(z)=(2\pi)^{-\frac{1}{2}}(\tfrac{1}{2}+\nu)^{-1}(z^2-1)^{-\frac{1}{4}}\left[\{z+(z^2-1)^{\frac{1}{2}}\}^{\frac{1}{2}+\nu}-\{z+(z^2-1)^{\frac{1}{2}}\}^{-\frac{1}{2}-\nu}\right]$$

$$q_\nu^{-\frac{1}{2}}(z)=-i(\tfrac{1}{2}\pi)^{\frac{1}{2}}(\tfrac{1}{2}+\nu)^{-1}(z^2-1)^{-\frac{1}{4}}\{z+(z^2-1)^{\frac{1}{2}}\}^{-\nu-\frac{1}{2}}$$

$$P_\nu^{\frac{1}{2}}(x)=(\tfrac{1}{2}\pi)^{-\frac{1}{2}}(1-x^2)^{-\frac{1}{4}}\cos\{(\nu+\tfrac{1}{2})\arccos x\}$$

$$Q_\nu^{\frac{1}{2}}(x)=-(\tfrac{1}{2}\pi)^{\frac{1}{2}}(1-x^2)^{-\frac{1}{4}}\sin\{(\nu+\tfrac{1}{2})\arccos x\}$$

$$P_\nu^{-\frac{1}{2}}(x)=(\tfrac{1}{2}\pi)^{-\frac{1}{2}}(\tfrac{1}{2}+\nu)^{-1}(1-x^2)^{-\frac{1}{4}}\sin\{(\nu+\tfrac{1}{2})\arccos x\}$$

$$Q_\nu^{-\frac{1}{2}}(x)=(\tfrac{1}{2}\pi)^{\frac{1}{2}}(\tfrac{1}{2}+\nu)^{-1}(1-x^2)^{-\frac{1}{4}}\cos\{(\nu+\tfrac{1}{2})\arccos x)$$

$$p_\nu^{\frac{1}{2}}(\cosh x)=(\tfrac{1}{2}\pi)^{-\frac{1}{2}}(\sinh x)^{-\frac{1}{2}}\cosh\{(+\tfrac{1}{2})x\}$$

$$q^{\frac{1}{2}}(\cosh x)=i(\tfrac{1}{2}\pi)^{\frac{1}{2}}\sinh x)^{-\frac{1}{2}}\exp\{-(\nu+\tfrac{1}{2})x\}\qquad\qquad x>0$$

$$p_\nu^{-\frac{1}{2}}(\cosh x)=(\tfrac{1}{2}\pi)^{-\frac{1}{2}}(\tfrac{1}{2}+\nu)^{-1}(\sinh x)^{-\frac{1}{2}}\sinh\{(\nu+\tfrac{1}{2})x\}$$

$$q_\nu^{-\frac{1}{2}}(\cosh\ x)=-i\,(\tfrac{1}{2}\pi)^{\frac{1}{2}}(\tfrac{1}{2}+\nu)^{-1}(\sinh\ x)^{-\frac{1}{2}}\exp\{-(\nu+\tfrac{1}{2})x\}$$

$$P_\nu^{\frac{1}{2}}(\cos\delta)=(-\tfrac{1}{2}\pi)^{-\frac{1}{2}}(\sin\delta)^{-\frac{1}{2}}\cos\{(\nu+\tfrac{1}{2})\delta\}$$

$$Q_\nu^{\frac{1}{2}}(\cos\delta)=-(\tfrac{1}{2}\pi)^{\frac{1}{2}}(\sin\delta)^{-\frac{1}{2}}\sin\{(\nu+\tfrac{1}{2})\delta\}\qquad\qquad 0<\delta<\pi$$

$$P_\nu^{-\frac{1}{2}}(\cos\delta)=(\tfrac{1}{2}\pi)^{-\frac{1}{2}}(\tfrac{1}{2}+\nu)^{-1}(\sin\delta)^{-\frac{1}{2}}\sin\{(\nu+\tfrac{1}{2})\delta\}$$

$$Q_\pi^{-\frac{1}{2}}(\cos\delta)=(\tfrac{1}{2}\pi)^{\frac{1}{2}}(\tfrac{1}{2}+\nu)^{-1}(\sin\delta)^{-\frac{1}{2}}\cos\{(\nu+\tfrac{1}{2})\delta\}$$

$$p_{-\frac{1}{2}}(z)=2\pi^{-1}(\tfrac{1}{2}z+\tfrac{1}{2})^{-\frac{1}{2}}K\left(\{(z-1)/(z+1)\}^{\frac{1}{2}}\right)\qquad\qquad z>1$$

$$q_{-\frac{1}{2}}(z)=(\tfrac{1}{2}z+\tfrac{1}{2})^{-\frac{1}{2}}K\{(\tfrac{1}{2}z+\tfrac{1}{2})^{-\frac{1}{2}}\}\qquad\qquad z>1$$

$$P_{-\frac{1}{2}}(x)=2\pi^{-1}K\{(\tfrac{1}{2}-\tfrac{1}{2}x)^{\frac{1}{2}}\}\qquad\qquad -1<x<1$$

$$Q_{-\frac{1}{2}}(x)=K\{(\tfrac{1}{2}+\tfrac{1}{2}x)^{\frac{1}{2}}\}\qquad\qquad -1<x<1$$

5. Bessel- and Related Functions

$\tau_{\nu,1}$ and $\rho_{\nu,1}=$ first positive root of $J_\nu(x)$ and $Y_\nu(x)$

respectively. $J_\nu(z)=(\tfrac{1}{2}z)^\nu\displaystyle\sum_{n=o}^\infty(-1)^n\{n!\,\Gamma(\nu+n+1)\}^{-1}(\tfrac{1}{2}z)^{2n}$

$$Y_\nu(z)=\cot(\pi\nu)J_\nu(z)-\mathrm{cosec}(\pi\nu)J_{-\nu}(z)$$

$$J_{-\nu}(z)=J_\nu(z)\cos(\pi\nu)-Y_\nu(z)\sin(\pi\nu)$$

$$Y_{-\nu}(z)=J_\nu(z)\sin(\pi\nu)+Y_\nu(z)\cos(\pi\nu)$$

$$H_\nu^{(1)}(z)=J_\nu(z)+iY_\nu(z);\quad H_\nu^{(2)}(z)=J_\nu(z)-iY_\nu(z)$$

Modified Bessel Functions

$$I_\nu(z)=e^{-\frac{1}{2}i\pi\nu}J_\nu(ze^{\frac{1}{2}i\pi})=e^{\frac{1}{2}i\pi\nu}J_\nu(ze^{-\frac{1}{2}i\pi})$$

$$I_\nu(z)=(\tfrac{1}{2}z)^\nu\sum_{n=o}^\infty\{n!\,\Gamma(\nu+n+1)\}^{-1}(\tfrac{1}{2}z)^{2n}$$

$$K_\nu(z)=\tfrac{1}{2}\pi\,\mathrm{cosec}(\pi\nu)\{I_{-\nu}(z)-I_\nu(z)\}$$

$$K_\nu(z)=\tfrac{1}{2}i\pi e^{\frac{1}{2}i\pi\nu}H_\nu^{(1)}(ze^{\frac{1}{2}i\pi})=-\tfrac{1}{2}i\pi e^{-\frac{1}{2}i\pi\nu}H_\nu^{(2)}(ze^{-\frac{1}{2}i\pi})$$

$$K_\nu(ze^{\frac{1}{2}i\pi})=-\tfrac{1}{2}i\pi e^{-\frac{1}{2}i\pi\nu}H_\nu^{(2)}(z);\quad K_\nu(ze^{-\frac{1}{2}i\pi})=\tfrac{1}{2}i\pi e^{\frac{1}{2}i\pi\nu}H_\nu^{(1)}(z)$$

Anger-Weber Functions

$$J_\nu(z)=\pi^{-1}\int_o^\infty\cos(z\sin t-\nu t)dt;\quad E_\nu(z)=-\pi^{-1}\int_o^\infty\sin(z\sin t-\nu t)dt$$

$$J_\nu(z) + J_{-\nu}(z) = \cot(\tfrac{1}{2}\pi\nu)\{E_\nu(z) - E_{-\nu}(z)\}$$

$$J_\nu(z) - J_{-\nu}(z) = -\tan(\tfrac{1}{2}\pi\nu)\{E_\nu(z) + E_{-\nu}(z)\}$$

$$J_{\frac{1}{2}}(z) = (\tfrac{1}{2}\pi z)^{-\frac{1}{2}}\left[\{C(z) - S(z)\}\cos z + \{C(z) + S(z)\}\sin z\right] = -E_{-\frac{1}{2}}(z)$$

$$J_{-\frac{1}{2}}(z) = (\tfrac{1}{2}\pi z)^{-\frac{1}{2}}\left[\{C(z) + S(z)\}\cos z - \{C(z) - S(z)\}\sin z\right] = E_{\frac{1}{2}}(z)$$

$$J_{\pm n}(z) = (\pm 1)^n J_n(z) ; \quad J_{-n-\frac{1}{2}}(z) = (-1)^n E_{n+\frac{1}{2}}(z)$$

$$E_{-n}(z) = (-1)^n E_n(z) ; \quad E_{-n-\frac{1}{2}}(z) = -(-1)^n J_{n+\frac{1}{2}}(z) \qquad n = 0,1,2,\ldots$$

Struve Functions

$$H_\nu(z) = (\tfrac{1}{2}z)^{\nu+1} \sum_{n=0}^{\infty} (-1)^n \{\Gamma(\tfrac{3}{2}+n)\Gamma(\tfrac{3}{n}+\nu+n)\}^{-1}(\tfrac{1}{2}z)^{2n}$$

$$L_\nu(z) = (\tfrac{1}{2}z)^{\nu+1} \sum_{n=0}^{\infty} \{\Gamma(\tfrac{3}{n}+n)\Gamma(\tfrac{3}{2}+\nu+n)\}^{-1}(\tfrac{1}{2}z)^{2n} = -ie^{-\frac{1}{2}i\pi\nu}H_\nu(ze^{\frac{1}{2}i\pi})$$

$$H_{-n-\frac{1}{2}}(z) = (-1)^n J_{n+\frac{1}{2}}(z) ; \quad L_{-n-\frac{1}{2}}(z) = I_{n+\frac{1}{2}}(z) ; \qquad n = 0,1,2,\ldots$$

Lommel Functions

$$s_{\mu,\nu}(z) = \{(1+\mu-\nu)(1+\mu+\nu)\}^{-1} z^{\mu+1} \; {}_1F_2(1; \tfrac{3}{2}+\tfrac{1}{2}\mu-\tfrac{1}{2}\nu; \tfrac{3}{2}+\tfrac{1}{2}\mu+\tfrac{1}{2}\nu; -\tfrac{1}{4}z^2)$$

$$(\mu \pm \nu = -1,-2,-3,\ldots)$$

$$S_{\mu,\nu}(z) = s_{\mu,\nu}(z) + 2^{\mu-1}\Gamma(\tfrac{1}{2}+\tfrac{1}{2}\mu-\tfrac{1}{2}\nu)\Gamma(\tfrac{1}{2}+\tfrac{1}{2}\mu+\tfrac{1}{2}\nu)$$

$$\cdot\left[\sin(\tfrac{1}{2}\pi\mu-\tfrac{1}{2}\pi\nu)J_\nu(z) - \cos(\tfrac{1}{2}\pi\mu-\tfrac{1}{2}\pi\nu)Y_\nu(z)\right]$$

$$s_{\mu,\nu}(z) = s_{\mu,-\nu}(z) ; \quad S_{\mu,\nu}(z) = S_{\mu,-\nu}(z)$$

Special Cases

$$s_{\nu,\nu}(z) = \pi^{\frac{1}{2}}2^{\nu-1}\Gamma(\tfrac{1}{2}+\nu)H_\nu(z) ; \quad S_{\nu,\nu}(z) = \pi^{\frac{1}{2}}2^{\nu-1}\Gamma(\tfrac{1}{2}+\nu)\left[H_\nu(z) - Y_\nu(z)\right]$$

$$s_{\nu+1,\nu}(z) = z^\nu - 2^\nu \Gamma(1+\nu)J_\nu(z) ; \quad S_{\nu+1,\nu}(z) = z^\nu$$

$$\lim_{\mu=\nu}\{\Gamma(\mu-\nu)\}^{-1}s_{\mu-1,\nu}(z) = 2^{\nu-1}\Gamma(\nu)J_\nu(z)$$

$$s_{0,\nu}(z) = \tfrac{1}{2}\pi\,\mathrm{cosec}(\pi\nu)\{J_\nu(z) - J_{-\nu}(z)\} = -\tfrac{1}{4}\pi\sec^2(\tfrac{1}{2}\pi\nu)\{E_\nu(z) + E_{-\nu}(z)\}$$

$$s_{-1,\nu}(z) = -\tfrac{1}{2}\pi\nu^{-1}\mathrm{cosec}(\pi\nu)\{J_\nu(z) + J_{-\nu}(z)\} = \tfrac{\pi}{4\nu}\mathrm{cosec}^2(\tfrac{\pi}{2}\nu)\{E_\nu(z) - E_{-\nu}(z)\}$$

$$S_{0,\nu}(z) = \tfrac{1}{2}\pi\,\mathrm{cosec}(\pi\nu)\left[J_\nu(z) - J_{-\nu}(z) - J_\nu(z) + J_{-\nu}(z)\right]$$

$$S_{-1,\nu}(z) = \tfrac{1}{2}\pi\nu^{-1}\mathrm{cosec}(\pi\nu)\left[J_\nu(z) + J_{-\nu}(z) - J_\nu(z) - J_{-\nu}(z)\right]$$

$$s_{1,\nu}(z) = 1 + \nu^2 s_{-1,\nu}(z) ; \quad S_{1,\nu}(z) = 1 + \nu^2 S_{-1,\nu}(z)$$

$$S_{\frac{1}{2},\frac{1}{2}}(z)=z^{-\frac{1}{2}}; \ S_{3/2,\frac{1}{2}}(z)=z^{\frac{1}{2}}; \ S_{0,-1}(z)=z^{-1}$$

$$S_{-\frac{1}{2},\frac{1}{2}}(z)-z^{-\frac{1}{2}}\{\sin zCi(z)-\cos zsi(z)\};$$

$$S_{-3/2,\frac{1}{2}}(z)=-z^{-\frac{1}{2}}\{\sin zsi(z)+\cos zCi(z)\}$$

Kelvin's Functions

$$J_\nu\{z\ \exp(\pm 3i\pi/4)\}=ber_\nu(z)\pm i\ bei_\nu(z);$$

$$K_\nu\{z\ \exp(\pm i\pi/4)\}=ker_\nu(z)\pm i\ kei_\nu(z)$$

Bessel Integral Functions

$$Ji_\nu(x) = \int_x^\infty t^{-1}J_\nu(t)dt; \ Yi_\nu(x) = \int_x^\infty t^{-1}Y_\nu(t)dt; \ Ki_\nu(x) = \int_x^\infty t^{-1}K_\nu(t)dt$$

6. Hypergeometric Series

Generalized Series:

$$_pF_q(a_1,a_2,\ldots,a_p;b_1,\ldots,b_q;z) = \sum_{n=0}^\infty \frac{(a_1)_n\cdots(a_p)_n}{(b_1)_n\cdots(b_q)_n} z^n/n!$$

$|z|<1$ if $p=q+1$, $|z|<\infty$ if $p\leq q+1$; divergent otherwise

Gauss's Series: $_2F_1(a,b;c;z)=F(a,b;c;z)=\sum_{n=0}^\infty \frac{(a_n)(b)_n}{(c)_n} z^n/n!$; $z<1$

Kummer's Series: $_1F_1(a;c;z)=e^z{}_1F_1(c-a;c;-z)=\sum_{n=0}^\infty (a)_n/(c)_n z^n/n!$; $|z|<\infty$

7. Whittaker Functions and Special Cases

$$M_{\kappa,\nu}(z)=z^{\frac{1}{2}+\nu}e^{-\frac{1}{2}z}{}_1F_1(\tfrac{1}{2}+\nu-\kappa;2\nu+1;z)=z^{\frac{1}{2}+\nu}e^{\frac{1}{2}z}{}_1F_1(\tfrac{1}{2}+\nu+\kappa;2\nu+1;-z)$$

$$W_{\kappa,\nu}(z) = \frac{\Gamma(-2\nu)}{\Gamma(\frac{1}{2}-\nu-\kappa)}M_{\kappa,\nu}(z) + \frac{\Gamma(2\nu)}{\Gamma(\frac{1}{2}+\nu-\kappa)} M_{\kappa,-\nu}(z); \ W_{\kappa,\nu}(z)=W_{\kappa,-\nu}(z)$$

Special Cases:

$$M_{0,\frac{1}{2}}(z)=2\ \sinh(\tfrac{1}{2}z), \ W_{0,\frac{1}{2}}(z)=e^{-\frac{1}{2}z}; \ W_{-\frac{1}{2},0}(z)=-z^{\frac{1}{2}}e^{\frac{1}{2}z}Ei(-z)$$

$$M_{\frac{1}{4},\frac{1}{4}}(z) = -\tfrac{1}{2}i\pi^{\frac{1}{2}}z^{\frac{1}{4}}e^{-\frac{1}{2}z}Erf(iz^{\frac{1}{2}}), \ M_{-\frac{1}{4},\frac{1}{4}}(z) = \tfrac{1}{2}\pi^{\frac{1}{2}}z^{\frac{1}{4}}e^{\frac{1}{2}z}Erf(z^{\frac{1}{2}})$$

$$W_{\frac{1}{4},\frac{1}{4}}(z) = z^{\frac{1}{4}}e^{-\frac{1}{2}z}; \ W_{-\frac{1}{4},\frac{1}{4}}(z) = \pi^{\frac{1}{2}}z^{\frac{1}{4}}e^{\frac{1}{2}z}Erfc(z^{\frac{1}{2}})$$

$$M_{\kappa+\frac{1}{2},\kappa}(z) = W_{\kappa+\frac{1}{2},\kappa}(z) = z^{\frac{1}{2}+\kappa}e^{-\frac{1}{2}z}$$

$$M_{\kappa,\kappa+\frac{1}{2}}(z)=(2\kappa+1)z^{-\kappa}e^{\frac{1}{2}z}\gamma(2\kappa+1,z)\,;\quad W_{\kappa,\kappa+\frac{1}{2}}(z) = z^{-\kappa}e^{\frac{1}{2}z}\Gamma(2\kappa+1,z)$$

$$M_{0,\nu}(z) = 2^{2\nu}\Gamma(1+\nu)z^{\frac{1}{2}}I_{\nu}(\tfrac{1}{2}z)\,;\quad W_{0,\nu}(z) = (z/\pi)^{\frac{1}{2}}K_{\nu}(\tfrac{1}{2}z)$$

$$M_{\kappa,\frac{1}{4}}(z) = -\tfrac{1}{4}\Gamma(\tfrac{1}{4}-\kappa)\pi^{-\frac{1}{2}}2^{-\kappa}(2z)^{\frac{1}{4}}\left[D_{2\kappa-\frac{1}{2}}\{(2z)^{\frac{1}{2}}\}-D_{2\kappa-\frac{1}{2}}\{-(2z)^{\frac{1}{2}}\}\right]$$

$$M_{\kappa,-\frac{1}{4}}(z) = \tfrac{1}{2}\Gamma(\tfrac{3}{4}-\kappa)\pi^{-\frac{1}{2}}2^{-\kappa}(2z)^{\frac{1}{4}}\left[D_{2\kappa-\frac{1}{2}}\{(2z)^{\frac{1}{2}}\}+D_{2\kappa-\frac{1}{2}}\{-(2z)^{\frac{1}{2}}\}\right]$$

$$W_{\kappa,\frac{1}{4}}(z) = 2^{-\kappa}(2z)^{\frac{1}{4}}D_{2\kappa-\frac{1}{2}}\{(2z)^{\frac{1}{2}}\}$$

$$D_{\nu}(z) = 2^{\frac{1}{2}\nu}\exp(-\tfrac{1}{4}z^2)\pi^{\frac{1}{2}}\left(\{\Gamma(\tfrac{1}{2}-\tfrac{1}{2}\nu)\}^{-1}\,{}_1F_1(-\tfrac{1}{2}\nu;\tfrac{1}{2},\tfrac{1}{2}z^2)\right.$$

$$\left.-2^{\frac{1}{2}}z\{\Gamma(-\tfrac{1}{2}\nu)\}^{-1}\,{}_1F_1(\tfrac{1}{2}-\tfrac{1}{2}\nu;\tfrac{3}{2};\tfrac{1}{2}z^2)\right) = 2^{\frac{1}{4}+\frac{1}{2}\nu}z^{-\frac{1}{2}}W_{\frac{1}{4}+\frac{1}{2}\nu,\frac{1}{4}}(\tfrac{1}{2}z^2)$$

Special Cases of Parabolic Cylinder Functions

$$D_0(z)=\exp(-\tfrac{1}{4}z^2),\quad D_n(z)=\exp(-\tfrac{1}{4}z^2)He_n(z)\,;\quad n=0,1,2,\ldots$$

$$D_{\frac{1}{2}}(z)=\pi^{-\frac{1}{2}}(\tfrac{1}{2}z)^{3/2}\left[K_{\frac{1}{4}}(\tfrac{1}{4}z^2)+K_{3/4}(\tfrac{1}{4}z^2)\right]\,;\quad D_{-\frac{1}{2}}(z)=(\tfrac{1}{2}z/\pi)^{\frac{1}{2}}K_{\frac{1}{4}}(\tfrac{1}{4}z^2)$$

$$D_{-1}(z)=(\tfrac{1}{2}\pi)^{\frac{1}{2}}\exp(\tfrac{1}{4}z^2)\mathrm{Erfc}(2^{-\frac{1}{2}}z)\,;\quad D_{-2}(z)=\exp(\tfrac{1}{4}z^2)\left[1-(\tfrac{1}{2}\pi)^{\frac{1}{2}}z\mathrm{Erfc}(2^{-\frac{1}{2}}z)\right]$$

$$D_{-3}(z)=\tfrac{1}{2}\exp(\tfrac{1}{4}z^2)\left[(\tfrac{1}{2}\pi)^{\frac{1}{2}}(1+z^2)\mathrm{Erfc}(2^{-\frac{1}{2}}z)-z\right]$$

$$D_{-n-1}(z)=(\tfrac{1}{2}\pi)^{\frac{1}{2}}(-1)^n(n!)^{-1}\exp(-\tfrac{1}{4}z^2)\left[\frac{d^n}{dx^n}\exp(\tfrac{1}{4}z^2)\mathrm{Erfc}(2^{-\frac{1}{2}}z)\right],n=0,1,\ldots$$

Incomplete Gamma Functions

$$\gamma(\nu,z) = \int_0^z t^{\nu-1}e^{-t}dt=\nu^{-1}z^{\nu}\,{}_1F_1(\nu;1+\nu;-z)=\nu^{-1}z^{\frac{1}{2}\nu-\frac{1}{2}}e^{-\frac{1}{2}z}M_{\frac{1}{2}\nu-\frac{1}{2},\frac{1}{2}\nu}(z)$$

$$\Gamma(\nu,z) = \int_z^{\infty} t^{\nu-1}e^{-t}dt=z^{\frac{1}{2}\nu-\frac{1}{2}}e^{-\frac{1}{2}z}W_{\frac{1}{2}\nu-\frac{1}{2},\frac{1}{2}\nu}(z)\,,\quad \Gamma(\nu,z)+\gamma(\nu,z)=\Gamma(\nu)$$

Special Cases

$$\gamma(1,z)=1-e^{-z}\,;\quad \Gamma(1,z)=e^{-z}\,;\quad \gamma(\tfrac{1}{2},z)=\pi^{\frac{1}{2}}\mathrm{Erf}(z^{\frac{1}{2}})\,,\quad \Gamma(\tfrac{1}{2},z)=\pi^{\frac{1}{2}}\mathrm{Erfc}(z^{\frac{1}{2}})$$

$$\gamma(n,z)=(n-1)!\left[1-e^{-z}\sum_{k=0}^{n-1}z^k/k!\right]\,;\Gamma(n,z)=(n-1)!e^{-z}\sum_{k=0}^{n-1}z^k/k!\,;n=1,2,\ldots$$

$$\Gamma(-n,z)=\frac{(-1)^n}{n!}\left[-\mathrm{Ei}(-z)-e^{-z}\sum_{k=0}^{n-1}(-1)^k k! k^{-k-1}\right]\,;n=1,2,\ldots\,;\Gamma(0,z)=-\mathrm{Ei}(-z)$$

$$\gamma(\tfrac{1}{2}+n,z)=2^{-2n}\frac{(2n)!}{n!}\left[\pi^{\frac{1}{2}}\mathrm{Erf}(z^{\frac{1}{2}})-z^{-\frac{1}{2}}e^{-z}\sum_{k=1}^{n}(4z)^k k!/(2k!)\right]$$

$$\Gamma(\tfrac{1}{2}+n,z)=2^{-2n}\frac{(2n)!}{n!}\left[\pi^{\frac{1}{2}}\mathrm{Erfc}(z^{\frac{1}{2}})+z^{-\frac{1}{2}}e^{-z}\sum_{k=1}^{n}(4z)^{k}k!/(2k)!\right]$$

$$\gamma(\tfrac{1}{2}-n,z)=2^{2n}\frac{n!}{(2n)!}\left((-1)^{n}\pi^{\frac{1}{2}}\mathrm{Erf}(z^{\frac{1}{2}})-2^{2-2n}z^{\frac{1}{2}-n}e^{-z}\sum_{k=0}^{n-1}(-4z)^{k}\frac{(2n-2k-2)!}{(n-k-1)!}\right)$$

$$\Gamma(\tfrac{1}{2}-n,z)=2^{2n}\frac{n!}{(2n)!}\left((-1)^{n}\pi^{\frac{1}{2}}\mathrm{Erfc}(z^{\frac{1}{2}})+2^{2-2n}z^{\frac{1}{2}-n}e^{-z}\sum_{k=0}^{n-1}(-4z)^{k}\frac{(2n-2k-2)!}{(n-k-1)}\right)$$

In the last 4 formulas $n = 1,2,3,\ldots$

Error Integrals

$$\mathrm{Erf}(z)=2\pi^{-\frac{1}{2}}\int_{0}^{z}\exp(-t^{2})dt;\quad \mathrm{Erfc}(z)=2\pi^{-\frac{1}{2}}\int_{z}^{\infty}\exp(-t^{2}dt)=1-\mathrm{Erf}(z)$$

$$\mathrm{Erf}(z)=2\pi^{-\frac{1}{2}}z\ {}_1F_1(\tfrac{1}{2};\tfrac{3}{2};-z^{2})=2(\pi z)^{-\frac{1}{2}}\exp(-\tfrac{1}{2}z^{2})M_{-\frac{1}{4},\frac{1}{4}}(z^{2})$$

$$\mathrm{Erfc}(z)=(\pi z)^{-\frac{1}{2}}\exp(-\tfrac{1}{2}z^{2})W_{-\frac{1}{4},\frac{1}{4}}(z^{2})$$

$$\mathrm{Erf}(x^{\frac{1}{2}}e^{\pm\frac{1}{4}i\pi})=C(x)+S(x)\pm i\{C(x)-S(x)\}$$

$$\mathrm{Erfc}(x^{\frac{1}{2}}e^{\pm\frac{1}{4}i\pi})=1-C(x)-S(x)\mp i\{S(x)-C(x)\}$$

Fresnel's Integrals

$$C(x)=(2\pi)^{-\frac{1}{2}}\int_{0}^{x}t^{-\frac{1}{2}}\cos t\ dt=(2x/\pi)^{\frac{1}{2}}\sum_{n=0}^{\infty}(-1)^{n}\{(4n+1)(2n)!\}^{-1}x^{2n}$$

$$S(x)=(2\pi)^{-\frac{1}{2}}\int_{0}^{x}t^{-\frac{1}{2}}\sin t\ dt=(2x/\pi)^{\frac{1}{2}}\sum_{n=0}^{\infty}(-1)^{n}\{(4n+3)(2n+1)!\}^{-1}x^{2n+1}$$

Exponential Integrals

$$-\mathrm{Ei}(-x)=\int_{x}^{\infty}t^{-1}e^{-t}dt;\quad \overline{\mathrm{Ei}}(x)=-\!\!\!\int_{-x}^{\infty}t^{-1}e^{-t}dt,x>0;-\!\!\!\int f(t)dt = \text{principal value}$$

$$-\mathrm{Ei}(-z)=-\gamma-\log z-\sum_{0}^{\infty}(n\cdot n!)^{-1}(-z)^{n};\quad \overline{\mathrm{Ei}}(z)=-\gamma-\log z-\sum_{0}^{\infty}(n\cdot n!)^{-1}z^{n}$$

$$\overline{\mathrm{Ei}}(x)=\tfrac{1}{2}\{\mathrm{Ei}(-xe^{i\pi})+\mathrm{Ei}(-xe^{-i\pi})\},x>0;\quad \overline{\mathrm{Ei}}(ze^{\pm i\pi})=\pm i\pi+\mathrm{Ei}(-z)$$

$$\mathrm{Ei}(-ze^{\pm i\pi})=\pm i\pi+\overline{\mathrm{Ei}}(z),\quad \mathrm{Ei}(-ze^{\pm\frac{1}{2}i\pi})=\mathrm{Ci}(z)\mp i\ \mathrm{si}(z)$$

$$\overline{\mathrm{Ei}}(ze^{\pm\frac{1}{2}i\pi})=\mathrm{Ci}(z)\pm i\{\pi+\mathrm{si}(z)\}$$

Sine- and Cosine Integral

$$\mathrm{Si}(x)=\int_{0}^{x}t^{-1}\sin t\ dt;\quad \mathrm{Ci}(x)=-\int_{x}^{\infty}t^{-1}\cos t\ dt;\quad \mathrm{si}(x)=-\int_{x}^{\infty}t^{-1}\sin t\ dt;\ x>0$$

$$\mathrm{Si}(z)=\tfrac{1}{2}\pi+\mathrm{si}(z)=\sum_{n=0}^{\infty}(-1)^{n}\{(2n+1)(2n+1)!\}^{-1}z^{2n+1}$$

$$Ci(z) = \gamma + \log z + \sum_{n=1}^{\infty} (-1)^n \{(2n)(2n)!\}^{-1} z^{2n} = \gamma + \log z - \int_0^z t^{-1}(1-\cos t)dt$$

$$Ci(z) = \tfrac{1}{2}\{Ei(-ze^{\frac{1}{2}i\pi}) + Ei(-ze^{-\frac{1}{2}i\pi})\}; Si(z) = \tfrac{1}{2}\pi + \tfrac{1}{2}i\{Ei(-ze^{\frac{1}{2}i\pi}) - Ei(-ze^{-\frac{1}{2}i\pi})\}$$

8. Miscellaneous Functions

Complete Elliptic Integrals **K**(k) and **E**(k)

$$K(k) = \int_0^{\frac{1}{2}\pi} (1-k^2\sin^2 t)^{-\frac{1}{2}}dt = \tfrac{1}{2}\pi \; {}_2F_1(\tfrac{1}{2},\tfrac{1}{2};1;k^2)$$

$$E(k) = \int_0^{\frac{1}{2}\pi} (1-k^2\sin^2 t)^{\frac{1}{2}}dt = \tfrac{1}{2}\pi \; {}_2F_1(-\tfrac{1}{2},\tfrac{1}{2};1;k^2)$$

Riemann's Zeta Function

$$\zeta(s) = \sum_{n=1}^{\infty} n^{-s} = \zeta(s,1) \qquad\qquad\qquad Re\ s>1$$

Hurwitz's Zeta Function

$$\zeta(s,v) = \sum_{n=0}^{\infty} (n+v)^{-s} \qquad\qquad\qquad Re\ s>1$$

Lerch's Transcendent

$$\Psi(z,s,\alpha) = \sum_{k=0}^{\infty} (\alpha+k)^{-s} z^n, |z| < 1$$

List of Functions

Symbol	Name of the Function	Listed on Page
$B(x,y)$	Beta function	250
$C(x)$	Fresnel's integral	256
$Ci(x)$	Cosine integral	256
$C_n^{\nu}(x)$	Gegenbauer's polynomials	250
$D_{\nu}(z)$	Parabolic cylinder function	255
$\mathbf{E}(k)$	Complete elliptic integral	257
$Ei(-x)$ $Ei(x)$	Exponential integrals	256
$Erf(x)$ $Erfc(x)$	Error integrals	256
$\mathbf{E}_{\nu}(z)$	Anger-Weber function	252
${}_mF_n(x)$	Hypergeometric series	254
$H_n(x)$ $He_n(x)$	Hermite polynomials	250
$H_{\nu}^{(1)}$ $H_{\nu}^{(2)}(z)$	Hankel's functions	252
$\mathbf{H}_{\nu}(z)$	Struve's function	253
$I_{\nu}(z)$	Modified Bessel function	252
$J_{\nu}(z)$	Bessel function	252
$\mathbf{J}_{\nu}(z)$	Anger-Weber function	252
$\mathbf{K}(k)$	Complete elliptic integral	257
$K_{\nu}(z)$	Modified Hankel function	252
$\tau_{\nu,1}, \rho_{\nu,1}$	First positive roots of $J_{\nu}(x)$ and $Y_{\nu}(x)$ respectively	
$L_n^{\alpha}(x)$	Laguerre polynomials	250
$\mathbf{L}_{\nu}(z)$	Struve's function	253
$M_{\kappa,\nu}(z)$	Whittaker's function	254
$P_n(x)$	Legendre polynomials	250
$P_n^{(\alpha,\beta)}(x)$	Jacobi polynomials	250
$p_{\nu}^{\mu}(z)$	Legendre functions	251

Symbol	Name of the Function	Listed on Page
$q_\nu^\mu(z)$ $P_\nu^\mu(x)$ $Q_\nu^\mu(x)$	Legendre functions	251
$S(x)$	Fresnel's integral	256
$si(x)$ $Si(x)$	Sine integral	257
$s_{\mu,\nu}(z)$	Lommel's function	253
$T_n(x)$ $U_n(x)$	Chebyshev polynomials	250
$W_{\kappa,\nu}(z)$	Whittaker's function	254
$Y_\nu(z)$	Neumann function	252
$\Gamma(z)$	Gamma function	249
$\gamma(a,z)$ $\Gamma(a,z)$	Incomplete gamma function	255
$\psi(z)$	Psi function	249
$\zeta(z)$	Riemann's zeta function	257
$\zeta(z,v)$	Hurwitz's zeta function	257
$\Psi(z,s,\alpha)$	Lerch's transcendent	257

F. Oberhettinger, Oregon State University, Corvallis, Oregon

Tables of Mellin Transforms

1974. V, 275 pp. Softcover DM 49,–
ISBN 3-540-06942-9

Contents: Part 1: Mellin Transforms. –
Part 2: Inverse Mellin Transforms.

This book contains tables of integrals of the Mellin transform type, sometimes referred to as the two-sided Laplace transforms. The Mellin transform has found widespread and effective use, particularly in the application to problems arising in analytic number theory. The author has also included a list of major contributions in the area of Mellin transforms.

Springer-Verlag Berlin
Heidelberg New York London
Paris Tokyo Hong Kong

Springer